Hungry Planet

Stories of Plant Diseases

Gail L. Schumann
Marquette University
Milwaukee, Wisconsin

and

Cleora J. D'Arcy
University of Illinois at Urbana-Champaign

The American Phytopathological Society
St. Paul, Minnesota U.S.A.

Cover Photographs
Front—Top (corn field) and Middle (rice harvest): Courtesy of Shutterstock; Bottom (potato field): Courtesy of Susie Thompson.
Back—Top (coffee beans), Middle left (bananas), and Middle right (cassava): Courtesy of Shutterstock; Middle center (vegetables): Courtesy of Scott Bauer–USDA; Bottom (tomato): Courtesy of Earthstar Stock, Inc.

Library of Congress Control Number: 2011938425
ISBN: 978-0-89054-399-3

A previous edition, *Plant Diseases: Their Biology and Social Impact,* by Gail L. Schumann, was published in 1991 by The American Phytopathological Society.

Printed in the United States of America on acid-free paper.

The American Phytopathological Society
3340 Pilot Knob Road
St. Paul, Minnesota 55121, U.S.A.

To our students

Contents

Online Resources for *Hungry Planet: Stories of Plant Diseases*

For Students and General Readers

Free materials that supplement this book are organized by chapter and available to readers at www.apsnet.org/edcenter/intropp/hungryplanet:

- Podcasts
- Color Images
- Suggested Readings
- Internet Resources
- Words to Know (linked to APS*net* Education Center Illustrated Glossary)
- Questions for Discussion

For Instructors

Instructors who examine this textbook for course adoption will be given login and password information to access the online resources. Like the student materials, the instructor resources are organized by chapter:

- Teaching Resources
- Podcasts
- Demonstrations and Desktop Lab Exercises
- Group Discussions
- Short and Long Writing Assignments

Preface

Plant diseases have changed human history and culture, yet few people know much about plants or the pathogens that cause plant diseases. Most of the people who will read this book enjoy the luxuries of abundant food and relatively low food prices. However, for many people around the world, getting enough to eat is a daily challenge. Sadly, famines still occur.

We invite you to read stories of plant disease epidemics that illustrate the past and present vulnerability of some important plants that people use for food, for fibers and oils, and for green spaces. As you read, we hope you will then consider how to preserve precious plant gene pools and how to stay one step ahead of the numerous pathogens that can destroy plants.

A variety of scientific fields are touched on in this book. A plant disease can be understood only by studying the biology of both the vulnerable plant and the pathogens that threaten it. Thus, the science of plant pathology introduces the reader to botany and to the various sciences associated with the pathogens: bacteriology, mycology, nematology, and virology. Beginning with the study of the nature of disease, plant pathology now includes studies of epidemics and the ecology of agriculture. Genetic studies have led to the development of disease-resistant crops, as well as the ability to manipulate plant and pathogen genes through genetic engineering. Biological concepts from these scientific fields are introduced along the way, and more technical information is found in the Science Sidebars in various chapters.

Plant pathology is a science with important ties to human welfare. All citizens should be aware of the biological basis of legislation governing environmental protection, pesticide use, release of genetically engineered organisms, land preservation for agricultural use, and genetic reserves. How do we balance safety and the cost of producing food, fiber, and fuel? How do we foster plant health in a world dominated by people?

These are two of the many questions we ask you to consider in *Hungry Planet.* We wrote this book to update *Plant Diseases: Their Biology and Social Impact,* published in 1991. Both books reflect our experiences as instructors of general education courses designed to enhance the scientific and agricultural literacy of university students.

Acknowledgments

We would like to thank our many colleagues in plant pathology, whose work helps protect the world's plants. Many of their images illustrate our stories. Additional color images, Internet links, an illustrated glossary, and other resources to accompany this book are available at the APS*net* Education Center (www.apsnet.org/edcenter). Special thanks are extended to Dr. H. D. Thurston, who contributed numerous images.

We appreciate the skills of artist Nancy Haver, who produced many of the drawings, and plant pathologist and artist Vickie Brewster, who contributed and modified many additional illustrations.

We also thank Dr. Darin Eastburn for his content and technical expertise in the production of the chapter podcasts available in the APS*net* Education Center.

Finally, we thank our many general education students. Over the past 25 years, they have listened to our stories and considered the important roles of nonscientists in helping to solve the problems of a hungry planet.

About the Authors

GAIL L. SCHUMANN received her B.S. degree in botany from the University of Michigan and her M.S. and Ph.D. degrees in plant pathology from Cornell University. She is professor emerita from the University of Massachusetts–Amherst, where she taught several plant pathology courses and had research and extension responsibilities in turfgrass pathology. Dr. Schumann is co-author of *Essential Plant Pathology,* more than 50 peer-reviewed publications on teaching and turfgrass pathology, and numerous extension publications. She is a recipient of The American Phytopathological Society (APS) Excellence in Teaching Award, the Northeastern Regional USDA/NASULGC Teaching Award, the Award of Merit from the Northeastern Division of APS, and a Fellow of APS. She is currently an adjunct professor at Marquette University in Milwaukee, Wisconsin.

CLEORA J. D'ARCY received her A.B. degree in biology from Harvard University and her M.S. and Ph.D. degrees in plant pathology from the University of Wisconsin–Madison. She is professor emerita in the Department of Crop Sciences at the University of Illinois at Urbana–Champaign, where she taught courses on plant pathology, professionalism and ethics, and teaching methods for undergraduates, graduate students, and new faculty. Dr. D'Arcy has authored more than 75 peer-reviewed publications on plant virology and teaching and is co-author of *Essential Plant Pathology* and co-editor of *Barley Yellow Dwarf: 40 Years of Progress*. She is a past president of The American Phytopathological Society, a University of Illinois Distinguished Teacher Scholar, a recipient of the National USDA/NASULGC Food and Agricultural Sciences Excellence in Teaching Award, and a Fellow of The American Phytopathological Society and the American Association for the Advancement of Science.

CHAPTER 1

The Irish Potato Famine: The Birth of Plant Pathology

In the early summer of 1845 in Ireland, the days were sunny and the potato crop was growing well. There was no warning of the disaster that would strike, causing misery, suffering, and death. Then, the weather turned overcast and rainy for weeks, and the potato plants rotted as the Irish peasants watched helplessly. The horrors of the Irish potato famine are still remembered—driving one more wedge between the English and Irish and contributing to political conflicts that continue to this day.

The story of this disaster is an important one. It introduces many of the ideas to be presented in this book, including the political aspects of food supply, the risks of genetic uniformity and dependence on only a few food crops, and the distribution of new crops from their origins throughout the world. But these same concepts could be derived from nearly any agricultural failure. The Irish potato famine is of specific importance because the debate surrounding the study of the stinking mass of rotted potatoes gave birth to the science of plant pathology: the study of plant diseases. Infectious microorganisms were finally to be accepted as the causes of disease, rather than the results, predating even Louis Pasteur's work with bacteria. The theory of spontaneous generation of microorganisms from decaying tissues was soon to be replaced with the germ theory.

This idea is so commonly accepted today that we assume a disease is caused by "germs" unless proven otherwise. Today's mysterious problem diseases are those that do not appear to be caused by bacteria, fungi, viruses, and other recognized but less familiar pathogens. In the mid-1800s, however, most people viewed disease—in plants, animals, or people—as the result of bad weather, punishment from God, or perhaps just bad luck.

Potatoes: From South America to Europe

In the 19th century, the peasant farmers of Ireland were so dependent on the potato that its common name, the Irish potato, still reflects its connection with these poor people. The potato's origin was not the Emerald Isle, however, but the highlands of South America—specifically, the Lake Titicaca region between Peru and Bolivia (Figure 1.1). The Spanish conquistadors discovered this plant in the 16th century while searching for gold in the Andes Mountains of South America. However, the European conquerors did not at first recognize the value of the potatoes they found growing throughout this region. The starchy tubers were and continue to be an important food crop in South America, particularly at higher elevations, where corn does not grow well (Figure 1.2). Religious and agricultural records of the ancient Inca contain many references to potatoes, and there is archeological evidence of potatoes as a South American food crop dating as far back as 400 B.C.

Some potatoes were brought back to Europe aboard ships, although it is likely that many failed to survive the long voyage because of poor storage conditions. The early history of the crop is confused because of the similarity of the term *la batata*, which refers to the sweet potato (an unrelated plant), and *la papa*, the Spanish name for what we now call the "white" or "Irish" potato. The first potatoes to reach Europe probably arrived in Spain in about 1570. Each language now has its own common name for this valuable plant. The French call potatoes *pommes de terre*, and the Dutch call them *aardappeln*, both of which translate to "apples of the earth"—certainly, a poetic description.

The people of Europe were reluctant to accept the potato as a food crop. Many religious advisors discouraged production of a food that grew in soil, which they thought made potatoes more appropriate for animals than humans. Besides, there was no mention of the potato in the Bible, which suggested that perhaps it was evil or might instill sinful desires in its eaters. Slowly, however, the crop spread throughout Europe, as hungry people discovered the many advantages of a highly nutritious crop that could be grown beneath the ground, safe from the trampling feet of invading armies.

The exact time of arrival of the potato in Ireland is unclear, but it was a well-established food crop by 1800. Harvests were low in some years, but they generally produced large amounts of the nutritious food. The population of Ireland grew from approximately 4.5 million in 1800 to more than 8 million by 1845. When the harvest was good, Irish peasants often ate 8–14 pounds (3–6 kilograms) of potatoes a day and with little else, unless some milk was available from a cow. Although decidedly boring, this daily intake provided substantial nutrition, including protein, carbohydrates, and many vitamins and minerals—particularly, vitamin C.

Figure 1.1. Map of South America. The potato originated in the region around Lake Titicaca, between Peru and Bolivia.

Figure 1.2. Native women selling many different potato cultivars in a Colombian market.

Figure 1.3. Harvesting potatoes in Maine in the early 1930s. Potato plants produce an abundant and nutritious food supply.

The grain crops, which grew poorly in Ireland anyway, were needed to pay the rent to the landowners, most of whom lived in England. In contrast to the wealthy landowners, most peasants lived in one-room, windowless huts with dirt floors, holes for a door and in the roof to let smoke from the fire escape, and little furniture or other possessions. Fortunately, potatoes thrived in the cool, moist climate of Ireland, which was similar to that of the South American highlands, their place of origin. A family could grow enough potatoes to feed themselves on half the land required to produce the same number of calories in grains (Figure 1.3).

The Potato Plant: Valuable Tubers

A field of growing potato plants is a beautiful sight, especially in midsummer, when the plants begin to bloom. The leaves on the bushy plants are composed of a number of dark-green, leaflike sections called leaflets, which are connected to a central petiole and therefore called compound leaves (Figure 1.4). Each plant has a bud at the growing tip and lateral (side) buds, where the leaves attach to the main stem. If the top of the plant is pinched off, the lateral buds begin to grow. Pinching off the top is a common practice for producing bushier house plants and other ornamental plants. For the potato plant, the lateral buds serve as a "back-up plan," enabling it to continue to produce new growth even if the top section is damaged or eaten.

The flowers on potato plants may not be familiar to someone who has not seen potato fields. The flowers are similar to those of the tomato, a close relative, but they vary in color depending on the cultivar (meaning "cultivated variety") and can range from white to deep purple. After blooming, fertile plants produce green berries that are similar to small tomatoes and filled with numerous tiny seeds. Rejection of these small, bitter fruits as food was probably another reason for the slow acceptance of this crop in Europe.

The potato is in the botanical family Solanaceae, which also includes bell and hot peppers, eggplant, petunia, tobacco, tomato, and some poisonous members, such as deadly

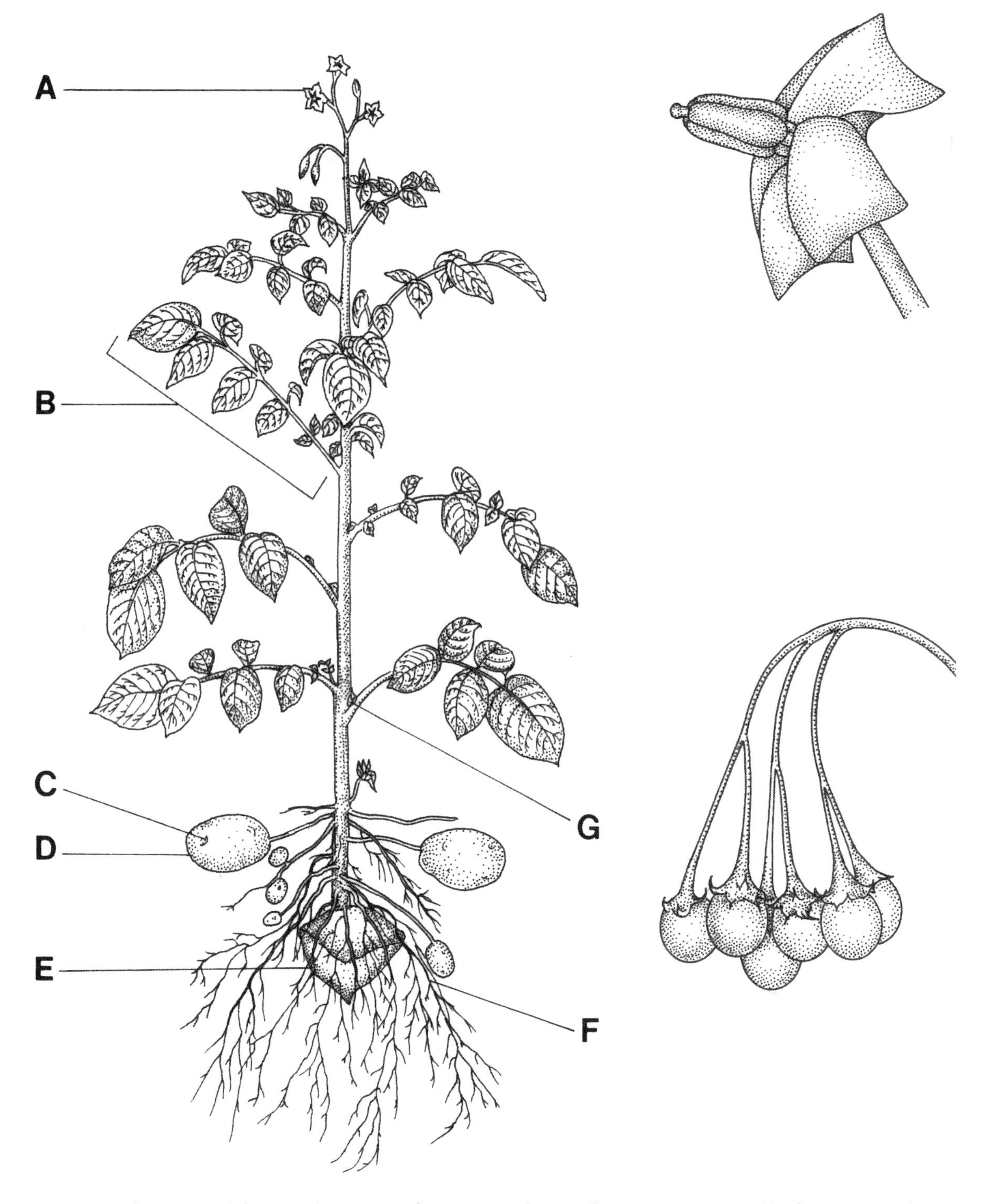

Figure 1.4. The potato, *Solanum tuberosum*. **Left,** an entire plant. **A,** flower; **B,** compound leaf; **C,** eye; **D,** tuber; **E,** tuber piece used for planting; **F,** roots; **G,** lateral bud. **Right top,** flower. **Right bottom,** fruit (berries).

nightshade. The Solanaceae family is worldwide in distribution, but some members—such as potato, tobacco, and tomato—are native to Central and South America and were unknown to Europeans before they began exploring the Western Hemisphere. The similarities in flower structure make the relations among these plants clear, even though they differ in other characteristics. The family name is derived from the Latin *solamen*, which means "comforting," reflecting the sedative effects of the alkaloids produced by some members of the family. Deadly members of the family slowed the acceptance of the edible members as food plants, but eventually, they were discovered to be safe. Some toxic alkaloids are present in the leaves of potato plants, causing digestive upsets to those who eat them. If potato tubers are left in the light, they too develop a green color and alkaloids and should not be eaten.

The familiar starchy potato is actually a tuber, as reflected in the plant's Latin or scientific name, *Solanum tuberosum.* All organisms are given a similar Latin binomial (two-word name) based on a system created in the 1700s by the famous Swedish naturalist, Carolus Linnaeus. Because tubers grow underground, they might appear to be roots, as carrots are. Closer examination reveals that the "eyes" of potatoes and other tubers are really buds in the axils of tiny scalelike leaves. Roots do not produce buds and leaves or turn green when left in the light. Tubers are actually underground stems adapted to the storage of nutrients. After several months in storage, the buds on potato tubers begin to sprout and are ready to grow into new plants.

In the field, at about the same time that potato plants begin to flower, small swellings start to develop at the ends of underground stems. These swellings are new tubers, which store the excess food the plant produces during photosynthesis. The starchy tubers may be harvested and eaten at any time. Small, freshly harvested "new potatoes" are tender and delicious, but to maximize yield, most growers wait until later in the season to harvest them. As the mature vines die aboveground, the tubers develop an outer cork layer for survival in the soil during the winter. This layer also protects them from desiccation and wounding when they are harvested and put into storage. Stored potatoes must be kept cool to prevent rotting by bacteria and fungi present on the tubers. They also need to be kept in the dark to prevent the "greening" of tuber tissues and the development of toxic alkaloids.

Most commercial potato growers and home gardeners do not plant the tiny seeds from the green berries produced after flowering. Instead, potatoes are grown by vegetative propagation; that is, small tubers or pieces of tubers are planted. To maximize the planting stock, farmers may cut each tuber into several pieces. Each piece can grow into a new

Science Sidebar

Scientific names

The Latin binomial was devised by students of Linnaeus, who were assigned to write down all the plants fed on by grazing animals. The students could not keep up with the animals when writing the long scientific names in use at the time, so they shortened the names to just two words. Until that time, there was no standard format for naming species, and the use of long, cumbersome definitions made communication between biologists difficult.

According to the Linnaean system, the first word is the genus: a name shared by closely related organisms. Other plants in the genus *Solanum* are closely related to the potato but sufficiently different to be considered different species. The second word is the specific epithet: a descriptive word that separates this species from all others in that genus.

Because the names follow the grammatical rules of Latin, they should be italicized or underlined when written. Once the genus has been written out in full, it can be abbreviated in subsequent appearances, using the uppercase first letter followed by a period and the specific epithet, as in *S. tuberosum.*

A species was once defined as "organisms that would produce fertile offspring if mated," but our expanding knowledge of the variety of life has made this definition difficult to apply in the case of many microorganisms. In such cases, species names change as our understanding of the relationship between organisms improves.

plant as long as an "eye," or bud, is present. Tuber pieces that do not have eyes are called "blind" and will not grow into new plants.

Vegetative propagation has several advantages. The tuber pieces, known as "seed" to farmers, contain substantial food reserves, so that a vigorous green shoot pushes up through the soil more quickly than a shoot from a tiny seed. In addition, each tuber piece grows into a plant that is genetically identical to the parent plant in size, time to maturity, and other important characteristics, including tuber taste and texture. True botanical seed, from the fruit of a plant, is the product of sexual reproduction, which results in genetic variation. Each seed produces a plant that is slightly different from its parents and all of its siblings. Genetic uniformity is a great advantage when uniformity has an economic benefit—as in flowers, fruits, ornamental plants, and some vegetables—so vegetative propagation has become a common practice in modern agriculture (Table 1.1). Tubers also provide an advantage to wild potato plants, because they help ensure survival in case the true seeds produced in the berries do not grow successfully. Like the lateral buds discussed previously, tubers provide another "back-up plan" for the plant.

There are, however, some disadvantages to vegetative reproduction of agricultural crops. The advantages of uniform characteristics may be outweighed by the disadvantage of uniform susceptibility to pests and pathogens. If one plant in a field is vulnerable to a particular disease, then so are all of its neighbors. Genetic uniformity increases the risk of loss. A second important risk lies in the large pieces of plant tissue that are planted during vegetative reproduction. Bulbs, cuttings, roots, tubers, and other relatively large plant pieces may be planted, rather than tiny seeds. These large plant pieces often carry pathogens, especially viruses and other systemic parasites, into the planting area at the very beginning of the season. Many of these pathogens would not be present in true seed, produced by sexual reproduction. The yield and quality of the crop may be greatly reduced by the introduction of pathogens on or in plant parts.

Of course, many kinds of parasites were not discovered or well understood until relatively recently, but farmers of the past observed the ravages of the diseases the parasites caused. Farmers suspected that reduced yields were related to the use of vegetative reproduction, which, despite its convenience, was considered "unnatural." Periodically, farmers allowed the plants to reproduce sexually, the "natural" way, and harvested the seeds to begin new selections for acceptable plants. They believed that sexual reproduction restored the vigor of their weakened plants, when, in fact, the seeds had simply escaped infection by the pathogens present in the parent plant.

Only small collections of potatoes were brought from the New World in the early days of European potato production, and from these tubers, further selections were made in an attempt to develop potato cultivars suitable for

Table 1.1. Top Food Crops of Developing Market Economies

Energy Production		Protein Production	
Crop	**Megajoules of Edible Energy**	**Crop**	**Kilgrams Produced per Hectare per Day**
Potatoes	216	Cabbages	2.0
Yams	182	Dry broad beans	1.6
Carrots	162	Potatoes	1.4
Maize	159	Dry peas	1.4
Cabbages	156	Eggplants	1.4
Sweet potatoes	152	Wheat	1.3
Rice	151	Lentils	1.3
Wheat	135	Tomatoes	1.2
Cassava	121	Chickpeas	1.1
Eggplants	120	Carrots	1.0

Source: From *Potato Atlas* (1985), published by International Potato Center (CIP). Used by permission of CIP.

consumption. As a result, the genetic variation among the potato cultivars was very limited. Crop losses occurred every year from various causes, and occasional food shortages were not unusual. When crops were good, tubers were plentiful in the months after the harvest, but as the winter months passed, supplies diminished and the tubers began sprouting, ready to plant for the next crop but no longer suitable for food. The summer months were a hungry time, usually requiring the Irish peasants to spend what few coins they had for grain to feed themselves until the next potato harvest.

The Late Blight Epidemic: Necessary Components

With this background, the stage was set for the impending disaster. What were the important components that led to the tragedy?

The human component should always be considered, and in this case, a large population depended on a single food crop. The Irish population had grown rapidly following the introduction of the potato, and no significant alternative foods existed when the crop failed. The grain crops the Irish planted were needed to pay the rent to the English landlords. If the rent was not paid, the people were thrown off their land and faced certain starvation. Perhaps this dependence on one crop for food sounds foolish or risky today, but the current human population, now approaching 7 billion, relies on essentially three species of plants—wheat, corn, and rice—to feed most of its people, an issue we will return to in Chapter 14.

In addition to the human component, we should consider the important biological components that combined to cause the blight of the potato crop. Plant pathologists have devised a memory aid called the disease triangle (Figure 1.5), which describes the three factors necessary for disease: a susceptible plant, a pathogen capable of causing disease in the plant, and environmental conditions favorable for disease development. The potato crop was derived from a small supply of tubers that survived the lengthy trip across the sea to Europe. The genetic variability of the crop had been further reduced by the Irish farmers' selection of tubers that grew best in their country. They particularly favored one type of potato called Lumper. From the name, it is obvious that this cultivar was not grown for the beauty of its tubers but rather for the large quantity of food produced. Thus, every plant in every field was nearly genetically identical—a desirable situation for agricultural characteristics, such as high yield to feed a large family, but a potentially dangerous situation for disease development.

For the early weeks of summer in 1845 in Ireland, records show hot and dry weather overall. Then the weather changed. Overcast skies dominated for 6 weeks, along with low temperatures. Records throughout Europe indicate a particularly cool and rainy period. In just a few weeks, the vigorous green potato vines became a blighted mass of decaying vegetation. When the tubers were dug from the ground, some were rotted, but many appeared to be sound. Later, however, these potatoes also rotted away in the storage bins. The potato crops failed throughout Europe, but the disaster was the worst in Ireland because of Irish peasants' nearly complete dependence on

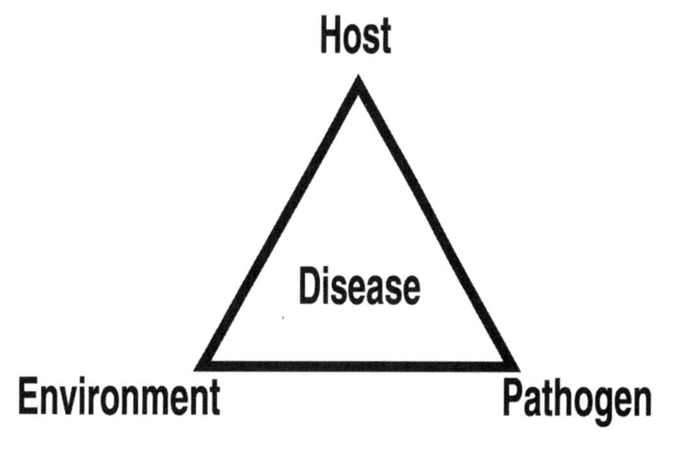

Figure 1.5. The disease triangle.

the potato for their food. Even though the cool, rainy weather affected all crop yields to some degree, the blight epidemic was specific to the potato. Something new and frightening was occurring in the potato fields, and the answer to the mystery lay in the third component necessary for disease: a virulent pathogen capable of infecting the potato crop (Figure 1.6).

At this point, it is interesting to consider the state of science in the mid-1800s in Europe. In general, only wealthy people had the leisure time and education to consider how the world operated, so most scientists were men trained in medicine or religion. The microscope had been invented nearly 200 years previously by Anton van Leeuwenhoek in the Netherlands. Robert Hooke, a physicist in England, first described the cells in plant tissues, although the concept that cells are the basic unit of plant and animal tissues was not proposed until 1838, nearly 150 years later. Hooke also first observed fungi through his microscope in 1667.

Spontaneous generation was the commonly accepted explanation for the presence of teeming populations of microbes in diseased or dead tissues. When, in 1845, a white "fungus" was found on the blighted potato vines, it was considered the result, rather than the cause, of decay. One common explanation for the rotted crop was that the plants took up too much water in the rainy weather. Numerous observations of diseased plants had been made over the centuries, however, and many thoughtful people had contributed clues to the solution of the puzzle of plant disease. The potato blight attracted the attention of a number of biologists and brought many of the clues together. Rainy seasons had occurred before without such losses. This new phenomenon required a new explanation: the role of a pathogen.

A pathogen is an agent capable of causing disease. In the case of the potato blight, the pathogen is an organism that parasitizes the potato plant. A parasite derives its food from another living organism and lives in intimate contact with its host. In nature, parasites are associated with nearly all living organisms. Familiar examples include the fleas, ticks, and intestinal worms that annoy most pets. A healthy pet can maintain reasonable populations of such parasites without being visibly harmed. Similarly, with wild plants and animals, parasites that do not cause noticeable disease should not be called pathogens.

The blight pathogen parasitizes potato plants and closely related plants in South America. Farmers there grow an enormous number of different potato cultivars that vary in their susceptibility to the parasite, so that even though some loss to the parasite occurs, devastating epidemics are avoided. The technology of modern agriculture—and of vegetatively propagated plants, in particular—leaves crops genetically uniform and thus uniformly susceptible to parasites. This situation can result in severe epidemics, in which the parasites not only derive food from their hosts but also increase so rapidly that they become pathogens, causing disease and loss.

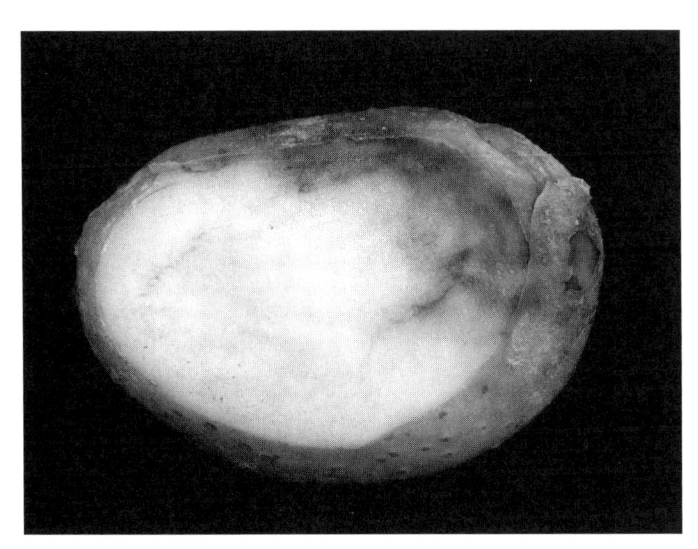

Figure 1.6. Potato tuber infected by *Phytophthora infestans.*

The blight pathogen arrived aboard a ship from the Americas, just as the potatoes had. It is likely that blighted tubers carrying the pathogen had been brought aboard many times previously but had then been eaten or rotted before arriving. Perhaps faster crossings allowed some of the tubers carrying the pathogen to survive and be planted in Europe. Perhaps, also, the environmental conditions were so favorable for infection by the pathogen in the mid-1800s that it was able to establish itself rapidly upon arrival. Certainly, the narrow genetic base of the existing potato plants left them vulnerable to the invading pathogen.

Until recently, the late blight pathogen was considered a member of the fungi. Modern studies of its genetic material, or deoxyribonucleic acid (DNA), have shown that it is really more closely related to certain algae and belongs to a group of funguslike organisms called oomycetes, or "water molds." The pathogen itself is not obvious on an infected potato plant. It appears as a fuzzy, whitish mildew usually on the lower surface of infected potato leaves and stems, where it emerges through the stomata (air-exchange pores) in humid weather. Microscopic treelike hyphae grow away from the plant tissue, producing lemon-shaped sporangia at their tips. A sporangium (plural: sporangia) is a structure that contains spores: the reproductive units of the pathogen. The sporangia are usually dispersed by air currents to neighboring plants and can travel easily to nearby fields when the air is moist. They dry out and die at high temperatures or after traveling long distances. At cool temperatures, such as those that prevailed in 1845, a change occurs in sporangia when they land on a wet potato leaf or stem. After a few hours, the cellular material inside the sporangium is converted into approximately eight wiggling zoospores, each with two flagella to aid in movement. The zoospores swim out of the sporangium and attach themselves to the leaf surface in an attack that is eightfold the assault from a single sporangium at warm temperatures. A small germination tube grows from the zoospore and invades the plant. From the tiny infection sites, an extensive network of threadlike hyphae pene-

Figure 1.7. Sporulation of *Phytophthora infestans* on potato leaves. Sporangia are produced on hyphae that emerge through stomata in the leaves.

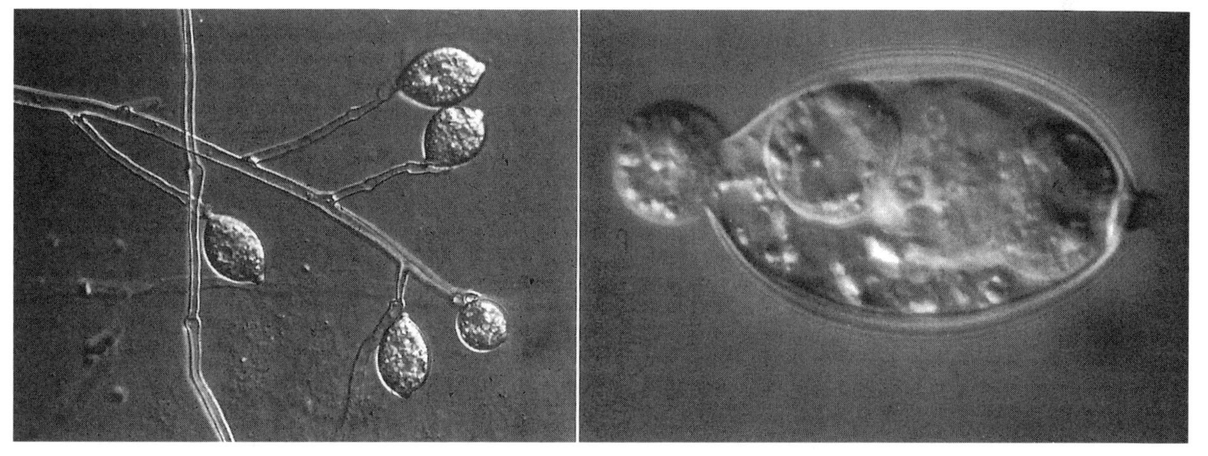

Figure 1.8. *Phytophthora infestans.* **Left,** lemon-shaped sporangia; **Right,** liberation of a zoospore from a sporangium (light micrographs).

trate between the cells, absorbing nutrients to feed the growing organism. Brown lesions of dead plant cells and newly produced sporangia begin to appear as soon as 3 to 5 days after infection. This pathogen can produce many generations of sporangia in a short time and rapidly colonize all available potato tissue. A whole potato plant can be turned into a slimy mass in less than 3 weeks (Figure 1.7).

Germination and infection by sporangia always require water on the plant surface. At higher temperatures (more than 68°F [20°C]), the sporangia germinate by producing a single germination tube, rather than zoospores (Figure 1.8). Thus, the number of potential infections from a single sporangium is greatly reduced in warmer environments. When temperatures are more than 68°F (20°C), disease is reduced but not eliminated.

Sporangia survive dispersal by air better when the air is humid, and water is necessary for infection. In addition, zoospores are produced at cooler temperatures, which maximize the number of potential infections possible from a single sporangium. The cool, wet environment of Ireland in 1845 was particularly favorable for this new disease, now known as late blight. In addition, the disease occurred late in the growing season, when the plants were large and the enclosed leaf canopies remained wet for a long time before drying out.

A final explanation is still needed: How do the tubers become infected by a pathogen that invades the leaves and stems? It is now known that sporangia from the leaves are

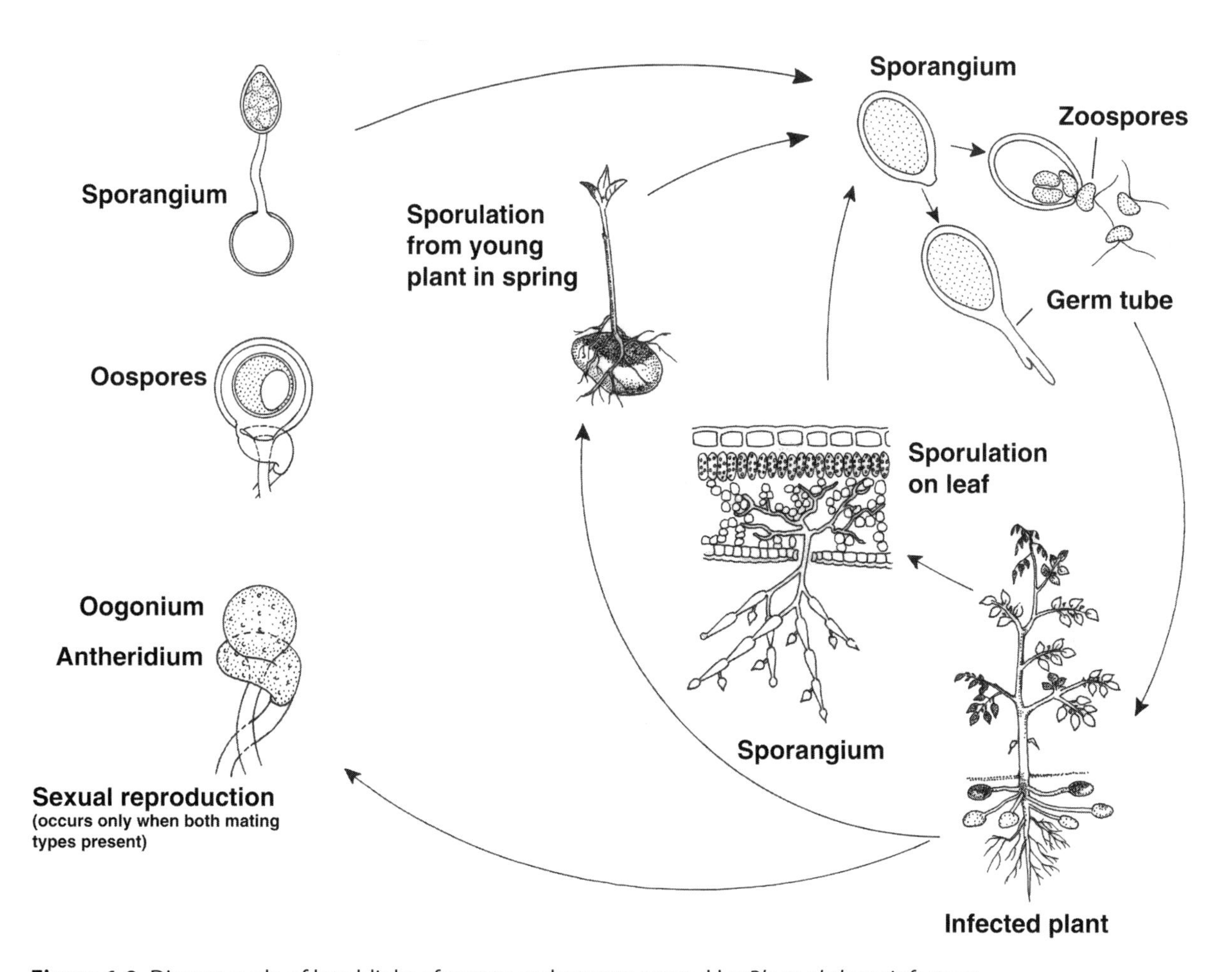

Figure 1.9. Disease cycle of late blight of potato and tomato caused by *Phytophthora infestans*.

washed by water into the soil, where they can infect the tubers. Potato growers have learned to "hill up" the soil around the stems of young potato plants to increase the depth of soil that sporangia have to traverse to attack the tubers. In addition, potato vines are usually killed before harvest with herbicides, with mechanical cutting, or by frost to protect newly dug tubers from exposure to sporangia on the leaves and stems. The pathogen is incapable of surviving temperate winters in the soil, but it spends the winter, well supplied with food, in infected tubers in storage bins and at the edges of fields in piles of discarded tubers, called cull piles. The cool environment of storage bins is conducive to growth of the pathogen in infected tubers, causing further losses. If conditions are moist in storage, water droplets may even spread sporangia to previously healthy tubers. Any infected tubers that survive the winter without being completely destroyed by the pathogen will serve as a source of sporangia to infect the new crop. And so begins a new late blight epidemic (Figure 1.9).

Consequences of the Late Blight Epidemic: The Irish Potato Famine

With the losses in the 1845 potato crop, the Irish peasants began to starve. Food was in short supply the first winter of the outbreak, leaving fewer seed potatoes to be planted the next season. The summer of 1846 brought new hope, as the plants began green and healthy, but the blight came again and destroyed any hope of a better winter. Help from England was slow to come. It was difficult to convince absentee landowners that this famine was any worse than the famines of the past, which had periodically left Ireland's poor a little hungrier than usual. Grain, as well as other food collected for rent, was exported from Ireland throughout the blight years. These cruelties and their political consequences are well described in *The Great Hunger*, by Cecil Woodham-Smith (Figure 1.10).

An important character in the political drama was Sir Robert Peel, the prime minister of the United Kingdom, who used the blight as an excuse to repeal protectionist trade laws by requesting the import of American maize (corn) to feed the hungry Irish. Part of Peel's agenda was to displace the landowners as the ruling class—an example of a political agenda playing into human suffering. The corn mash was cooked in open kitchens, often miles from the villages. Those too weak or sick to walk to the food had to share the portions of their stronger kin. The cooked mush also prevented the sale and distribution of the grain on the black market. The rations were dubbed "Peel's brimstone" by the less-than-delighted recipients, who were considered ungrateful, since food was not only provided but cooked for them as well.

Figure 1.10. "An Eviction." Tenants are thrown out of their homes in Ireland for not paying rent. Printed in the *Illustrated London News*, December 16, 1848.

The rainy weather and resulting blight continued throughout what came to be called the "Hungry Forties." In continental Europe, weather and political events contributed to political unrest and famine in the middle to late 1840s. Hundreds of thou-

Figure 1.11. A political cartoon printed in the United States at the time of the Irish potato famine.

sands died during these years from hunger or diseases related to hunger. The poorest people depended on potatoes and rye. Potatoes suffered from late blight and rye from another disease called rust, and as a result, the costs of both bread and potatoes rose. In addition to food shortages, an economic recession occurred around 1848, which led to riots and political debates about the role of government in helping the poor. National constitutions were established for the first time in several countries, and Marx and Engels's *The Communist Manifesto* was published.

Nowhere was the suffering as grim as in Ireland, however. In a 15-year period, the Irish population of 8 million decreased by at least 1 million as a result of disease and starvation and by an additional 1.5 million as a result of emigration by those individuals who found a way to leave the island—many traveling to the United States and Canada (Figure 1.11). Today, more than 10% of the U.S. population has Irish ancestry, largely as a result of emigration because of the potato famine.

In 1847, nearly 100,000 emigrants sailed for Quebec. They went first to a quarantine station at Grosse Île, an island in the St. Lawrence Seaway many miles from Quebec City, where they were inspected and quarantined if ill. More than 5,000 died at sea, and more than 5,000 died of typhus and dysentery without ever leaving the immigration station.

Canada has established a National Historic Site, Grosse Île and the Irish Memorial, where visitors can learn about quarantines and public health. Some of the buildings for patients have been restored, and the restorations have included the placement of large rectangular, hinged doors in the walls above all the beds. These doors were designed to allow the so-called miasma to escape from dying and dead patients. Miasma was an imagined cloudlike substance emitted from diseased bodies, which was believed to be responsible for the spread of a disease. In 1909, a large Celtic cross was placed on a high cliff overlooking the island by the Ancient Order of Hibernians. Mass graves have been marked and are accompanied by a panel of names of those who died in the typhus epidemic that spread through the immigrants, who were weakened by starvation during the Irish potato famine. *Ship Fever* is a novella by Andrea Barrett that describes life on Grosse Île among the Irish immigrants and the medical personnel who tried to save them. Memorials to the victims of the Irish potato famine also have been established in Ireland, Battery Park in New York City, and other sites touched by the epidemic.

The Birth of Plant Pathology: The Study of Plant Diseases

The history of the scientific study of the blight is well chronicled in E. C. Large's popular book *The Advance of the Fungi*. Some of the important characters in this drama should be mentioned.

The organism found on the potatoes was first described and named in 1845 by Dr. C. Montagne, a French physician in Napoleon's army. He shared his observations with the Rev. M. J. Berkeley in England, who recognized that this new organism was probably connected with the blighting in some way. Berkeley's rival was Dr. John Lindley, a botany professor at University College in London, who did not believe that the organism was the cause of the blight. Their arguments were published in *The Gardeners' Chronicle*, reflecting some of the most intense philosophical arguments among scientists throughout Europe. Were the rust, smut, and other fungi the products of diseased tissues, or could they be the causes of the diseases?

The scientific philosophers were unable to save the potato crops or prevent the blight in succeeding years, but from this disaster came an important new understanding of plant disease. In about 1860, Anton deBary, a German botanist, performed the experiments that proved the role of the white funguslike organism in late blight (Figure 1.12). He followed the established procedures of the scientific method, in which scientists make observations that lead to a hypothesis or question. deBary observed the white growth of a funguslike organism on potato plants and speculated that it was the cause of the disease. He hypothesized that the sporangia that he had observed would infect the plants and cause the disease. Other researchers had hypothesized that excess water caused the spontaneous production of the white growth. deBary realized that he had to refute this idea by putting two sets of potato plants in the same environmental conditions. Even though both sets of plants were exposed to the same favorable wet environment, only the plants inoculated with the sporangia became blighted. The noninoculated plants served as a control, in which all conditions are the same except for the factor (inoculation with sporangia) to be tested. It was clear that the plants did not rot away because they absorbed too much water. In the absence of the pathogen, no disease occurred.

It is important to appreciate the intellectual significance of deBary's experiments. They contributed to major advances in the study not only of plant diseases but also of human and animal diseases, which eventually led to a new understanding of the importance of contaminated food and water and unsterilized medical instruments in infectious and contagious diseases. In science, all experiments are open to challenge by skeptics. Many scientists were involved in the testing of various organisms, mostly fungi and bacteria, as possible causes of disease. For 200 years, people had been observing and even naming disease-causing organisms, but these had mostly been seen as the result, rather than the cause, of disease. Scientists were now prepared to abandon the theory of spontaneous generation for Louis Pasteur's germ theory of disease, which he proposed in 1863. A scientific theory is a generalized truth accepted by the scientific community after many experiments.

Figure 1.12. Anton deBary.

The cause of late blight was considered a fungus in the 1800s and recognized as new to Europe. Like all other organisms, it was given a Latin binomial. It was first named *Botrytis infestans* by Montagne, but its name was changed to *Phytophthora infestans* by deBary.

Science Sidebar

Scientific name authorities

In the formal designation of a scientific name, a name or initial (called the authority) follows the Latin binomial to allow finding the description of the organism that was officially recorded when it was named. Many plant and animal names are followed by an L., as in *Solanum tuberosum* L., because Linnaeus first created the system and named many of the commonly known plants and animals. Most of those names were well accepted, because the organisms were already familiar and well studied.

Names of less familiar organisms are often changed, as their biology and relationships to other organisms become better understood. This is particularly true of microorganisms because of the rapid improvements in the technology for their culture and study. In such cases, the specific epithet is usually retained, but the genus is changed. The name of the original describer is then put in parentheses, followed by the name of the new describer. For example, the potato late blight pathogen is now called *Phytophthora infestans* (Mont.) deBary. Montagne's specific epithet, *infestans,* was retained, but deBary realized that the organism was not really similar to other species in the genus *Botrytis,* so he created the genus name *Phytophthora* to describe the blight pathogen.

Linnaeus used Latin to describe organisms, because that was the international language of educated Europeans. Many of the names can be translated from their Greek and Latin roots to reveal the ideas of the describers. *Phytophthora* comes from the Greek (*phyto* means "plant," and *phthora* means "destroyer"), and the species name, *infestans,* suggests the devastating infestation. People use common names for local plants and animals in their various languages, but scientists communicate internationally using the Latin names to avoid confusion. Although there are common names for the diseases that pathogenic microorganisms cause, most of these microorganisms do not have common names, because usually, only scientists observe and talk about them.

deBary's work led many scientists to observe diseased plants and describe the parasitic fungi associated with them. Of course, the presence of an organism in diseased tissue is not proof that it is the primary causal agent in the disease. In fact, most fungi are incapable of parasitizing healthy plant tissue and can invade only dead plant tissue. Some means was necessary to determine whether a fungus or funguslike organism present on a diseased plant had caused the disease or was simply an opportunistic saprophyte—that is, an organism that obtains its nourishment from dead organic matter.

The methodology for a proof of pathogenicity was derived from studies in the 1880s by German microbiologist Robert Koch, who worked with anthrax disease of sheep. By chance, the bacterium that causes anthrax has particularly large particles that were visible even with 19th-century microscopes. Koch created a process to convince himself and others that the bacterium in the blood of sheep and cattle actually caused the anthrax disease. Koch's postulates, which can be applied to diseases of plants as well as animals, include the following steps:

1. The symptoms and any evidence of the pathogen in the diseased host are carefully described.
2. The suspected pathogen is isolated from the diseased host and from all other contaminating microorganisms, usually on a nutrient medium on which the organism can grow and reproduce. A description is made of the suspected pathogen.
3. A healthy host is inoculated with the suspected pathogen. It is later observed for symptoms, which must be identical to those described in step 1.
4. The pathogen is isolated from the inoculated host and must be identical to the organism described in step 2.

When all four steps have been completed, the proof of pathogenicity is established.

Recently, Barry Marshall and J. Robin Warren became convinced that the bacterium *Helicobacter pylori* was the cause of stomach ulcers, but other scientists were skeptical. With unusual dedication, Marshall performed step 3 of Koch's postulates on himself. He drank a culture of the bacterium, developed an ulcer, and cured himself with antibiotics. Although this self-endangerment is not a generally recommended procedure, Marshall and Warren received a Nobel Prize in 2005. Koch's postulates continue to guide the study of new diseases, although modifications must be made for pathogens that cannot be cultured outside a living host.

Following deBary's work with *P. infestans*, other scientists began to report numerous fungi capable of infecting plants and causing disease. In the early years, bacteria seemed associated mostly with animal and human diseases, and an inaccurate separation of pathogens was created by scientists who jumped to conclusions because of their preconception that huge differences existed between plants and animals. Near the end of the 19th century, however, bacteria and viruses also were shown to be plant pathogens. By 1900, plant pathology—the study of plant diseases—had become an important focus of scientific study.

Protecting Potatoes from the Blight: A Never-Ending Battle

Scientists had come to a consensus about the cause of the blight, but they had not yet produced a solution to the problem of saving the potato crop from invasion by the pathogen. Protective copper fungicides still lay in the future, so the poorest Europeans went hungry in the blight years that occurred periodically. According to Carefoot and Sprott in their interesting book *Famine on the Wind,* the last major famine caused by *P. infestans* occurred in 1916, during World War I. It resulted in the deaths of 700,000 German civilians, who were unable to protect their potato crop because copper was needed to produce bullets, rather than fungicides. Even today, more than 170 years after the Irish epidemic, frequent applications of fungicides are necessary to grow potatoes in moist climates, and losses occur even in dry areas, such as Israel and the western United States. Potatoes remain a fungicide-intensive crop, despite more than 150 years of study of *P. infestans* and the disease it causes.

Figure 1.13. Potatoes with resistance to *Phytophthora infestans* survive an epidemic better than the blighted susceptible cultivar in the foreground.

Identifying potato cultivars that are resistant to infection by *P. infestans* has been an important goal. For years, farmers have selected potato plants that survived blight better than others (Figure 1.13). Potatoes also have been collected in South America in an attempt to find blight-resistant plants. Although some cultivars have been more resistant than others, no potatoes have been able to survive cool, rainy blight years without treatment with fungicides.

In the first half of the 20th century, plant breeders were able to transfer genes that conferred resistance to *P. infestans* to potato from a closely related plant, *Solanum demissum,* which had been collected in Mexico. The breeders cross-pollinated the plants, harvested the seeds, and selected the most blight-resistant plants. Eventually, new potato cultivars were created that remained blight free even when deliberately inoculated with the pathogen. Some plant pathologists believed that late blight had been eliminated. Unfortunately, this was not the case.

Within a few years, new genetic variants of *P. infestans*, called races, appeared that were capable of infecting the resistant plants. Farmers reported that the resistance in the plants had "broken down," but in reality, the pathogen had evolved to overcome the resistance through genetic change. For some years, *S. demissum* remained an important source of resistance genes, but each time a potato cultivar with a new resistance gene was planted, it was only a matter of time before the pathogen developed a new race and caused a disastrous epidemic. This boom-and-bust cycle in the use of resistance genes for various diseases has been observed in many important crops. Plants with resistance to nearly every important disease have been produced in breeding programs, but the production of plants with durable resistance remains a difficult challenge. This topic will be considered more fully in Chapter 5.

Recent Late Blight Epidemics: It's Not Just History

Crop losses resulting from late blight increased dramatically in Europe, the Middle East, the Near East, and North America in the late 1980s and early 1990s. Large quantities of potatoes were shipped from Mexico to Europe in 1976–1977 because of drought-related shortages. Exotic strains of the pathogen were inadvertently carried in the tubers. Exotic strains also were introduced into North America, probably via infected tomato fruit from Mexico. The tomato, a close relative of the potato, is also susceptible to late blight. Some of the strains brought to North America belonged to the second mating type of *Phytophthora infestans* that is required for sexual reproduction.

Until now, we have focused on the asexual sporangia and zoospores and their role in late blight. In parallel to the life cycle of the potato, the zoospores have a reproductive function similar to that of the asexual tubers and involve no genetic change. In Figure 1.9, which illustrates the *P. infestans* life cycle, the sexual stage (oospores) can be seen on the left side. The sexual stage cannot occur unless both mating types are present. The early outbreaks of late blight that occurred in Ireland in the 1840s, and later in all potato-growing areas of the world, involved only one mating type, so oospores were not involved in those epidemics. Oospores provide a means of survival in the soil and also provide genetic variation in the organism, just as seeds provide genetic variation for the potato plant. The importance of both asexual and sexual reproduction for fungi and funguslike organisms is discussed more in Chapter 2.

Genetic variation can increase the ability of *P. infestans* to overcome resistance genes and to become resistant to certain fungicides. Some of the strains that were introduced to North America after 1975 were already resistant to some key late blight fungicides. Farmers who thought they were protecting their potato crops suddenly watched the plants melt away, just as had happened during the Irish potato famine. In 1992, losses to late blight came to $1 billion in the United States and were especially severe in the Northeast and Northwest.

In 2009, cool, wet weather dominated much of the northeastern United States throughout most of the summer. As shown by the disease triangle, weather alone does not necessarily lead to a late blight epidemic. However, that spring, infected tomato plants were sold to many home gardeners at several large chain stores. What better way to distribute a pathogen? The sporangia from the tomato plants completed the disease triangle and led to severe late blight for the home gardeners. The sporangia also were able to travel by wind to commercial tomato and potato fields. Late blight was identified on Long Island, New York, on June 23, 2009, and was widespread in New York State by mid-July. Late blight reports continued throughout the northeastern United States and in several southern and midwestern states as the summer progressed. By the end of the growing season, late blight was reported on potatoes and tomatoes in more than 25 states. Complete losses were reported from many gardeners and organic farmers; fungicides protected the crops of conventional growers.

Late blight is of international importance. A cold wave and heavy fog caused late blight epidemics in Bangladesh and western India in the fall of 2009 and into 2010. Many farmers could not afford fungicides. Some researchers suspect that the seed potatoes used for planting may have been carrying the pathogen. Losses of more than 35% were reported in many areas.

Late blight remains a threat to potatoes, which are an important and nutritious crop grown in nearly every country. According to the Food and Agriculture Organization of the United Nations, potatoes are now produced in more than 150 countries around the world, with China and India the largest producers. The required fungicide protection of the plants in most areas adds to the cost of production. Annual losses to late blight in developing countries, where fungicides are often not available, are estimated at $3 billion.

Lessons from the Irish Potato Famine: People and Plant Diseases

From its beginnings in the Irish potato famine, plant pathology has faced a clear set of important problems. Human beings wish to produce food with the greatest efficiency and the least effort. Even in the earliest agriculture, this has meant growing large numbers of the same species of plants close together in the same area. In addition, the growers prefer that the crop be fairly uniform, meeting standards of taste, color, yield, and even maturation date, so that harvesting is efficient. In many cases, genetic uniformity, via vegetative propagation, is highly desirable for the production of a particularly good apple or an unusual color of tulip. Unfortunately, these horticultural demands create an environment at risk for disease epidemics. In addition, dependence on only one or even a few food crops was risky in the 1800s and remains risky today. Thus, an important goal of the plant pathologist is to determine ways to balance the needs of agricultural production with the need for plant protection in a world filled with highly adaptable pathogens.

Human beings have been responsible for most of the major plant disease epidemics in history, often because of having moved plants and pathogens from their points of origin to every corner of the earth. Plants are most vulnerable to attack by a pathogen to which they have never been exposed. Plants and pathogens in their places of origin have had millions of years in which to coevolve. Two important diseases of trees—chestnut blight and Dutch elm disease—are examples of the devastation caused by introduced pathogens in Europe and North America (see Chapter 12). Similar problems have arisen when foreign insect pests, such as the Japanese beetle and the gypsy moth, have been introduced.

Tragically, many Native Americans died of "childhood" diseases introduced by Europeans, such as measles and chicken pox, when first exposed to these viruses. More recently, astronauts were kept in isolation following early space explorations to ensure that they had not brought back some pathogen capable of destroying Earth's human population.

A similar situation can exist when a plant is taken to a new environment, where it is exposed to new pathogens. This was demonstrated by the severe pear losses from fire blight in North America (see Chapter 5). Plants and their associated pathogens were carried all over the world without restriction in the mid-1800s. Plant quarantine laws, first passed in the United States in 1912, could not be created until the basic principles of plant pathology were understood. However, tremendous plant losses could have been prevented with earlier restrictions. Even today, most people who travel do not understand the continuing threat of foreign pests and pathogens.

The disease triangle demonstrates that three components must be considered in the study of any plant disease: the plant, the pathogen, and the environment. The disease triangle also warns that significant change in any one of the components should be investigated. The unusual weather conditions of the 1840s certainly contributed to the devastation of the potato crops in Ireland but only because *P. infestans* had arrived from the Western Hemisphere. Cool, rainy weather was not a serious threat before the pathogen was introduced.

From the tragedy, however, grew a new perspective of the natural world. Disease was no longer viewed as a dark, magical force but as a biological activity involving tiny pathogens that invaded the plants and caused their decay. The new science of plant pathology involved the study of plants and their pathogens. Scientists could use Koch's postulates to identify the pathogen responsible for a particular disease. Crop protection could move from ancient religious rituals to a scientifically based understanding of the means by which farmers and growers can favor crop growth and discourage infection by pathogens. People could devise ways to prevent pathogens from infecting plants, using chemicals, resistant plants, or environmental management.

It soon became clear that every crop is subject to many different diseases. Most of the more than 75,000 described species of fungi are incapable of infecting living plants, and even parasitic fungi are usually able to infect only a few closely related species of plants. What is the basis of this specificity, and how are plants able to defend themselves from the multitude of microorganisms in their environment? These fascinating questions have not yet been fully answered, but current research continues to work on answering them, as well as on solving the applied problems of crop protection.

The human component of plant pathology was probably never as clear as in Ireland in the "Hungry Forties." Many foods are important economic commodities, and politicians become involved in their production and distribution. In the 1840s, politics played a role in the Irish potato famine, and today, politics can still determine who will be well fed and who will go hungry. As the human population increases, food production needs increase. At the same time, the amount of arable land decreases each year, as does the number of living species of plants. The centers of origin of the major food crops, which are an important source of the genetic diversity needed for continued plant protection, lie mostly in the areas of the world with the greatest population pressure. An important principle of plant pathology, learned since the potato famine, is that we will never eliminate the pathogens of plants. That means we must learn to coexist with them and still produce and efficiently distribute enough food for an ever-growing human population.

The study of the epidemic caused by *P. infestans* also introduces the important points of a disease cycle. To understand any plant disease, we must know how the pathogen survives unfavorable conditions (such as winter or hot, dry weather), how it is dispersed to plants, and how environmental conditions affect the various stages of its life cycle. The study of plant diseases also requires some knowledge of the biology of the pathogens. Since fungi cause many important plant diseases, the next chapter introduces the fungi and their role as plant pathogens, including *P. infestans* and some other organisms that are now considered funguslike organisms.

CHAPTER 2

The Most Important Plant Pathogens: Fungi and Oomycetes

Human beings constantly try to categorize the components of the world around them. By relating new objects to familiar objects, we can more easily discern their similarities and differences.

Early biologists tried to divide the living organisms of the world into plants and animals. At first, this attempt at categorization presented few problems, because most large organisms fell easily into one of these two groups. Most plants are green and sedentary, and their cells possess a cell wall in addition to the cell membrane that encloses animal cells. The mushrooms and other fungi, so commonly seen on decaying vegetation in fields and forests, presented something of a problem to the early biologists. These organisms do not move, and their cells have cell walls, but they do not possess the green pigments for photosynthesis so characteristic of plants. In fact, their whole biology presented an array of mysteries, such as the sudden appearance of mushrooms after rain and the growth of fungi in dark recesses, where plants cannot normally live.

Fungi were certainly not animals, so botanists adopted this confusing group. The study of fungi has generally remained within the realm of botanical scientists, frequently ignored by zoologists and even many microbiologists. Molecular studies now separate the kingdom Fungi from true plants, and as will be discussed shortly, current evidence suggests that fungi are actually more closely related to animals than plants (Figure 2.1). In addition, certain organisms that had been called fungi for decades—such as the potato late blight pathogen, *Phytophthora infestans*—have been reclassified into other groups of organisms.

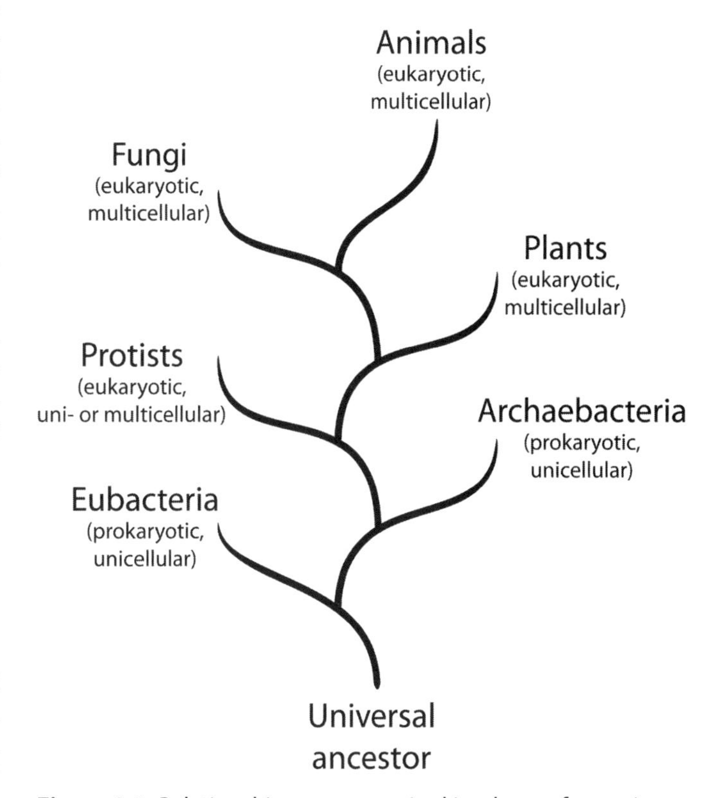

Figure 2.1. Relationships among major kingdoms of organisms.

Characteristics of Fungi

How Do You Know a Fungus When You See One?

Most fungi consist of a mass of living filaments called hyphae (Figures 2.2 and 2.3). A single filament is a hypha, a term originating from the Greek word for "web." Many scientific terms are written according to the linguistic rules of Latin, including the means of making a word plural. Thus, there is a single hypha, but there are several hyphae. The word fungus (plural: fungi) also is from Latin. The mass of hyphae that comprises the vegetative, or nonreproductive, part of a fungus is called the mycelium, using the Greek root *mykes,* which refers to fungi. Similarly, the study of fungi is called mycology. The words hyphae and mycelium are often used interchangeably.

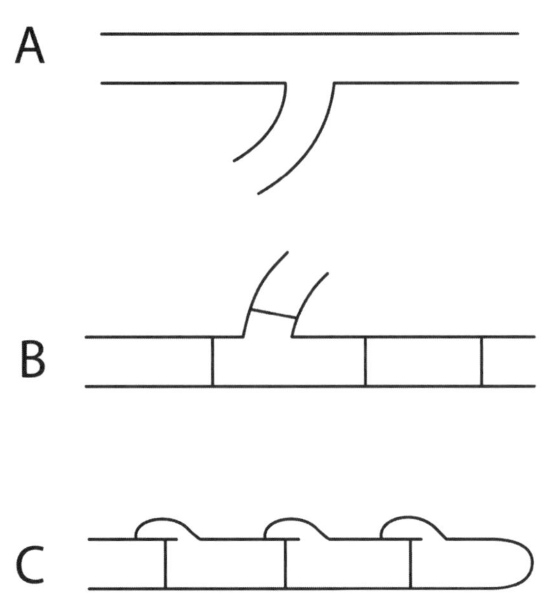

Figure 2.2. Various forms of fungal hyphae. **A,** without septa (crosswalls); **B,** with septa; and **C,** with septa and clamp connections.

The fragile, threadlike hyphae are quite vulnerable to destruction. They are easily damaged by physical pressure, temperature extremes, and desiccation. Hyphae also are not suitable for dispersal in the environment, except perhaps in water or in plant parts less prone to desiccation, such as bulbs, roots, and tubers. If a nutrient source is not available, the mycelium usually dies quickly.

Despite having a simple structure, fungi are highly evolved organisms. Their cellular organization is similar to that of all higher organisms (Figure 2.4), including such features as a membrane-enclosed nucleus containing organized chromosomes and organelles, called mitochondria, for respiration. Such organisms are called eukaryotes (from

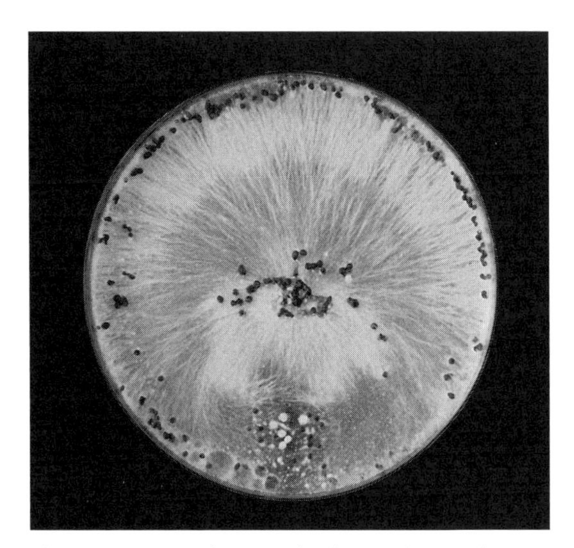

Figure 2.3. Mycelium and sclerotia (survival structures) of a fungus growing on potato dextrose agar.

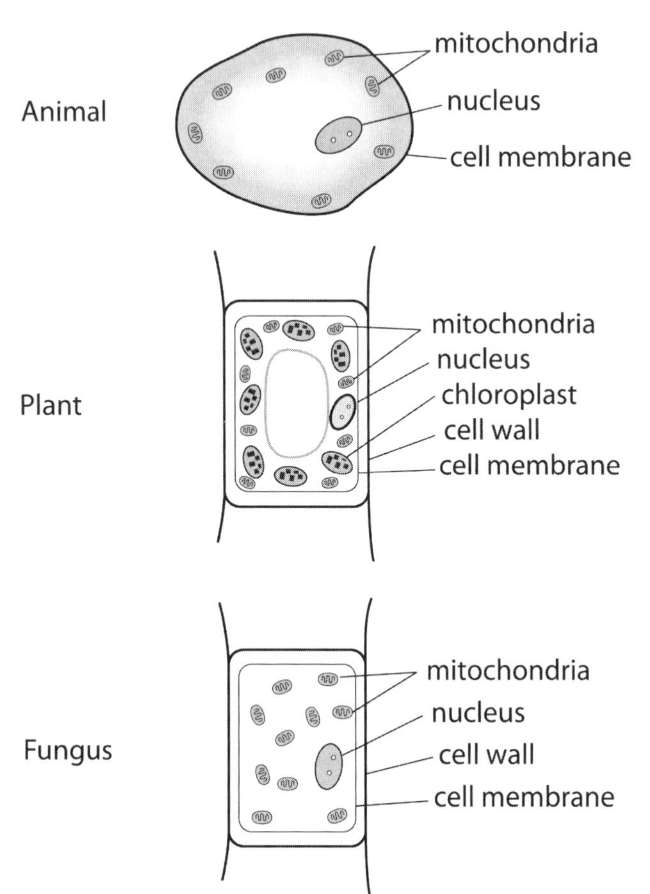

Figure 2.4. Cell structures of various eukaryotic organisms: animal, plant, and fungus.

the Greek *eu* for "true" and *karyon* for "nucleus"), because they have true nuclei. They are unlike the prokaryotes, such as bacteria, which lack nuclei and other membrane-bound organelles. All plants, animals, and fungi are eukaryotes. A fungal cell is surrounded by a cell membrane and an outer cell wall, just as a plant cell is. Plant cell walls are composed primarily of cellulose, however, and fungal cell walls are composed primarily of chitin, a compound found in crab and lobster shells and insect exoskeletons. Animal cells have cell membranes but not cell walls.

All fungi are heterotrophic, which means they require an external source of organic compounds for food. They are different from autotrophs, such as green plants, which make their own food by photosynthesis or chemosynthesis. Simple compounds can be absorbed through the fungal cell wall, but most potential food sources consist of more complex molecules that are too large to be absorbed. Fungi then produce various enzymes that degrade, or break down, large molecules into smaller ones that can be absorbed. Thus, fungi have external digestion, whereas animals have internal digestion.

Fungi grow by adding new cells to their hyphal tips, resulting in outward radial growth of the mycelium. This pattern of growth helps a fungus discover possible food sources. As long as a nutrient source and moisture are available, the hyphae continue to grow. This pattern of growth is unusual compared with that of plants and animals, which produce structures of familiar shapes and sizes. Fungi can be thought of as masses of living filaments that continue to grow outward in search of new food sources. They form various shapes, such as mushrooms, to enhance their chance for survival or to distribute spores.

The fragile mycelium is subject to desiccation and starvation if environmental conditions become unsuitable, so the fungus must be prepared to survive inhospitable circumstances or be able to disperse to a more suitable place. Fungi produce various reproductive structures that function as dispersal and survival devices. Many fungi commonly produce various types of spores to serve these functions. A spore is a package of fungus that contains at least one nucleus and the cellular contents necessary to germinate into a new, growing hypha in a new location. Spores may have thickened walls for survival and may be dispersed in various ways, including by animal vectors, water, and wind. They may be thought of as the "seeds" of the kingdom Fungi.

Growth and Asexual Reproduction: Mitosis of the Eukaryotic Cell

The nucleus in a typical eukaryotic cell contains the genetic information necessary for cellular control. This genetic information is stored in the chromosomes, which are composed of deoxyribonucleic acid (DNA). The structure of DNA and the genetic code embedded within it are common to all living organisms. These topics are discussed in Chapter 6. Cells must have processes to allow this genetic information to be passed on to new cells during growth and to succeeding generations during reproduction. Whether in a human being or a fungus, these processes are the same. At the cellular and molecular levels, we can see the biological links that connect all forms of life, despite vast differences in size and shape.

The study of plant pathology allows us to especially appreciate these connections, because the relationship between a pathogen and a host is a dynamic genetic interaction. Genetic changes in the host plant correspond to genetic changes in the pathogen, as an ever-changing balance is pursued. Effective agricultural practices require a clear understanding of the interactions of the genetics and life cycles of hosts and pathogens. Plant pathology studies have demonstrated how similar these functions are in all life forms.

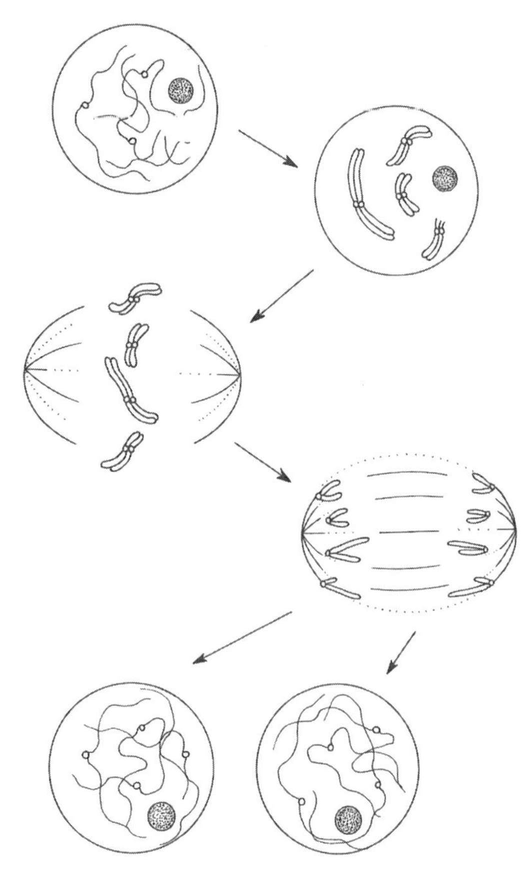

Figure 2.5. Mitosis: production of two identical nuclei for growth or asexual spore production.

Each time a cell divides, the DNA in the nucleus must be replicated and divided between the resulting "daughter" cells, so that both cells receive an identical set of the genetic "blueprints" necessary for cellular control. The carefully controlled replication and subsequent division of the nuclear material of a cell is called mitosis (Figure 2.5). When cells prepare to divide, the chromosomes double and become visible. The chromosome pairs become arranged along an equatorial plane and are then drawn apart, so that identical sets of chromosomes collect at both ends of the dividing cell. New nuclear envelopes surround the two sets of chromosomes, and mitosis is complete.

The nuclei of all eukaryotes, including fungi, undergo mitosis during cell division. This provides a nucleus for each new cell produced during normal growth and during production of asexual fungal spores. The daughter cells produced during mitosis are genetically identical to the mother cell, resulting in genetic stability both within a single organism and across generations when asexual reproduction occurs. The new hyphae that grow from asexual spores are essentially clones of the mycelium that produced them.

Genetic Recombination Through Sexual Reproduction: Meiosis of the Eukaryotic Cell

Despite the tremendous variety of life forms on Earth, all eukaryotes share the sexual processes by which genetic information is passed from one generation to the next. Maintenance of a species requires genetic stability between generations but enough variability to allow the species to survive in a changing environment. Sexual reproduction serves both of these functions. Genetic information is contributed from two different nuclei (from two different "parents"), and the genes are recombined to produce new, genetically different nuclei. The resulting individuals are quite similar but not genetically identical to the parent organisms.

How does this genetic mixing occur? In sexual reproduction, each parent nucleus contributes one set of chromosomes to the offspring nucleus. There are actually two steps in this process (Figure 2.6). In the first step, the two nuclei come together in the same cell (plasmogamy). In the second step, the two nuclei fuse

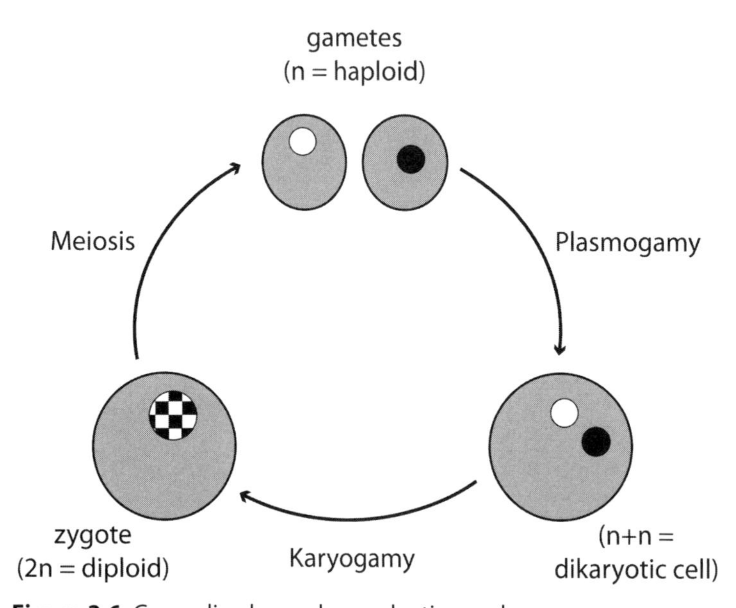

Figure 2.6. Generalized sexual reproduction cycle.

(karyogamy). The resulting nucleus consists of pairs of chromosomes and is therefore termed diploid. (The parent nuclei that contain only one of each kind of chromosome are haploid.) If this process were to continue, the number of chromosomes would double with each succeeding generation. To avoid this problem, an additional division occurs in the diploid nucleus that reduces the nuclei back to the haploid state. In this way, each parent contributes part of its genetic makeup to the offspring, but the final number of chromosomes remains stable for the species.

This two-division process, called meiosis (Figure 2.7), begins in a manner similar to mitosis. The diploid nucleus contains pairs of all the types of chromosomes. All the chromosomes double and divide, as in mitosis, but the process continues through a second division. Four haploid nuclei are produced from the original diploid nucleus, each con-

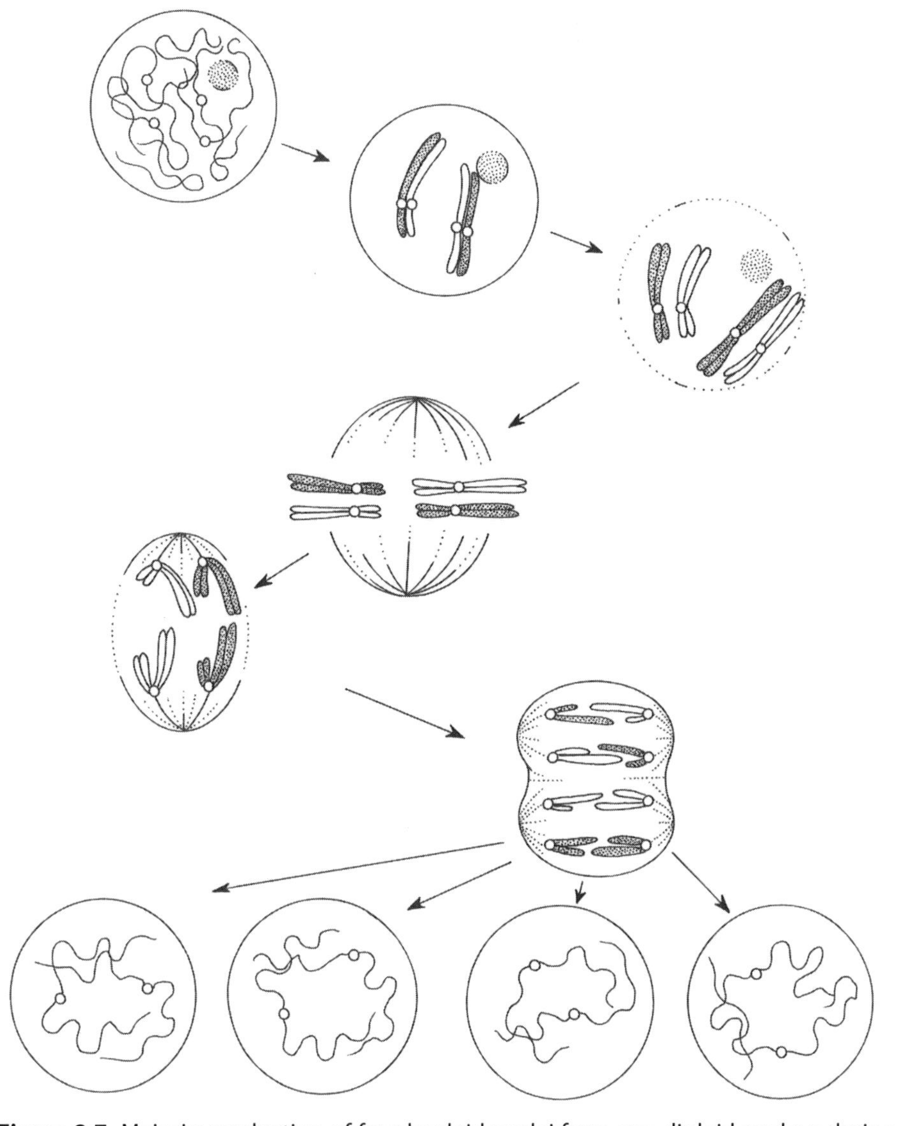

Figure 2.7. Meiosis: production of four haploid nuclei from one diploid nucleus during sexual reproduction, resulting in genetic recombination while keeping the chromosome number stable.

taining only one chromosome of each pair. During meiosis, there is a random distribution of chromosome pairs, so each haploid nucleus contains a different assortment of chromosomes. Each of the four haploid nuclei is genetically different but still contains a complete set of the genetic blueprints for that species. For example, human beings receive a complete set of genetic information from each of their parents, but the combination of genes received from each parent may be different for each individual.

In animals, the products of meiosis are the haploid, single-celled gametes: the sperm and eggs. During fertilization, each sperm and egg contributes one set of chromosomes to the new nucleus. After these gametes fuse (plasmogamy), the nuclei also fuse (karyogamy), and the diploid state is restored. The new diploid nucleus contains one set of chromosomes from each parent. The resulting fertilized egg divides and grows, creating a diploid nucleus for each new cell by mitosis. All the cells in an animal's body are diploid except the haploid eggs and sperm. Thus, sexual reproduction maintains a stable chromosome number in the species but allows a reshuffling of the genetic material to produce variability in the next generation. Each new animal produced by sexual reproduction is genetically similar but not identical to its parents.

The processes of meiosis, plasmogamy, and karyogamy characterize sexual reproduction in all eukaryotic organisms, but the predominance of the diploid body does not. For instance, plants produce two different vegetative bodies: one composed of haploid cells and another composed of diploid cells. In many simple plants, such as mosses, the haploid body predominates. Haploid spores germinate and grow into the familiar small plants found in moist and shady places. The little green plants produce sperm and eggs. After fertilization, a tiny diploid plant grows from the tip of the haploid plant and depends on it for nutrients. Following meiosis, the diploid plant produces haploid spores that begin the cycle again.

The diploid stage predominates in large, vascular plants, as it does in animals. In the male organs of flowering plants, called anthers, meiosis is followed by mitosis to form myriad haploid pollen grains, which may travel on the wind or on an insect to a female part of a flower, called the stigma. A pollen grain grows through a tube down from the stigma to reach the haploid embryo sac of the flower, where a single cell functions as the egg. A cell in the pollen grain produces sperm by mitosis. The egg and sperm nuclei fuse to produce a diploid cell that is protected and nourished in the seed until germination. In this case, all the cells of the new plant contain diploid nuclei. However, some plants have more than two sets of chromosomes and are known as triploids, tetraploids, and even hexaploids. These complexities are of agricultural importance and are discussed in Chapter 6.

Life Cycles of Fungi: Survival and Reproduction

The estimated 1.5 million fungi are a diverse group of organisms with complex life cycles. Like animals and plants, most fungi undergo sexual reproduction. To understand a life cycle, it is necessary to make detailed studies of chromosomes in cells at various stages and to determine chromosome numbers. With that information, it is possible to determine which cells are diploid versus haploid and thus the points at which plasmogamy, karyogamy, and meiosis occur. Conducting such studies may be difficult with fungi, because their chromosomes are often very small and difficult to count accurately. Some important fungi grow only in host tissue, rather than on nutrient media in a laboratory, where they are more easily studied. In addition, many fungi require different mating types, special nutrients, or precise environmental conditions to reproduce sexually. It is known, however, that the nuclei in the vegetative mycelium of fungi are haploid.

Why is knowledge of life cycles and genetic recombination so important? Recall from Chapter 1 discussion of the interaction of *P. infestans* and the potato. The genetic uniformity of the potato crop resulted in the disastrous epidemics in Ireland, and the potato remains vulnerable to late blight today. Early attempts to create resistant potato cultivars using single resistance genes failed because of the pathogen's ability to quickly develop new genetic races that were able to overcome this host resistance. If we wish to protect crops with genetic resistance, we must understand the genetic interaction between pathogens and their hosts.

Many factors determine the success or failure of genetic resistance in a plant. Important factors related to the plant include whether it is an annual (replanted each year) or a perennial (grows back each year), whether it is vegetatively propagated, and what genetic diversity is available to plant breeders. Factors related to the life cycle of the pathogen include the number of generations produced during a growing season, the means by which the pathogen is dispersed to other plants, and the inherent genetic variability of the organism. Many fungi do not commonly reproduce sexually but still find ways to overcome resistance in plants. However, since sexual reproduction is often an important means of genetic variation for fungi, we need to understand the role of sexual reproduction in the pathogen's life cycle if we intend to find ways to manage the disease it causes.

A second important reason to study fungal life cycles is that spores produced during sexual reproduction are sometimes survival structures themselves but are more often produced in or on survival structures that protect the fungus from adverse conditions. Thus, sexual spores often initiate epidemics after a dry or cold period or after the prolonged absence of a susceptible host plant. Asexual spores contain only mitotic nuclei but often serve the important function of rapid multiplication during an epidemic. In temperate climates, they are sometimes called "summer spores."

The fungi are a group of organisms that evolved from a common ancestor. Their evolutionary relationships have been difficult to study, because hyphae decay easily and have not been well preserved in the fossil record. Some of the best fungal fossils are of parasitic species preserved in or on their host plants. The relationships between the major fungal groups have become better understood with modern genetic and molecular studies.

The ability of fungi to be plant pathogens is governed by their biological characteristics. An array of adaptations to host plant species and plant tissues can be found within each of the major fungal groups. For example, leaf pathogens and root rot species can be found in all the major fungal groups, but the conditions under which various species predominate are governed by their basic biology. The following sections introduce two major fungal groups that include important plant pathogens and the group of funguslike organisms to which *P. infestans* belongs.

Science Sidebar

-mycetes versus -mycota

The classification of organisms is in constant flux, because scientists continually learn more about the relationships among organisms. Each organism is given a species name and then grouped with similar organisms in a genus, family, order, class, phylum, and kingdom (see Figure 2.1). The suffix -mycetes indicates the taxonomic rank of class. The terms Ascomycetes, Basidiomycetes, and Oomycetes were valid taxonomic subdivisions until the 1990s, when mycologists changed fungal taxonomy. They did so by raising the divisions among these groups one taxonomic level to that of phylum. The suffix for that rank is -mycota, so the groups were renamed Ascomycota, Basidiomycota, and Oomycota.

Although these are the correct current terms for these three groups of organisms, the informal terms in lowercase letters (ascomycetes, basidiomycetes, oomycetes) are used in many publications and by many instructors and mycophiles. We will use these common but archaic terms informally throughout this book.

Science Sidebar

Ascospore formation

In the early stages of sexual reproduction, male (antheridia) and female (ascogonia) structures are produced. Nuclei migrate from the male to the female (plasmogamy), and a limited dikaryotic mycelium develops, in which each cell contains two different haploid nuclei. The two nuclei divide by mitosis, in tandem, as the hyphae grow. Eventually, the two nuclei fuse in a special terminal cell (karyogamy), and the resulting diploid nucleus undergoes meiosis almost immediately to create four haploid nuclei. These four nuclei often undergo one mitotic division, resulting in the production of eight haploid nuclei. A spore wall forms around each nucleus to produce ascospores.

The yeasts do not have the mycelium of typical fungi. Instead, they have a vegetative structure that is reduced to single cells that bud to produce new cells. Sexual reproduction of yeasts involves the fusion of the haploid nuclei of two single cells followed by meiosis, resulting in an ascus filled with ascospores.

Ascomycota: Sac Fungi

Fungi in the group Ascomycota, less formally called the ascomycetes, spend most of their life cycle with haploid nuclei. The hyphae are usually septate (having crosswalls), but the haploid nuclei and other cellular structures can move through the crosswalls, so it is not possible to identify a single "cell" of a hypha. The common name for this group is "sac fungi," because the sexual ascospores, usually eight, are always produced in a saclike ascus. Ascospores germinate and form the new vegetative mycelium with haploid nuclei. Thus, in the ascomycetes, the haploid state predominates, and the diploid state is limited to the single cell that produces each ascus (plural: asci) and its ascospores (Figure 2.8).

Ascomycetes, like other fungi, also produce asexual spores of various types. The most common are called conidia, a term derived from the Greek word for "dust." Masses of these spores are sometimes seen in bathroom showers or on moldy bread and cheese. Each conidium contains at least one haploid nucleus that was produced by mitosis and enough cytoplasm to grow into a new hypha. No genetic change occurs when conidia are produced.

Many important plant pathogens are ascomycetes, including species that cause powdery mildews and some leaf spot, canker, wilt, and root rot diseases. In addition to plant pathogens, the ascomycete group includes many other well-known fungi. Probably the best-known ascomycetes are the yeasts, which are important in the production of bread and alcoholic beverages (Figure 2.9). Brightly colored, cup-shaped saprophytic fungi on decaying

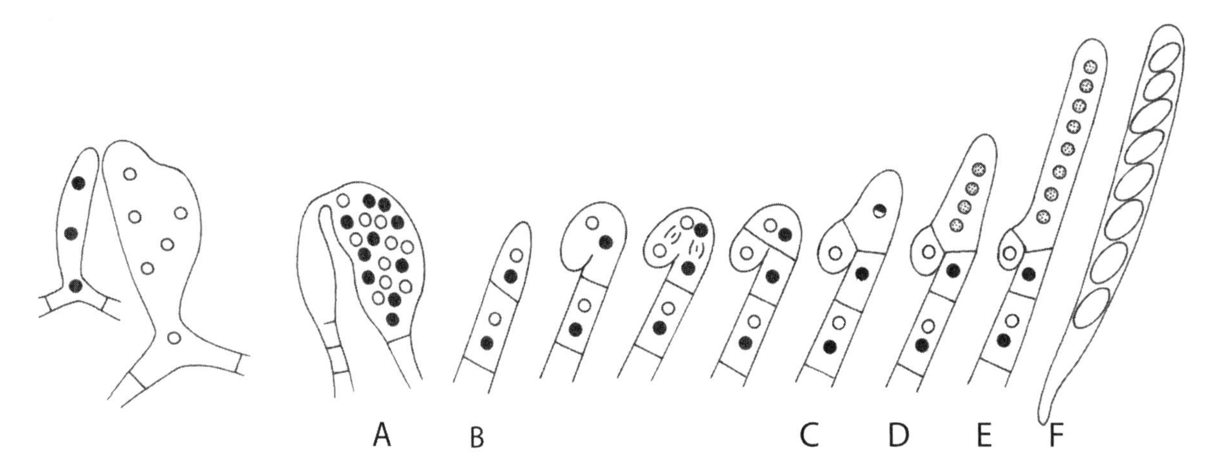

Figure 2.8. Sexual reproduction in the ascomycetes. **A,** plasmogamy; **B,** dikaryotic cells; **C,** karyogamy; **D,** meiosis; **E,** mitosis; **F,** ascospores in an ascus.

wood and the morels of gastronomic fame are the reproductive structures of other familiar ascomycetes. Microscopic examination of their surfaces reveals rows of asci containing ascospores, the characteristic spores of this group of fungi.

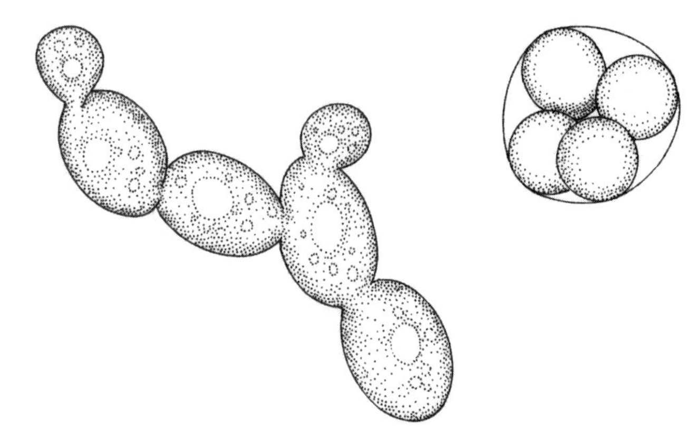

Figure 2.9. Yeast. **Left,** vegetative cells bud to increase their numbers; **Right,** in sexual reproduction, an ascus containing four ascospores is produced.

Basidiomycota: The Mushroom Fungi

A second major fungal group is Basidiomycota, less formally called the basidiomycetes. Fungi in this group spend most of their life cycles with dikaryotic hyphae, in which each cell contains two genetically compatible haploid nuclei. Sexual reproduction occurs later in specialized cells, in which the two nuclei fuse to form a diploid nucleus followed by meiosis to produce four haploid nuclei for the externally produced basidiospores (Figure 2.10). Thus, diploid nuclei have a limited existence in both the ascomycetes and the basidiomycetes. Basidiospores are used for dispersal to new host plants.

The basidiomycetes include most of the large fungi seen in the woods (mushrooms and conks, or "shelf fungi") and some significant plant pathogens of various sizes. For example, the rust fungi and smut fungi are basidiomycetes. Rust fungi are named for the

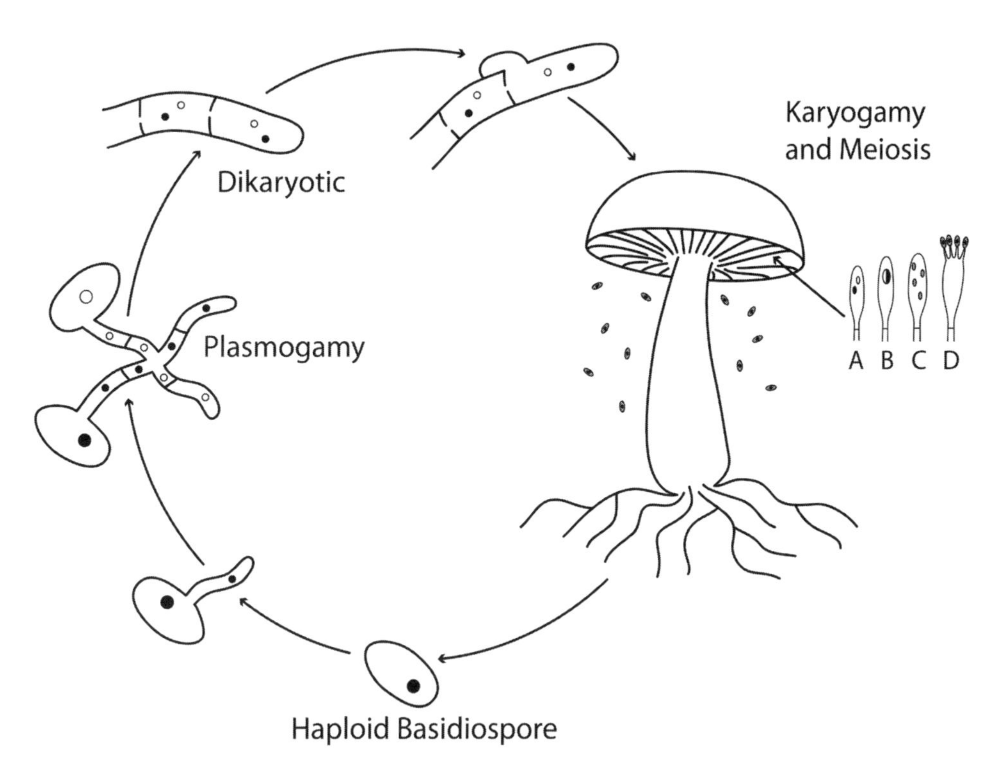

Figure 2.10. Sexual reproduction in the basidiomycetes. **A,** dikaryotic cell; **B,** karyogamy (diploid nucleus); **C,** meiosis; **D,** basidiospores with haploid nuclei.

Science Sidebar

Basidiospore formation

The formation of basidiospores in mushrooms, conks, rust fungi, and smut fungi is quite variable, but it is the feature all basidiomycetes share. Many basidiomycetes produce large fruiting structures (mushrooms, conks) composed of dikaryotic hyphae. Their various gills and pores are lined with specialized cells (basidia), in which karyogamy and meiosis occur to produce haploid nuclei that migrate into the developing basidiospores.

When mature, basidiospores are usually forcibly discharged, fall from the gills and pores, and become dispersed by wind. Basidiospores can be observed as "spore prints" from a mushroom cap placed with its gills or pores down on a piece of paper and covered with a bowl overnight. The haploid basidiospores germinate and produce a limited haploid mycelium, which must fuse with a different, genetically compatible mycelium for development to continue. If this fusion does not occur, the haploid mycelia die. When the two mycelia fuse, plasmogamy brings the nuclei into the same cell, but karyogamy is delayed, so the predominant vegetative (nonreproductive) mycelium is dikaryotic. The hyphae are septate, and each cell of the hyphae contains two genetically different nuclei, which divide in tandem as the hyphae grow. In some cases, clamp connections form at each septum. These specialized structures aid in the proper distribution of the pairs of nuclei during mitosis and hyphal growth.

orange, powdery spores that are often produced during their complex life cycles, which can include up to five spore stages and two unrelated host plants. Rust fungi are discussed in detail in Chapter 11. Smut fungi are named for the black, powdery spores that they produce. They are discussed in more detail in Chapter 10. Many mushrooms found near trees are actually beneficial symbionts called mycorrhizae, but other mushrooms and conks that grow on the trunks of dying trees are the fruiting bodies of wood decay or root rot pathogens. Mushrooms and conks are discussed in more detail in Chapter 12.

Oomycota: The Funguslike "Water Molds"

The group Oomycota, also called the oomycetes, consists of more than 500 species. Because of the importance of water in the life cycles of oomycetes, the common name for this group is "water molds." Water molds can easily be found in almost any surface water. A simple means of observing them is to float a popped kernel of popcorn in a cup of pond water. In a few days, a growth of white mycelium will become visible, and examination with a microscope will reveal hyphae, sporangia, zoospores, and perhaps even oospores. Some water molds are pathogens of aquatic animals, including fish and crayfish, and cause significant losses in aquaculture systems, such as pond-grown salmon. This group also includes many important plant pathogens.

The oomycetes used to be considered fungi, but they have long been recognized as different from true fungi. For example, oomycetes have diploid nuclei in their hyphae, and their cell walls do not contain chitin. Unlike the septate hyphae of ascomycetes and basidiomycetes, the filamentous mycelium of oomycetes is multinucleate. Each individual hypha is an undivided tube of cytoplasm containing a number of nuclei. Meiosis typically occurs in the reproductive structures just before fertilization.

Genetic studies have shown that oomycetes are most closely related to brown and golden algae and diatoms. Some scientists have placed them in the kingdom Stramenopila. The production of motile zoospores is a prominent characteristic of this group but is not a characteristic of true fungi. Oomycetes are similar to true fungi in the production of a mycelium and the production of spores. Thus, in terms of their adaptations as plant pathogens and how we manage the diseases they cause, they are more similar to the fungi than to the other biotic pathogens of plants: bacteria, nematodes, and viruses.

Chapter 1 addressed some details of the life cycle of *P. infestans* in relation to the late blight epidemics that it can cause. Zoospores are produced from sporangia, or sporangia may germinate and infect directly, bypassing the zoospore stage. Germination and infection by both sporangia and zoospores require the presence of water. Water must be absorbed by the sporangium for germination, and zoospores require water for movement and infection of the host plant.

Clearly, reproduction and dispersal with sporangia is rapid and effective in a late blight epidemic, but this means of reproduction has some limitations. When sporangia are produced from the mycelium of *P. infestans,* reproduction is asexual. The sporangia simply enclose the nuclei created during normal nuclear mitosis. The nuclei of the vegetative mycelium, the sporangia, and the zoospores are all diploid. Because these structures are not products of sexual reproduction, they have no opportunity for genetic recombination, and variability is thus limited. Another limitation is that these asexual structures are not involved in long-term survival of the pathogen.

Sexual Reproduction in Oomycetes

Is there sexual reproduction in *P. infestans?* In the 1800s, this question puzzled botanist Anton deBary, who was familiar with other oomycetes and their sexual reproduction (Figure 2.11). He expected to find the thick-walled, diploid oospores characteristic of the oomycetes. Yet he never found oospores of *P. infestans* in his studies.

These oospores were finally found in nature in the Toluca Valley of Mexico during the 1950s. Unlike many oomycetes, *P. infestans* requires two different mating types for sexual reproduction. Both types are commonly found in the Toluca Valley, which explains why oospores are regularly produced there. The center of diversity of *P. infestans* appears to be in Mexico, just as the center of diversity of the potato plant is in the Andes Mountains of South America. Apparently, only one mating type of *P. infestans* was transported to the

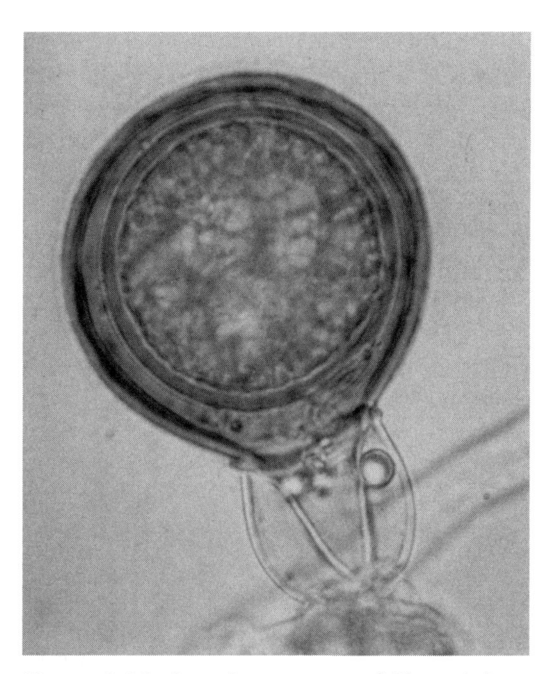

Figure 2.11. Sexual structures of *Phytophthora* species. The collarlike antheridium (male) is below the round oogonium (female), which contains the thick-walled oospore (light micrograph).

Science Sidebar

Oospore formation

The mycelium of an oomycete produces male and female structures, termed antheridia and oogonia, respectively. The structures were named by botanists who observed that their functions were analogous to those of the anthers and ovules of flowers. Meiosis occurs in the antheridia and oogonia. A haploid nucleus then travels from the antheridium into the oogonium, which also contains a haploid nucleus, and karyogamy restores the diploid state. The resulting diploid oospore has a thick wall that resists adverse environmental conditions. The term oospore reflects the egglike appearance of the spore, which gives this group of organisms the name "egg fungi," or oomycetes.

United States and Europe in the 1840s, which means oospores could not have been produced in these newly infested areas.

Without oospores, *P. infestans* cannot survive in the soil during temperate winters. Farmers can take advantage of this weakness in the life cycle of *P. infestans* by practicing careful sanitation in their fields. Infected tubers are destroyed, fields are carefully cleaned of volunteer plants from tubers missed in the previous harvest, and tuber pieces for planting are carefully inspected to prevent introduction of the pathogen to the field along with the "seed." Even with these precautions, *P. infestans* can sometimes escape detection. It is a threat to crops each year because of its explosive ability to multiply rapidly during favorable weather conditions.

As described in Chapter 1, both mating types were discovered in Europe and North America during the 1980s and 1990s. Since then, the second mating type has been detected around the world. The impact of oospore production outside Mexico is still undetermined, however, since relatively little time has passed. In many places, the second mating type seems to have displaced the first mating type. Today, only the second mating type is usually detected in the United States, as it was during the 2009 epidemic that began on tomato plants in home gardens.

What are the possible long-term consequences of the introduction of the second mating type of *P. infestans?* First, the sexual spores may allow the pathogen to survive in the soil outside tuber tissue. Thus, sanitary practices to remove infected tubers may become less effective. The pathogen may be able to survive in the soil for years as thick-walled oospores waiting to attack the next potato or tomato crop. In addition, the genetic recombination that occurs during sexual reproduction may increase the genetic diversity of *P. infestans.* One important result may be changes in the ability of the pathogen to infect potatoes. Another concern is that the pathogen may become resistant to the new systemic fungicides even more quickly than it does now, an important problem discussed in Chapter 8.

Even without sexual reproduction, *P. infestans* has exhibited considerable genetic variation. How is this possible? Sexual reproduction is not the sole source of genetic variability. Mistakes occur during the replication of DNA that change the genetic information in the chromosomes. These mistakes, called mutations, are relatively rare. They are estimated to occur in any particular gene at a rate of once in every 200,000 to 1,000,000 cell divisions. Despite their low frequency, mutations are an important and basic source of genetic change in the evolution of all organisms. In an organism that produces many millions of sporangia every few days during an epidemic, mutation can be a significant source of genetic variation. The tremendous reproductive capacity of microscopic pathogens greatly increases their ability to change genetically, relative to the much slower rate of reproduction in their host plants. Thus, even without sexual reproduction, *P. infestans* has been able to overcome specific resistance genes in potato plants after only a few growing seasons.

The Impact of Oomycetes Today

Clearly, the battle with late blight is by no means over, and in addition, other oomycetes are threatening other plant species. *Phytophthora ramorum* was discovered in Germany and the Netherlands in 1993 and has been found in 15 more countries since then.

In the United States, a new disease caused by this pathogen was noticed in Marin County, California, in 1995, and it subsequently spread to additional counties in the northern coastal area of the state (Figure 2.12). Because the disease was initially found mostly in oak trees, it was named sudden oak death. It is now commonly called ramorum blight, because the pathogen causes diseases in more than 70 species of plants. The hosts of *P. ramorum* include hardwoods, softwoods (coast redwood), landscape plants (camellia and rhododendron), and some herbaceous plants.

P. ramorum thrives in cool, wet conditions and survives primarily in the form of thick-walled asexual spores. The pathogen can be spread in windblown rain, irrigation water, soil, or potting mix, and it can infect the trunk, branches, or leaves of a tree. It was originally a sexually reproducing species but now exists as three asexual lines that are unable to produce oospores. The movement of nursery stock presents the greatest risk for long-distance travel of the pathogen. Because this pathogen threatens plants in many parts of the United States, government inspection of host plants is required before shipping them across state lines in California, Oregon, and Washington. Eradication of the pathogen from currently infested nurseries also is an important goal.

Many other *Phytophthora* species are dangerous plant pathogens. Some have narrow host ranges, whereas others are pathogenic to numerous plant species. Many of the pathogens are difficult to manage because they attack roots, and no means are available for adequately protecting underground plant parts. *Phytophthora* species also infect plants in the so-called splash zone, when sporangia are carried to an infection site in a splashing drop from falling rain or an overhead irrigation system. It is even a challenge to protect host plants from the airborne sporangia of species such as *P. infestans* and *P. ramorum.*

Another important genus of oomycetes is *Pythium.* The life cycles and appearances of species in this genus are so similar to those of *Phytophthora* species that some taxonomists

Figure 2.12. Sudden oak death, or ramorum blight. A view of a mixed evergreen forest in Humboldt County, California. Some of the trees are dying, and others have died rapidly with the leaves still on the trees. Affected species include coast live oak, canyon live oak, and tanoak.

would like to see the genera combined. Nearly every outdoor soil contains some species of the genus *Pythium.* Most species can survive equally well as saprophytes or oospores.

In cold, poorly drained soils, most seeds germinate slowly. Under these conditions, *Pythium* species can be an important cause of damping-off—a term used to describe the deaths of seedlings at various stages of growth. Germinating seedlings exude nutrients that attract the zoospores. Either the infected seeds rot in the soil, or young plants topple over at the soil line. Usually, *Pythium* species infect only slowly growing young seedlings or plants that are wounded.

Because *Pythium* species are ubiquitous, natural soils should be heat treated before being used to pot houseplants or garden seedlings. In heavy soils and poorly drained gardens, raised beds improve drainage and make plants less vulnerable to attack by these water molds. Allowing soils to warm in the spring before planting seedlings also hastens germination and reduces losses to *Pythium* species. Many seeds are coated with a fungicide to provide short-term protection for germinating seedlings. Treated seed is easily detectable, because it is often colored bright red or purple to make sure it is not mistaken for food and eaten. Usually, plants become less susceptible to attack as they mature. *Pythium* species can potentially cause damping-off of nearly every species of plant. Thus, genetic resistance is less likely to be a suitable management strategy than careful attention to environmental conditions at planting time.

Downy mildews are another important group of diseases caused by oomycetes. The life cycles of most downy mildew pathogens, like those of other oomycetes, involve sporangia that produce zoospores and sexual oospores for survival. In contrast to most of the previously discussed soilborne oomycetes, downy mildews infect aboveground plant parts and are well adapted for air dispersal. The common name for these oomycetes reflects the downy masses of hyphae and sporangia that appear on the undersides of infected leaves during wet weather. The sporangia are produced on treelike, branched hyphae, where they can easily become airborne on a passing breeze.

Although the sporangia of most downy mildew pathogens produce zoospores, some form asexual spores that never produce zoospores. This is considered an advantageous adaptation to an aboveground environment, because the organism no longer depends on a zoospore stage, which requires water for locomotion and infection. Regardless of whether zoospores are produced, the sporangia still require water for germination and are quite susceptible to desiccation during dispersal. Thus, downy mildews, like other diseases caused by oomycetes, are most severe during wet weather.

Unlike *Phytophthora* and *Pythium* species, downy mildew pathogens are obligate parasites. This means they are unable to exist as saprophytes and rely completely on oospores for survival in the absence of living host plant tissue. As the host tissue dies, oospores form, allowing the downy mildew pathogens to survive in plant debris and soil until new host tissue becomes available.

Downy mildew pathogens are host specific. Thus, one species causes disease only in grapes, another only in tobacco, another only in spinach, and so on. Many downy mildew pathogens cause diseases of closely related plants in one botanical family. Some of these diseases have had important economic impact. For example, hop production in the northeastern United States was virtually eliminated by downy mildew caused by *Pseudoperonospora humuli,* which appeared in 1928. Hop production is now predominantly found in the dry western states. In 1979, Cuba lost most of its tobacco crop to blue mold, a downy mildew caused by *Peronospora tabacini.* The economic hardship that resulted contributed to the emigration of many Cubans to the United States in the "Freedom Flotilla" of 1980.

Yet another downy mildew disease led to the demonstration of the first effective foliar fungicide in France in 1885. At that time, French vineyards were being decimated by a downy mildew disease caused by *Plasmopara viticola* (Figure 2.13). Where had this new plague come from? The scenario was a familiar one: the movement of plants and therefore pathogens by people. This downy mildew disease was actually the third significant disease or insect problem to affect European grapevines during the 1800s, and all of them were brought to Europe from North America.

The first new disease appeared in the French vineyards around 1845 and was caused by a powdery mildew fungus. Powdery mildews are caused by ascomycetes that grow profusely on the surfaces of plants. These whitish fungi remain exposed to the environment and are therefore susceptible to the toxic effects of sulfur when it is dusted on leaves. By 1854, the powdery mildew epidemic had reduced French wine production by approximately 80%, but after extensive applications of sulfur, production began to rise again.

An alternative to this chemical management strategy was to select grape cultivars resistant to powdery mildew from U.S. grapes. Unfortunately, a second devastating pest was inadvertently brought into France along with these plants around 1860. The new pest was an aphidlike insect of the genus *Phylloxera* that lived and fed underground on the grape roots. It had not been an important problem in the United States. The U.S. grapes had become resistant to this native pest, because the two had coevolved over many years. However, the insect was new to the European grapes, which had no resistance to it, and disaster struck the French vineyards again. To solve this problem, most of the vineyards were replanted with plants made by grafting *Phylloxera*-resistant North American rootstocks to shoots from French grape vines. The grafted plants produced the desirable French grapes, while the roots tolerated the insects.

Figure 2.13. Downy mildew of grape (*Plasmopara viticola*) infects foliage and destroys berries. The downy appearance of the hyphae and sporangia give the fungus its common name.

During the importation of these rootstocks, the downy mildew pathogen arrived. It caused the third and most devastating grape plague around 1878. The thick-walled sexual oospores were able to survive in plant debris and soil. Sporangia were produced from the oospores and splashed onto the foliage. Zoospores produced in the sporangia swam out to cause new infections. Shortly after infection, new sporangia were produced through stomata for air dispersal to other plants, causing extensive damage to both leaves and grapes.

Grapevines in France and elsewhere in Europe suffered from these overlapping plagues for nearly 40 years, affected first by powdery mildew, then by *Phylloxera,* and finally succumbing to downy mildew. The monetary loss that resulted from downy mildew amounted to nearly $50 billion, as European wine production was far less in the 1880s and 1890s than it had been in the 1840s.

From this agricultural and economic disaster began the era of chemical pesticides. As the story is told, Pierre-Marie-Alexis Millardet, a French professor of botany, was walking down a lane and observing the grapes infected with downy mildew when he noticed that some grapes were covered with a bluish-white wash. He also noticed that the leaves of these plants were healthy, whereas those of the neighboring plants were badly diseased. When Millardet questioned the farmer to whom the grapes belonged, he was told that the grapes along the road had been sprayed with a mixture of lime and copper sulfate ($CuSO_4$) to discourage pilferers.

This event marked the accidental discovery of what is now called Bordeaux mixture, named for the area of France where the discovery was made. Millardet experimented with the mixture and first applied it to prevent downy mildew on grapes in 1885. The copper ions of the mixture were toxic to the pathogen, and the lime reduced the phytotoxicity of plain copper sulfate. Bordeaux mixture is effective against bacteria, most fungi, and oomycetes, and it is quite inexpensive. Even today, more than 100 years later, it is the most widely used fungicide in the world.

The previous examples of diseases caused by oomycetes illustrate the variety that exists among these organisms, despite the biological similarity of their life cycles. The mycelia, sporangia, and zoospores of oomycetes all contain diploid nuclei. Meiosis occurs in the sexual reproductive structures, and the diploid state is restored with the production of the thick-walled oospore.

All diseases caused by oomycetes are associated with wet environmental conditions. Water is necessary to prevent desiccation of sporangia, to allow movement of zoospores, and to cause germination of oospores, sporangia, and zoospores. Some pathogenic oomycetes are strictly soilborne, causing infections only on underground or splash-zone plant parts. Others, such as the downy mildew pathogens, are airborne and infect aboveground plant parts. Some species have wide host ranges, such as the species of *Pythium* that attack nearly any vulnerable seedling, but they generally exist as soil saprophytes. At the other end of the spectrum are the host-specific downy mildew pathogens, which are obligate parasites incapable of saprophytic existence. The oomycetes are a diverse and fascinating group of organisms.

Conclusions

The examples in this chapter demonstrate that the ability of organisms to be pathogens is always conditioned by their basic biology. If a pathogen is known to belong to a certain fungal group (ascomycetes or basidiomycetes) or funguslike group (oomycetes), then some important facts about its life cycle are immediately known. To better understand the disease it causes, it is necessary to determine the specific adaptations the pathogen has made to its hosts and its environment.

Life cycle patterns provide the basic framework of plant pathology. The individual life cycles of the many thousands of pathogenic fungi and funguslike organisms can be divided into just a few categories, allowing quick discernment of the important aspects of each disease. The search for such patterns is an important component of the science of plant pathology.

At this point, you know what fungi and funguslike organisms are and how they differ from other forms of life. These organisms have developed ways to reproduce both asexually (to maintain genetic stability and to spread quickly) and sexually (to increase variability and to survive adverse conditions). Because they are the most economically important group of plant pathogens, you will read about other fungal diseases in the chapters that follow. As you learn about each new disease, consider the plant disease triangle and the life cycle of the specific pathogen. Consider, as well, how this basic information can be used to devise management strategies to reduce the impact of plant diseases on people around the world.

CHAPTER 3

Coffee and Rubber: Monocultures and Quarantines

When the potato, *Solanum tuberosum,* was transported from its native home in South America to North America and Europe, a new and important food crop began its journey to all parts of the world. Today, the potato is the fourth most important world food crop—behind rice, wheat, and maize (corn)—and the rate of increase in its production is greater than that of any other crop. The potato is often grown in climates similar to the climate of the cool South American highlands, but some cultivars are adapted to warmer climates, thus extending the crop's range to tropical lowlands as well. The potato is grown in each state in the United States and in almost all of the world's countries.

Approximately 200 years after the potato's introduction to Europe, one of its important plant pathogens inadvertently crossed the ocean in infected tubers, causing late blight epidemics and human starvation. Since no one understood the nature of plant disease at that time, no safeguards could be put in place to prevent such an occurrence. *Phytophthora infestans* is now an established pathogen in all potato-growing areas of the world, as are the various other fungi, bacteria, viruses, nematodes, and other pathogens that traveled with the tubers of their host plant.

How to move useful and important plants around the world without transferring dangerous pathogens and pests is one of the great difficulties facing the human race. We know today that foreign pests and pathogens represent a dangerous threat to native plants. If an introduced pathogen is capable of attacking a plant, the initial exposure can be disastrous, because the plant is not likely to have defenses against the attack. Well-known examples of disastrous outcomes are late blight in the mid-19th century and Dutch elm disease and chestnut blight in the 20th century. After years of selection pressure, plants evolve various means of resistance to their attackers, but at the time of first exposure, they have not had the opportunity to coevolve with the pathogen to create a genetic balance. Without well-developed resistance mechanisms, plant losses can be great. We see the same problem with insect pests. In certain parts of the United States, the emerald ash borer, gypsy moth, and Japanese beetle—all introduced insect pests—have caused tremendous plant injury.

The danger of initial exposure to pathogens is a threat to animals and human beings too. When the European colonists first arrived in the Western Hemisphere, they brought with them diseases—chicken pox, measles, mumps, smallpox, and whooping cough—that caused enormous numbers of deaths among the native people (far more than were caused by guns). The Europeans had greater resistance to many of these diseases, because they and generations before them had already been exposed to the pathogens. Today, the world is experiencing the effects of the human immunodeficiency virus (HIV) epidemic, a new virus introduced into the human population that can produce acquired immune deficiency syndrome (AIDS). A small number of people have natural resistance to the virus, but many more are dying or surviving only through the use of demanding and expensive drug regimens.

In the case of human disease, people recognized that keeping sick people separate from healthy people helped to slow the spread of the disease. In fact, the word quarantine is derived from the Latin root for the number 40. The word originated during the 1300s, when people were kept on ships for 40 days to be sure they did not have plague (the "Black Death"). Presumably, they would have died or recovered during this period. More recently, the Apollo astronauts were kept in isolation for 3 weeks after returning from their trips to the moon to make sure they did not bring a "lunar plague" back to Earth. After many space flights and no evidence of a potential threat, isolation is no longer required upon the astronauts' return. Future astronauts who travel to Mars will be quarantined until more is known about potential life on that planet. Today, health departments have the power to quarantine people with certain diseases or to prevent their entry into the country at border stations.

Also vulnerable are the various introduced plant species that have been removed from the selection pressure of pests or pathogens in their native lands. This occurred with the potato in Europe and many other species that have been moved to new continents. During this protected time, the plants usually are selected to be quite genetically uniform to meet the agricultural requirements of farmers, and in the absence of the pest or pathogen, they have no means of selectively maintaining resistance. When a pest or pathogen is then accidentally introduced into the new land, the threat of disaster is great because of reduced resistance and reduced genetic diversity.

As human travel increased and the time required to move a long distance decreased, plants—and, unfortunately, many of their pathogens—were collected and distributed to new lands. In fact, plant collecting was actively pursued in the United States and many other countries, especially in the first half of the 19th century. Many nonnative Western Hemisphere plants had already been introduced into what was to become the United States by Native Americans in their travel and trade with people in the Caribbean and Central and South America.

The first European settlers planted wheat immediately upon their arrival in North America. They also brought with them seeds and cuttings of many other crops, such as apples, cabbage, carrots, melons, onions, oranges, peaches, and pears. When Benjamin Franklin visited Europe beginning in the 1760s, he sent back to the United States many different kinds of seeds. So did Thomas Jefferson when he served as minister to France. In 1790, Secretary of State Jefferson wrote, "The greatest service which can be rendered any country is to add a useful plant to its culture." In 1827, President John Quincy Adams made an official proclamation encouraging U.S. consuls around the world to collect and introduce useful trees and plants. He wrote, "The President is desirous of causing to be introduced into the United States all such trees and plants from other countries not heretofore known in the United States as may give promise, under proper cultivation, of flourishing and becoming useful." The consuls were given instructions on how to pack and ship these living materials, and the Secretary of the Navy ordered ships to carry the collections back to the United States. A number of familiar plants were received, including alfalfa, camphor, cork oak, cotton, sorghum, and tea.

In 1862, the U.S. Department of Agriculture (USDA) was organized and the land-grant college system was established. One of the duties of the new department was to collect, test, and distribute new seeds and plants, and the land-grant colleges helped evaluate and maintain new breeding lines. Since 1898, more than 500,000 plant accessions have been made, and today, the National Plant Germplasm System oversees this effort.

With the birth of the science of plant pathology in the 1840s, people began to understand the role of introduced pathogens in some of the significant epidemics of the time. They also began to see how some traditional agricultural practices make crops more vulnerable to plant diseases. Many of these lessons were accompanied by tremendous economic losses and important changes in the agriculture of the areas in which the epidemics occurred. Major changes in agriculture always have political and economic consequences.

This chapter describes some historical examples of how pathogen movement and agricultural practices led to epidemics. It also addresses how quarantine legislation helps protect plants within political borders against invasion by dangerous pests and pathogens.

The Story of Coffee Rust: Why the British Drink Tea

Coffee is a tropical crop, surpassed only by oil in its value as a world commodity. For centuries, it has been imported into Europe and recognized as economically important to the European countries that ruled tropical colonies. Today, it remains an important crop to the independent nations that emerged from those colonies.

The first coffeehouses were established in the 1500s in Arabia, Egypt, and Turkey. The scientific name of the most important coffee species, *Coffea arabica,* reflects these origins (Figure 3.1). Coffee became a popular drink in Europe in the 1600s, when contaminated drinking water limited people to consuming fermented beverages and those made with boiled water, such as tea and coffee. Tea was a more economical drink, because more cups could be made from a pound of tea leaves than from a pound of coffee beans, but coffee was more popular. In the 1600s, coffeehouses were major social centers in England, and by 1675, more than 3,000 had been established. Each coffeehouse had its own clientele, often based on occupation or political persuasion. In fact, the king of England often called for the coffeehouses to be temporarily shut down, because they were considered centers of political unrest.

The Dutch were the first major European coffee importers, transporting coffee from their colonial plantations in Ceylon (now Sri Lanka), Java, and Sumatra. The small, non-deciduous tree *C. arabica* produces red berries that contain the seeds, or beans, that are roasted and then brewed into a potent caffeinated drink. The trees grow best in cool, humid climates but cannot survive frost and are thus limited to tropical highlands. Like cacao (chocolate) and bananas, this crop is grown only in the tropics but is consumed in great quantities in countries with temperate climates.

Figure 3.1. Coffee (*Coffea arabica*) with the berries that contain the seeds, or "beans," that are harvested, dried, and roasted.

During Napoleon's time, much of the world's coffee-producing region was lost by the Dutch to the British. In 1825, the British began to develop their property in Ceylon, and every suitable piece of land was turned into a coffee plantation. Thousands of farm workers were brought over from southern India to augment the native workforce. By 1870, Ceylon had become the world's greatest producer of coffee. "Java" remains a slang term for coffee, reflecting the time when coffee production was centered in that part of the world. Today, however, most of the world's coffee comes from the tropical Western Hemisphere, although production has recently increased in Southeast Asia, especially Vietnam. Sri Lanka is now well known for its tea production, and having a cup of tea, rather than coffee, is a familiar element of English culture. As with the Irish potato famine, these changes were caused by a plant pathogen but abetted by agricultural practices.

The fungal pathogen responsible for these changes probably arose in southern Ethiopia, the origin of the coffee plant (Figure 3.2). It is a basidiomycete (see Chapter 2) and belongs to a subgroup known as the rust fungi. Some of the most important rust fungi are unique in requiring up to five spore stages and two unrelated plants to complete their life cycles. The spores required for infection of a second host plant species (teliospores and basidiospores) are produced on coffee trees, but so far, no alternate host has been found. If an alternate host is someday discovered, it will probably be where coffee originated in Ethiopia. In coffee rust, the important spore stage is the urediniospore. Urediniospores are called the repeating stage of a rust fungus, because they are the only spore stage that can repeatedly infect the same host species on which they were produced. Other spores in these complex life cycles function to move the pathogens from one host to the next.

In a coffee rust epidemic, the pathogen rapidly infects and reproduces on the foliage of the coffee tree (Figure 3.3). A single, tiny rust pustule on a coffee tree leaf can produce 150,000 urediniospores, and a single leaf can contain hundreds of pustules. When the coffee rust fungus, *Hemileia vastatrix,* reached Ceylon in 1875, nearly 400,000 acres (160,000 hectares) of the island were covered with coffee trees. No effective chemical fungicide was available to protect the foliage, so the fungus was able to colonize the leaves until nearly all the trees had been defoliated. Unlike late blight of the potato, the coffee rust pathogen does not directly kill the tree. Rather, it causes the tree to decline over several years because of lost foliage, until it ultimately dies.

Coffee trees are understory trees in their native environment and are widely spaced. The dense foliage surrounding the trees reduces the probability that a coffee rust spore will find its way from one tree to another. In a plantation, however, there is no blocking foliage of nonhost plants, and coffee plants exist in high density. This creates the perfect en-

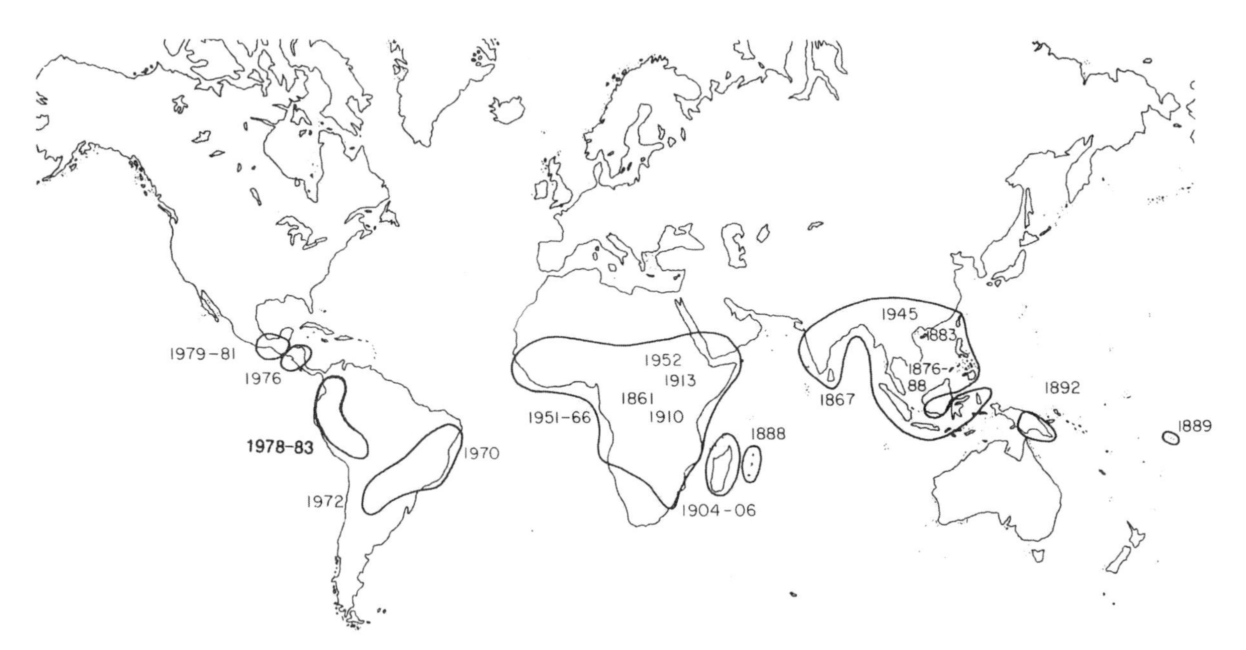

Figure 3.2. World distribution of the coffee rust fungus, with the dates it was first discovered.

vironment for an epidemic, especially in the continuous warmth and moisture of a tropical climate. Thus, it is not surprising that the previously minor coffee rust parasite became a deadly pathogen in the plantations of Ceylon.

The urediniospores produced on the leaves are quite resistant to desiccation, unlike the sporangia of the potato late blight pathogen, *P. infestans,* and are capable of long-distance movement in a viable state. They moved easily through the acres of coffee trees, feasting on the banquet prepared by unsuspecting plantation owners. In 1870, Ceylon exported 100 million pounds (45 million kilograms) of coffee. By 1889, production was down to 5 million pounds (2.3 million kilograms). In less than 20 years, many coffee plantations had been destroyed, and production had essentially ceased.

What did politicians and scientists do to ameliorate this disaster? The British government responded to the coffee growers' pleas for assistance by sending 25-year-old H. Marshall Ward, fresh from studies with Anton deBary at Cambridge, to Ceylon in 1875 (Figure 3.4). Even though Ward failed to save the coffee plantations, he presented the infant sci-

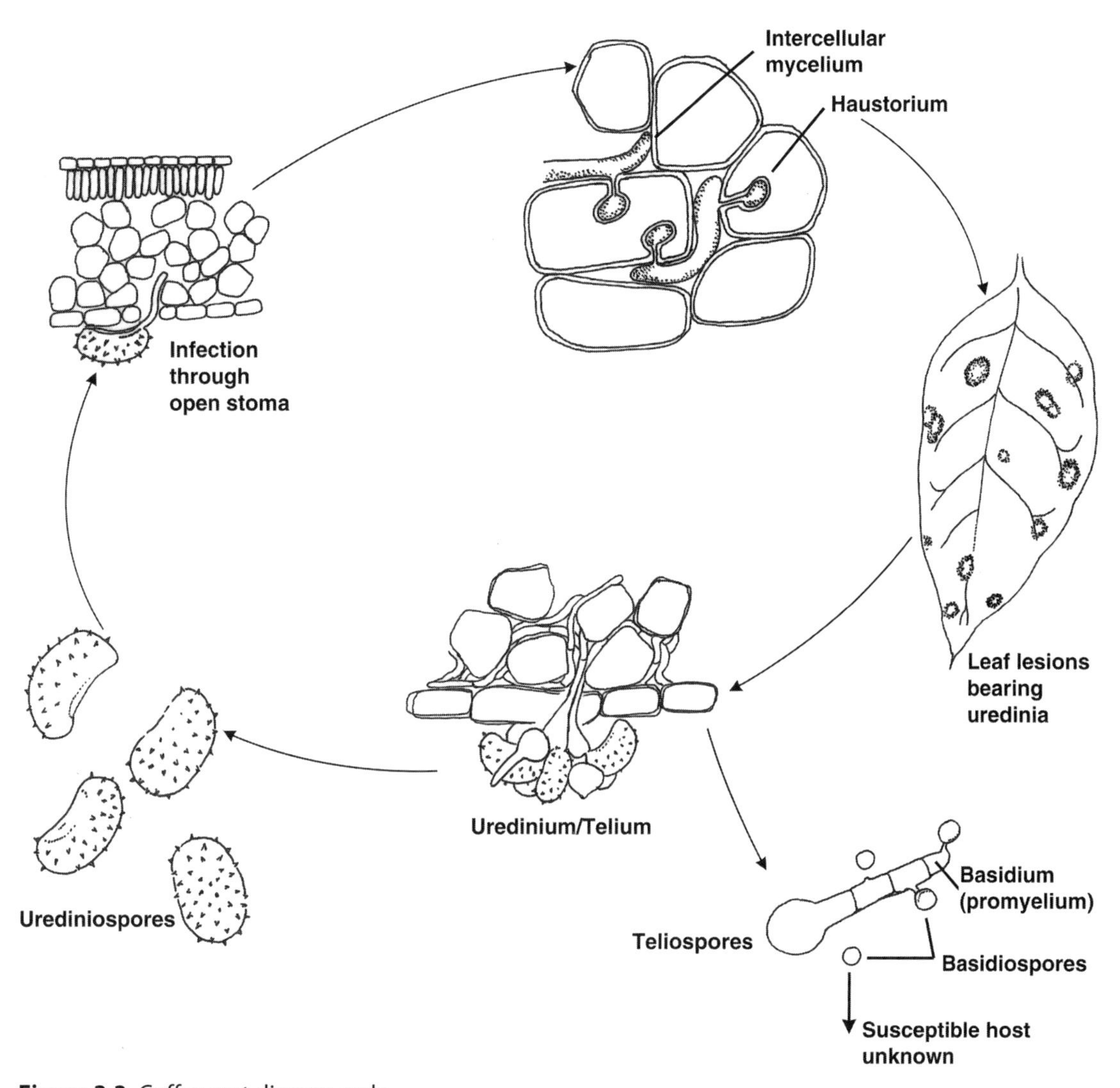

Figure 3.3. Coffee rust disease cycle.

ence of plant pathology with two important concepts that are still fundamental to plant protection.

Ward's studies of the life cycle of the rust fungus convinced him that the germinating spores represented a vulnerable stage for attack. To effectively protect the plant from invasion, Ward recommended applying some newly discovered fungicides as a protective coating on the leaves before the spores arrived. Once infection had occurred, the hyphae inside the leaf tissue were no longer vulnerable to the fungicide. Thus, it was important to anticipate the disease and not wait for its symptoms to appear before spraying. Unfortunately, the sulfur fungicides of that time were neither readily available nor very effective, and the rust epidemic was too well established to save the coffee trees.

Ward also warned about the dangers of monoculture—planting only one species in an area. He recognized that the widespread planting of coffee trees over the island had created the perfect environment for a fungal epidemic. Rust fungi, like downy mildew pathogens, are obligate parasites and require living host tissue for their growth and reproduction. Thus, the rapid spread of the coffee rust was facilitated by the availability of many acres of the host plant. As the coffee trees were dying, however, the plantation owners noticed that the thousand or so acres of tea bushes were still healthy. Ignoring Ward's warning about monoculture, the owners replaced most of the dead coffee trees with tea bushes. By 1875, more than 1 billion tea seedlings had been planted on 300,000 acres (120,000 hectares)—an amazing increase from the acreage planted only a few years earlier.

Figure 3.4. H. Marshall Ward, who was sent to Ceylon (now Sri Lanka) by the British government to save the coffee plantations from the rust epidemic.

It is difficult to imagine the economic chaos that occurred during these years. Like the Irish peasants who fled the potato famine, thousands of Indians left Ceylon, returning home with the help of the Royal Navy. The native workers that remained tried to establish the tea plantations, but the tea bushes took time to reach production size. Plantation owners also had to find a market for all of the tea they hoped to sell. Luckily, no fungus invaded the tea crop immediately, and newly discovered fungicides were soon available to protect the tea from its fungal pathogens.

In an attempt to escape the rust disease, coffee growers moved production to the Western Hemisphere. Coffee had been grown in the Caribbean Islands since the 1700s, but planting quickly spread to the tropical highlands of Brazil, Colombia, and Central America. Today, Brazil and Colombia are major coffee producers. Coffee production became centered in the tropical Americas, because coffee rust was successfully excluded through the use of carefully enforced quarantines.

The quarantines were successful for more than 100 years, but in 1970, coffee rust was discovered in Brazil. How the fungus arrived there is not completely understood. Intercontinental movement of the rust spores from coffee plantations in East Africa is one possibility. Another is that the dustlike spores were carried on airplanes, luggage, people, or plants, which continuously move between the continents. Eradication of the infected trees failed to eliminate the pathogen, which spread throughout the coffee-growing areas, moving into Colombia and the countries of Central America in less than a decade (Figure 3.5). The spread was delayed by careful quarantines between many of the countries, but political unrest and human travel, along with natural dispersal of the spores by wind, allowed the fungus to circumvent the quarantines.

What consequences have resulted from the introduction of such a dangerous pathogen? Frequent fungicide applications are necessary to protect the highly susceptible cultivars of *C. arabica,* which produce the best-quality coffee (Figure 3.6). Chemical inputs are particularly demanding for growers with small planatations, who must purchase fungicides and spraying equipment. In addition, trees must be grown at a lower density to allow complete fungicide coverage of the susceptible foliage. Wider spacing of trees also increases air movement between them, reducing the leaf-drying time. When the foliage dries more quickly, infections are reduced, since rust spores, like almost all fungal spores, require water for germination. Many farmers with small crops prefer to grow coffee in the forest understory—the crop's natural habitat. Doing so reduces dew formation but makes fungicide protection more difficult. Chemical inputs and changes in planting practices increase the costs of production and hence the price of coffee for consumers.

All of the coffee trees growing in the Western Hemisphere were derived from a single, rust-susceptible tree, so genetic variation is minimal. Although the details are sketchy, coffee trees were presented as a royal gift to the French King Louis XIV from the Dutch. A Frenchman stole one of the coffee trees from the royal gardens and planted it in the Caribbean in about 1723, beginning coffee production in the Western Hemisphere. There are rust-resistant cultivars of *C. arabica* (arabica coffee) and other species, such as *C. canephora* (robusta coffee), but the crops are lower in quality (Figure 3.7). Plant breeders often struggle with the problem of combining desirable genetic traits for crop quality with genes for resistance in the same plant. Furthermore, rust fungi are capable of producing many genetically different strains, which can be differentiated by their ability to cause disease on different host cultivars. More than 40 of these so-called races have been detected for *H. vastatrix.*

Figure 3.5. A billboard near the airport of Bogotá, Colombia, warning about the danger of importing the coffee rust fungus before its introduction to South America.

Figure 3.6. Coffee plantations on a steep hillside in Colombia.

Figure 3.7. *Coffea arabica* (left) defoliated by coffee rust, compared to a resistant hybrid of *C. arabica* and *C. canephora* (right).

Finding durable resistance to a pathogen is particularly difficult when the crop is perennial (growing year after year) and growing in a frost-free environment. The pathogen population is not reduced by winter stresses, and replanting with new cultivars is expensive and infrequent. Resistance that is effective against all races of the pathogen remains the long-term goal. Most of the genetic resources of arabica coffee lie in its center of origin: Ethiopia. There is great concern within the coffee industry that many of the areas where coffee trees grow in the wild are threatened by farming and population growth. The Ethiopian government is involved in protecting these resources for the future.

Since the 1970 outbreak of coffee rust, fungicide applications have become routine on most coffee plantations in the Western Hemisphere. Coffee rust has been found in almost all coffee-producing countries, but it has not yet affected the state of Hawaii. Quarantines against coffee rust in that state date back to 1888, when King David Kalakaua declared a quarantine on imported coffee to prevent the introduction of diseases. In many countries, alternative production methods for coffee include shade growing, which provides the understory environment of coffee in its natural habitat. This method of growing coffee may be more sustainable, and it favors biodiversity because it does not involve a single-species plantation. Yields are lower, but coffee grown without fungicides can be sold for higher prices. In addition, a number of groups have established a marketing policy referred to as fair trade, which advocates purchasing coffee directly from the owners of small farms to increase their income and to encourage the use of sustainable production methods.

The Story of South American Leaf Blight of Rubber: Henry Ford's Failure

Whereas the British switched to tea production following the coffee rust epidemic, the Dutch converted many of their dying coffee plantations in Southeast Asia to rubber plantations. Like the potato, the rubber plant had its origins in South America. And like the coffee rust epidemic, the decline of the rubber crop was caused by a fungal pathogen and exacerbated by unsound agricultural practices.

The rubber plant, *Hevea brasiliensis,* produces natural rubber in a latex sap contained in a system of tubes throughout its trunk. More than 12,000 species of plants exude a milky latex, including such familiar species as dandelion, lettuce, milkweed, and poinsettia. The exact function of the latex is not known, but in some cases, it contains compounds that protect against insect pests. The latex of many of these plants contains the polymer compounds known as rubber, and the latex of *H. brasiliensis* is a particularly efficient source, containing 40–50% rubber. In the days before the abundant variety of plastics and other polymers were derived from oil, there were many industrial, medical, and household needs for a durable and flexible waterproof substance. Unfortunately, natural rubber melts when the temperature is high and cracks when the temperature is low. Charles Goodyear spent years trying to solve this problem, and in 1840, he accidentally discovered the process of vulcanization: the addition of sulfur that allows rubber to remain firm but flexible over a wide range of temperatures.

In the 1700s, Europeans considered rubber quite remarkable. The material was named by Joseph Priestley, who, in 1770, observed that it could be used to rub out pencil marks on paper. With the increased prevalence of motorized vehicles in the late 1800s and early 1900s, the demand for rubber greatly increased for use in producing tires. Obtaining enough rubber proved difficult, however, because rubber trees grew in the jungles of South America at a very low density—only a few trees per acre. Rubber was collected by native people, who cut down whole trees and collected the mass of latex that bled from each one. This process became increasingly inefficient, as the collectors were forced to travel farther and farther into the jungle, searching for rubber trees. The early history of the rubber industry is tainted by the cruel treatment and even enslavement of the workers who collected the rubber from different plant species in both South America and Africa.

Seeds of the valuable rubber plant were smuggled out of South America by an Englishman, H. A. Wickham, in the 1870s. Plantation production began in Asia after a tapping system was developed that allowed the latex to be harvested repeatedly from the trees without killing them (Figures 3.8 and 3.9). Plantation production followed in South America, as jungle harvesting became more and more difficult. Following World War I, there was particular pressure to develop rubber plantations in South America to provide secure sources of rubber for the United States during times of war.

In 1929, Henry Ford developed 8,000 acres (3,200 hectares) for rubber production in the rain forest of Brazil (modestly called "Fordlandia"). In addition, Ford attempted to apply his business philosophy from the U.S. auto industry and created a town for the mostly native workers similar to one that might be found in Michigan—the center of the U.S.

Figure 3.8. A rubber plantation in the Philippines.

Figure 3.9. Tapping rubber. The milky latex leaks out of a shallow cut in the bark and collects in a cup.

industry. Some aspects of his approach were successful, such as paying fair wages, but others were not, such as implementing an 8-hour workday in a tropical climate. Ford's enterprise ultimately failed. As had occurred elsewhere, a microscopic fungus killed all the trees that had been planted. The trees planted in a second attempt were killed, as well, by a disease called South American leaf blight, which resulted from infection by the fungus *Microcyclus ulei.*

M. ulei is a member of the ascomycetes, a major fungal group mentioned in Chapter 2. As shown in Figure 3.10, which illustrates the disease cycle, it produces both sexual and asexual spores. The asexual spores, called conidia, are important in the rapid reproduction of the fungus but represent no genetic variation except that produced by mutation. Conidia of *M. ulei* are produced on the surface of infected leaf tissue, where they are easily dispersed by air to other plants. Only a few days after infection, conidia begin to be produced in numbers too large to imagine, continually infecting the tender new tissue of young leaves. Interestingly, *M. ulei* produces a second type of asexual spore (pycnidiospores) in fruiting bodies (pycnidia), but this spore does not appear to play a significant role in the disease cycle.

M. ulei also produces a survival stage through sexual reproduction, resulting in the production of ascospores in asci. The asci are produced in a protective fruiting body that usually consists of dark-colored hyphae. Ascopores are not only able to survive adverse conditions, but they also represent a genetically diverse new generation, because they are a product of meiosis and genetic recombination. The ascospores are forcibly discharged through an opening at the top of the fruiting body by a puffing mechanism caused by a change in humidity, as drier air passes by the liquid-filled asci. This small propulsion is sufficient to send ascospores on a journey to other rubber plantations or to renew an epidemic in the home plantation.

Why did *M. ulei* destroy the South American rubber plantations but not the rubber trees in the jungles? As in the case of coffee rust, the answer lies primarily in the density of the plantings. All plants are parasitized by relatively harmless leaf spot fungi. If a rubber plant in the jungle becomes infected by *M. ulei* and spores are produced, it is highly unlikely that those spores will land on another rubber plant. The low density of the rubber trees, combined with the dense foliage of surrounding nonhost plants, protects rubber plants from frequent infection. As happened with coffee rust, this previously minor parasite became a deadly pathogen in plantations, where many rubber trees were planted closely together.

The plantations of Southeast Asia were and still are just as vulnerable to destruction as those in Fordlandia, but they have been able to exist on a continent without the leaf blight fungus—a mirror image of coffee production. Workers on rubber plantations are trained to recognize the lesions of *M. ulei,* and their continual vigilance, combined with restrictions on travel directly from South America to Southeast Asia, have kept the Asian plantations blight free so far.

In Brazil, only repeated fungicide applications have been effective in protecting the rubber leaves against the continuous supply of spores in a plantation environment. Genetic

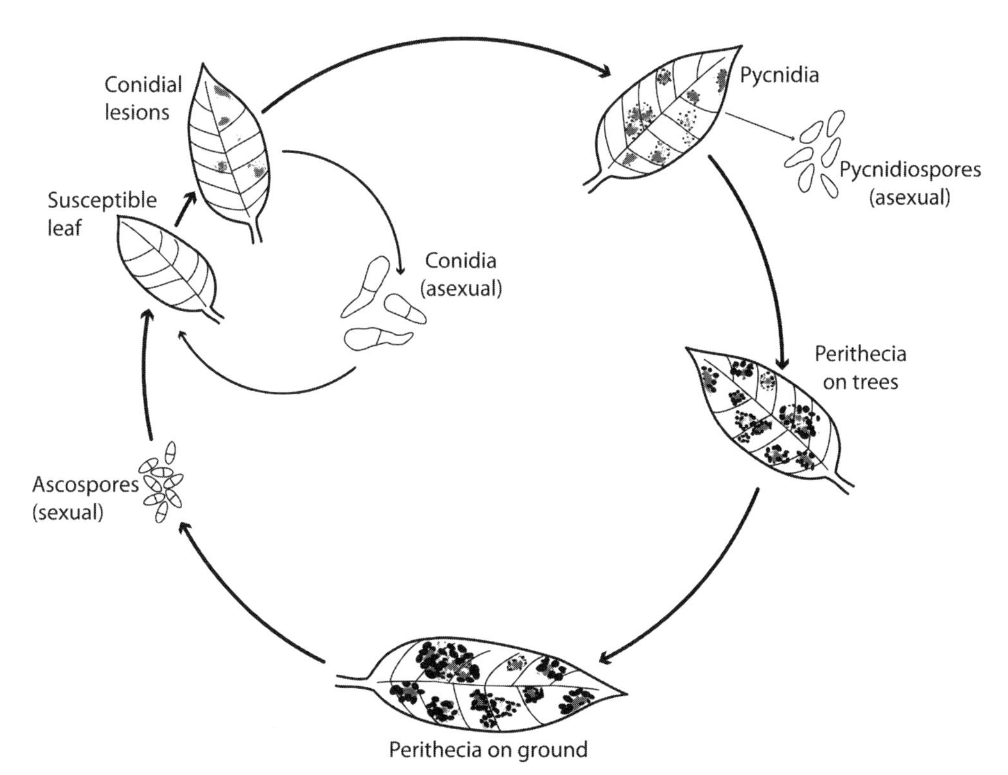

Figure 3.10. Disease cycle of South American leaf blight. Conidia are produced on the surfaces of infected leaves. Later, pycnidia (asexual fruiting bodies) form in the leaves and release pycnidiospores, which have no apparent role in the disease cycle. Still later, perithecia (sexual fruiting bodies) form and release ascospores.

resistance has been investigated, but trees that are resistant to the fungus yield less rubber than those that are susceptible. Disease-resistant tops can be grafted onto high-yielding root stocks, but the process is costly. Despite being the native home of the rubber tree, Brazil is now an importer of natural rubber. Today, 90% of the world's natural rubber is produced by Asian plantations, created from the fungus-free seeds smuggled out of South America. Thus far, only a strict quarantine and many miles of ocean have prevented devastation of the Asian plantations.

The Need for Quarantines: They Buy Us Time

The stories of coffee rust and South American leaf blight of rubber epidemics serve to reemphasize the lessons learned from the potato late blight epidemic. First, each story demonstrates how monoculture agriculture increases the likelihood of a major epidemic developing. This is true for staple crops such as potatoes, luxury crops such as coffee, and commodity crops such as rubber. In today's agriculture, most crops are in a similarly precarious situation. A field or orchard planted with hundreds or thousands of plants of a single species offers a generous invitation to any pathogen. In perennial crop cultures in tropical environments, the danger is even greater, because susceptible plant tissue may be continually available, and winter stresses that reduce the pathogen population are absent. For both rubber and coffee production, relocation to another continent was the most effective initial solution. Rubber remains safe from leaf blight in Asia, but coffee producers must learn to coexist with the rust fungus throughout the world.

A second lesson reemphasized by these two epidemics is that disaster can result when people move crops from their areas of origin to new regions. Potatoes, coffee, and rubber were all moved and cultivated for many decades away from the pathogens that eventually decimated them. In fact, the time of separation caused increased susceptibility in the host plants, both from natural evolution and from human selection of cultivars with desirable characteristics, such as higher yields.

A third lesson to be learned from the coffee rust and South American leaf blight stories is that pathogens that are insignificant in a natural setting, where their hosts are spaced widely apart, can become devastating when those same hosts are planted closely together. Both monoculture and dense planting increase the efficiency of agricultural production. The dangers introduced by these common practices must be recognized.

Another lesson to be learned from the two epidemics is that the use of quarantines provides a valuable tool in the management of plant diseases. The earliest quarantines were put in place out of fear of exposure to diseased people, even before the nature of contagious or infectious disease was understood. Plant quarantines began in Europe in the late 1800s to prevent introduction of the Colorado potato beetle from North America. Three important plant disease epidemics that occurred in the early part of the 20th century led to the first U.S. quarantine legislation in 1912. In two of the epidemics, native plants were killed by introduced pathogens to which they had little resistance. The American chestnut tree, *Castanea dentata,* a dominant species throughout the Appalachian Mountains, was nearly eliminated by a fungus from Asia that causes chestnut blight, and the eastern white pine, *Pinus strobus,* suffered tremendous losses from a fungus introduced from Europe that originated in Asia and causes white pine blister rust. In the third epidemic, a serious bacterial disease, citrus canker, decimated U.S. citrus orchards soon after citrus fruits were introduced into the country. All three diseases have interesting histories and will be discussed in greater detail in later chapters.

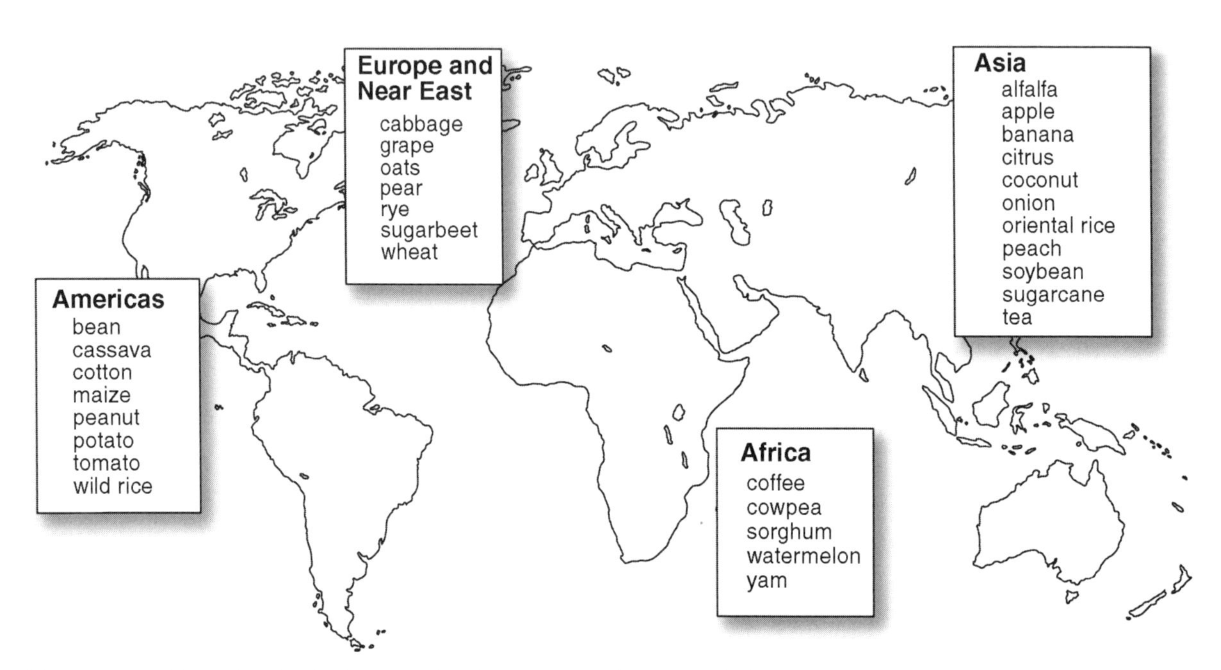

Figure 3.11. Sites of early plant domestication, revealing how people have moved many important crops around the world.

As a result of these epidemics, it became clear that important native and agricultural plants were threatened by open borders and unrestricted importation of plants and plant products, such as fruits, nuts, and lumber. It also became clear that the continued importation of plants and plant products was necessary for agriculture and the economy. Examination of a world map showing the origins of many food crops makes it obvious that U.S. agricultural production would be excessively limited if it were restricted to plants native to this specific country or region (Figure 3.11).

Microscopic pathogens in plant tissue are difficult to detect, unless the infection is so well established that rotted areas, lesions of dead cells, or other symptoms of disease are evident. Early infections and the presence of microscopic spores and other resting structures of pathogens are impossible to detect visually. Some infected plants remain symptomless, and others, such as woody plants infected with certain viruses, may take years to exhibit symptoms. In addition, only a small percentage of the plant material imported in large shipments can be inspected at ports of entry in the United States. Thus, the successful detection of pathogens by government plant inspectors at borders falls short of the level necessary to completely exclude threatening pathogens.

How are pathogens being stopped at the borders when they cannot be detected quickly and economically? One simple solution in place is to forbid the importation of plant material by travelers. Everyone entering the United States is required to dispose of all plant material. Individuals may even be checked by USDA dogs specially trained to detect plant material. Dogs in the "Beagle Brigade" are selected to be docile around the milling crowds of travelers waiting to pass through U.S. Customs stations (Figure 3.12). Although these dogs are amazingly effective, they sometimes make mistakes. For example, a dog trained to sniff out citrus fruit to prevent the reintroduction of the citrus canker bacterium was confused by lemon-lime scented shaving cream. Like all laws, mandatory border quarantines require the cooperation of educated travelers, not just strict law enforcement, to be successful. Each year, uninformed travelers threaten U.S. agriculture by bringing home botanical souvenirs (Figure 3.13).

Figure 3.12. The U.S. Department of Agriculture trains beagles to help detect hidden plant and animal products carried by travelers.

Figure 3.13. Some of the material confiscated at JFK International Airport in New York in a single day.

People working in agriculture may import plants, seeds, and plant products by special permit, according to the restrictions of quarantine legislation. Three main categories have been established for the importation of plants.

The broadest category is labeled "restricted" and includes most plants and seeds. Items in this category must be inspected and possibly subjected to chemical treatment before being released into the United States.

A second category includes many vegetatively propagated crops, such as flowers, fruit trees, and woody ornamentals. These items must be maintained for a postentry period of approximately two years before their general release. During this period, the plants are carefully inspected and subjected to laboratory tests to determine whether they carry pathogens that could threaten plants already present in the United States. Before release, the plants are often maintained in agricultural experiment stations designated as entry portals. The plants are grown in specially designed greenhouses or growth chambers that exclude insects. Any pots, soil, or water from these plants that may be contaminated is sterilized.

A third category includes plant material that is prohibited from import, because it may carry pathogens that pose a serious hazard to U.S. plants. This category is generally the most controversial, because growers sometimes feel they are unable to obtain plant material necessary for competition in world markets. Before material is prohibited, careful surveys are done to determine that the pest or pathogen is not already present in the United States, and the threat is periodically reevaluated. Sometimes, a U.S. crop or plant is barred from another country, because the potential pathogen exists in the United States but not in the importing country. For example, after Karnal bunt of wheat, a seedborne disease, was discovered in the southwestern United States in 1996, China placed an embargo on U.S. wheat imports.

Many changes have occurred since the first U.S. quarantine legislation was passed in 1912. The Federal Plant Pest Act of 1957 includes restrictions on importation of a broad range of plant pests and pathogens. It allows the federal government to restrict interstate movement of plant material and permits destruction of diseased or infested material in emergency situations. The USDA's Animal and Plant Health Inspection Service (APHIS) monitors imported plants and the distribution of pests and pathogens within the United States. Individual states may have their own protective legislation as well.

Quarantines are enacted by federal legislation and enforced by APHIS. For example, because the sudden oak death pathogen (*Phytophthora ramorum*) infects a wide variety of woody plants, strict quarantines were established in 2004 in nurseries in California, Oregon, and Washington to prevent distribution of this pathogen to other parts of the United States. *Plum pox virus,* which infects stone fruits (such as cherries, peaches, plums), was accidentally introduced from Europe and arrived in the United States in 1999, leading to implementation of a quarantine in Pennsylvania. The quarantine was recently lifted after successful eradication of the virus. Soybean rust, caused by *Phakopsora pachyrhizi,* originated in Asia but spread to South America, where a large crop of soybeans is grown, and then arrived in the United States in 2004, despite a national quarantine.

Since the 1990s, grapes and many other California crops have been threatened by the introduction of an efficient vector (the glassy-winged sharpshooter) of a bacterium *(Xylella fastidiosa)* that infects the xylem (water-conducting tissue) of plants. Thus, in some cases, quarantines must focus on vectors as well as pathogens. In general, the globalization of trade in recent years has greatly increased the exposure of crops and native plants to exotic pathogens and their vectors. This is an often unacknowledged result of the global economy.

In countries that are geographically isolated, such as Australia, plants have been biologically isolated from outside organisms and are therefore particularly vulnerable. Arriving travelers may be sprayed with insecticide while still on the airplane and questioned in detail by border officials about carrying plant and animal products and having recently visited farms. Border control is easier in countries bordered by oceans than in land-locked countries. Countries that share borders and climates often have relaxed restrictions with one another, because the natural dispersal of pests and pathogens is inevitable and will circumvent any human attempts to restrict it.

A major limitation of quarantines is that they are based on political borders, which may or may not correspond with the biological borders that determine the spread of pathogens and vectors. The International Plant Protection Convention (IPPC), sponsored by the Food and Agriculture Organization (FAO) of the United Nations, provides member countries with phytosanitary export permits for eight regional organizations. The region that includes the United States is the North American Plant Protection Organization (NAPPO), which maintains a website where quarantine reports can be found. Between neighboring countries, such as the United States and Canada, restrictions are usually confined to soil or machinery that might carry soilborne pests. In many countries, it is difficult to protect borders adequately from invading pathogens. Language differences, prohibitive costs, and lack of adequate training of border guards all contribute to the rapid spread of new pathogens. Because of these problems, more than three-fourths of all countries have no restrictions on plant material carried by travelers.

At best, quarantines result in a delay in the spread of a pathogen. Restrictions are least effective for pathogens that will likely be dispersed naturally by air movement. They are most effective for pathogens that are strictly soilborne or that occur in isolated areas separated by a large expanse of ocean or mountains. Still, a delay can be useful for plant protection. It gives plant pathologists time to determine the best practices for reducing an epidemic and, more importantly, to select disease-resistant cultivars before the pathogen arrives.

Bioterrorism: Movement with Intent to Harm

Agricultural crops are particularly vulnerable to bioterrorism for the same reasons they are vulnerable to exotic pathogens and pests. Namely, most crops are grown in genetically uniform monocultures. In addition, most crops are easily accessible. Large fields are not even fenced, in most cases.

Research into biological warfare was part of some official U.S. government programs before World War II and particularly during the Cold War. Some of the diseases investigated included brown spot of rice, late blight of potato, rice blast, and stem rust of wheat. Plant pathologists studied pathogens as potential agents of biological warfare at Fort Detrick, Maryland. They considered possible ways to distribute spores in foreign territories and ways to defend homeland crops. As part of this program, huge quantities of wheat stem rust spores were produced and stockpiled.

In 1969, President Richard Nixon unilaterally ended the U.S. biological warfare program and ordered that all biological weapons, including the rust spores, be destroyed. In 1972, the international Biological and Toxin Weapons Convention (BTWC) was approved, and today, it is considered a key tool in controlling weapons of mass destruction. In the early 2000s, Saddam Hussein of Iraq was accused of stockpiling such weapons, including stinking smut spores for possible use against Iran's wheat crop.

In the current climate of concern about bioterrorism, the U.S. Department of Homeland Security has provided funds and created directives for improved crop biosecurity. To date, security measures have included improved crop monitoring, enhanced communication among plant disease diagnosticians, and development of emergency response plans. Having additional funding for crop biosecurity has allowed plant diagnosticians to improve their laboratories and communication systems. Regional plant diagnostic networks maintain websites and conduct so-called first-detector education and training for people who work in agriculture and are likely to notice unusual occurrences of plant disease. Many new pathogens have invaded areas of the world by natural dispersal and inadvertently via imported plant materials, packing materials, and soil. Improved monitoring for new diseases and more trained personnel watching for unusual outbreaks will help safeguard U.S. food and fiber production.

Monitoring and preventing the movement of plant pathogens is a global challenge and thus requires international cooperation. Although pathogens can move naturally, human beings have been responsible for most of their long-distance travel. People move plants, pathogens, and vectors to new areas despite legal restrictions and may do so in the future with the intention of inciting devastating plant disease epidemics. As new pathogen strains evolve in one area, members of the agricultural community must be on guard for their arrival in other areas, where the host plants are produced. Those who live in temperate regions also must be aware of diseases found only on tropical crops, because many of these crops, such as coffee and rubber, contribute to the international economy. Consumers would greatly miss these products if they were no longer abundant and affordable.

CHAPTER 4

To Grow a Healthy Plant: Soil, Water, and Air

Plants have evolved in a world of potential pathogens. Some features of their anatomy, such as woody tissues and waxy surfaces, help protect them from these potential killers. Other features, such as openings for air and water movement, increase plants' vulnerability to pathogen attack. Before plants dominated the landscape, the world was a very different place. The atmosphere was anaerobic (without oxygen), and the lack of an ozone layer made life on land quite precarious because of the intense ultraviolet radiation.

Today, we benefit from plant life in many ways. In fact, plants are critical to our existence on Earth. It has been said that we are guests on this planet of green plants, because they provide food (via photosynthesis), oxygen in the atmosphere, and protection of the ozone layer. In these ways, plants make possible all animal life, including our own.

It Begins with Soil

Plant life on land depends on the presence of soil. The soils that dominate the landscape in many parts of the world were created slowly over many millions of years. On the rocky surfaces of land masses, erosion by heating, freezing, glacial ice, rain, and wind broke up the large rocks into small pieces, releasing some minerals. Primitive land plants contributed to the breakdown of the rock, as they colonized the cracks and depressions. The mineral particles became mixed with decaying organic material from animals, microbes, and plants that had once been alive. Thus, every soil is influenced from its origin by the underlying inorganic bedrock and rocky material carried by glaciers and water, as well as the organic matter contributed by the organisms that lived on the surface.

If you slice straight down into most soils in grasslands and forests, you will find a typical profile composed of several layers (Figure 4.1). The relatively thin layer at the top contains dead and decaying organic matter. Stubble and leaves from a previous crop may be present on agricultural lands, or this layer may be plowed into the soil underneath. The next layer is a combination of dead and decaying organic material (10% of the solids) and fine pieces of mineral material (90% of the solids). This layer is where most plant roots are found. Below the root zone is a layer containing much less decayed organic material and mostly small pieces of minerals. The final layer in a soil profile, just above the bedrock, consists of large pieces of the mineral components but little or no organic matter. Our very survival depends mostly on the vulnerable top layers, which contain most of the organic material.

The size and relative proportion of the mineral particles present determine the soil type (Figure 4.2). Mineral soil particles vary in diameter, ranging from sand (0.05–2.00 millimeters) to silt (0.002–0.050 millimeters) to clay (less than 0.002 millimeters). Where sand predominates, the soil contains relatively large spaces between particles, allowing rapid drainage after a rain. In clay soils, the smallest particles predominate, and because the

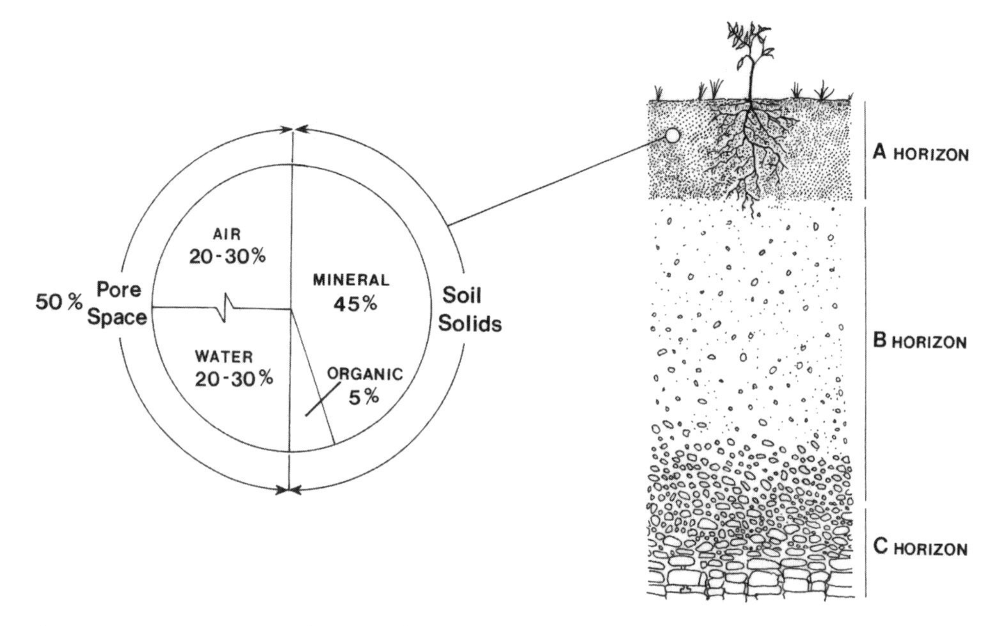

Figure 4.1. Composition of soils. **Left,** silt loam soils, which are good for plant growth, are approximately half pore space and half soil solids. In well-drained soil, the pore space is filled half with water and half with air. **Right,** the three primary horizons, or soil layers, in a typical soil. The **A** horizon contains relatively high amounts of organic matter and plant roots. The **B** horizon contains less organic matter. The **C** horizon is composed primarily of inorganic particles of the underlying bedrock.

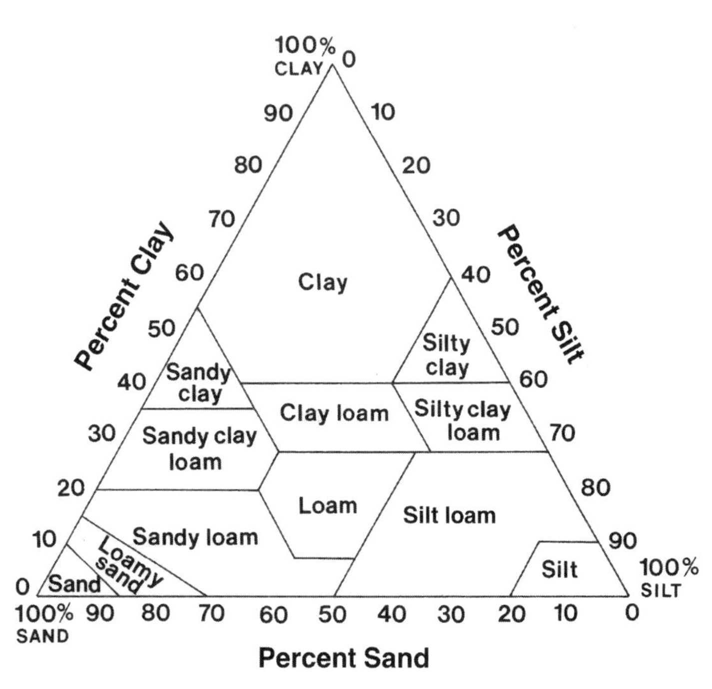

Figure 4.2. Soil classification. The class name of a soil is based on the distribution of particle sizes. Silt loam soils are excellent for agricultural uses.

spaces between particles are tiny, drainage may be slow. Some of the best agricultural soils are loams, in which there is a balance among sand, silt, and clay. This balance results in variation among the sizes of spaces between soil particles.

Water is present in soils in a thin layer on the surface of soil particles and also as free water in the spaces between soil particles. Because water molecules tend to cling together, the water in the tiny spaces between soil particles, although drawn downward by gravity, also is held in the pore spaces. This same phenomenon can be seen when water drips slowly from a faucet. The drop stretches and enlarges until the force of gravity finally pulls it down.

In a silt loam soil, approximately half of the volume of the soil consists of pore spaces between soil particles. When soil moisture is optimal, half of the pore spaces are filled with air and half with water. Thus, a silt loam soil can maintain soil moisture but also allow good drainage after rain. The air spaces are critical, because oxygen (O_2) must be available to most soil organisms, as well as to plant roots. When drainage is blocked or very slow, soil becomes anaerobic, which causes plant roots and other aerobic organisms (those requiring O_2) to die.

Carbon Cycling: The Basis of Life

Life on Earth is based on the element carbon. The mineral soils just described contain 1–10% decaying life forms, called organic matter. The decay of organisms after death is an important part of the carbon cycle (Figure 4.3). This cycle refers to the continual processing of carbon atoms in carbon dioxide (CO_2) into organic molecules by photosynthesis and the release of these same carbon atoms back to CO_2 by cellular respiration in specialized organelles called mitochondria. These processes are essential to life, because the chemical bonds of an organic (carbon-containing) molecule contain energy that can be released in a controlled manner during cellular respiration and used in the life processes of organisms. In addition, the carbon in organic molecules is available to organisms for building the organic molecules necessary for growth: carbohydrates, lipids, nucleic acids, and proteins.

The original source of all this energy is the sun. Photosynthesis is the process by which the energy of the sun is captured and stored in the chemical bonds of organic molecules produced from CO_2 and water (H_2O). In plants, these compounds are simple sugars, which

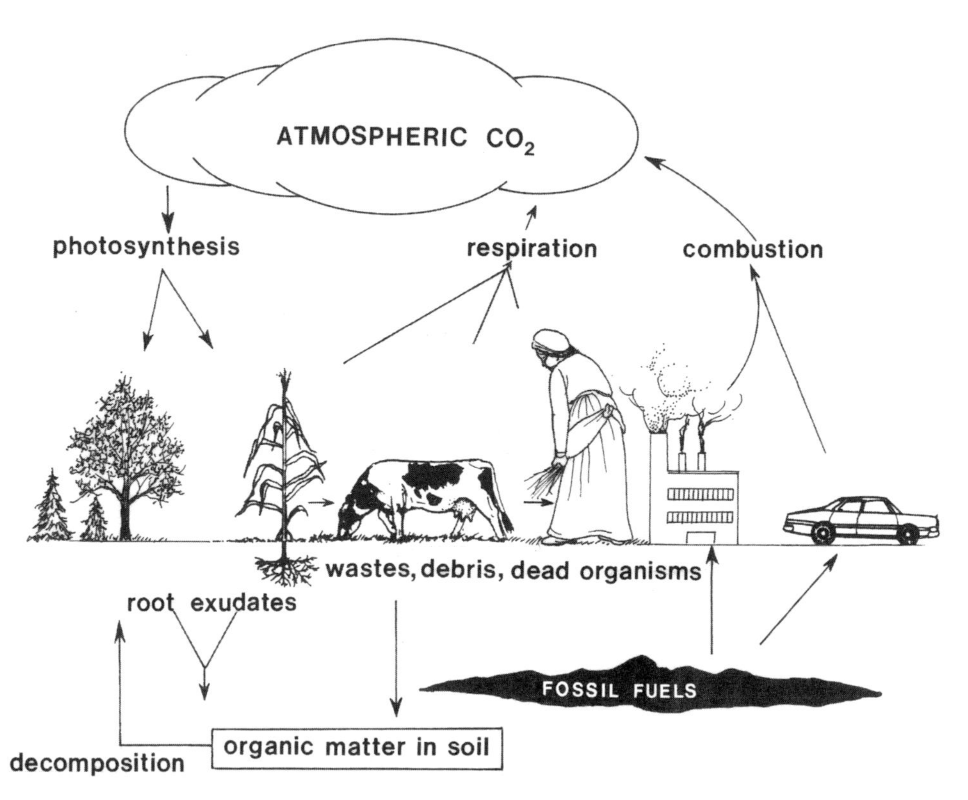

Figure 4.3. The carbon cycle. Photosynthesis fixes inorganic carbon into organic forms. Respiration and combustion return organic carbon to its inorganic form.

are often combined and stored as starch. In eukaryotic cells, photosynthesis takes place in specialized organelles called chloroplasts, which trap sunlight and fix carbon atoms from CO_2 into organic compounds, releasing O_2. The absorption of CO_2 from the air results in most of the dry weight of a plant, accompanied by small amounts of minerals absorbed through the roots.

The sun's energy is trapped in organic molecules (sugar) by chloroplasts during photosynthesis. Plants obtain energy from the sugar through the process of cellular respiration, which takes place in the mitochondria. Cellular respiration requires O_2 and releases CO_2, the opposite of photosynthesis. When humans and other animals eat plants, the complex compounds are broken down during digestion into simpler compounds. Then, through cellular respiration, humans and animals are able to use the energy of the chemical bonds to fuel the metabolic processes necessary for their existence. Just as in plants, in humans and animals, O_2 is required by the mitochondria and CO_2 is released. Because inhaling brings O_2 into the body for cellular respiration and exhaling expels CO_2, breathing is sometimes called respiration. Upon death, all types of organisms decay, serving as food sources for bacteria and fungi, which use the organic molecules of the decaying organisms as sources of carbon and energy. All organic molecules eventually degrade to CO_2 and H_2O as part of the continuous carbon cycle.

The carbon cycle demonstrates our total dependence on photosynthesis to provide energy in the form of organic molecules. The cycle is completed with the eventual return of carbon atoms in organic matter to carbon in the form of CO_2 through the processes of respiration and decay. In addition, photosynthesis is the only source of atmospheric O_2, which means this process not only accomplishes carbon fixation but also provides the O_2 necessary for aerobic respiration. Until photosynthesis evolved, all life on Earth was anaerobic, functioning without free O_2. Aerobic respiration is more efficient than anaerobic respiration, because it derives more energy from the same organic molecules. Thus, aerobic organisms dominate Earth today.

The presence of O_2 in the atmosphere also led to the formation of the ozone (O_3) layer in the upper atmosphere, which shields Earth from much of the ultraviolet light coming from the sun. By reducing mutagenic ultraviolet radiation, photosynthesis had a second major effect: It allowed life forms to leave the protection provided by deep water to colonize the land. Ozone in the lower atmosphere is considered a pollutant that can harm both animals and plants (see Chapter 14), but the ozone layer high above Earth is critical for life on land.

Carbon Cycling and Climate Change

Atmospheric CO_2 levels have been declining over the past 200 million years, for several reasons. Much of the carbon has become fixed in organic molecules in the living organisms of the world's ecosystems and in the organic matter in the soil. Organic carbon also is stored as coal, natural gas, and oil. Some carbon is dissolved as carbonic acid in the oceans. In contrast to this long-term historical decline, the level of CO_2 in the atmosphere began to rise at the onset of the Industrial Revolution. Today, ice cores from Antarctica indicate that atmospheric CO_2 levels are the highest they have been in 600,000 years.

People have become an important component in the carbon cycle because of their rapid degradation of many organic molecules through the burning of fossil fuels and the development of industry. Fossil fuels—including coal, gas, and oil—are biological in origin, created mostly from ancient plant materials. The combustion of these fuels and of wood, leaves, and other organic matter results in the release of energy. Thus, combustion has the same goal as respiration: to obtain energy. Like respiration, combustion contributes CO_2 to the atmosphere. Energy sources that do not contribute CO_2 include nuclear, solar, water, and wind power. People also have destroyed large areas of photosynthetic activity through deforestation for agriculture, fuel, lumber, and urbanization. Large areas that were previously covered with plant life now contain buildings, parking lots, and roads. Plowing, or tillage, is an important part of agriculture, but it also increases the decomposition of organic matter, releasing CO_2.

There is no doubt that these changes already affect the environment. CO_2 is a "greenhouse gas," which means it helps trap heat from the sun in the atmosphere, just as the glass panels of a greenhouse raise the temperature inside the structure on a sunny day. Many gases can function as greenhouse gases, including water vapor and methane. People who live in winter climates know that on very cold nights, the sky is usually clear, but the presence of clouds, which contain water vapor, can moderate how low the temperature drops. A certain amount of the greenhouse effect is necessary for our survival on Earth. Without the atmosphere, the planet would be cold and unlivable, like the moon.

Science Sidebar

Atmospheric CO_2 and the ozone layer

People are concerned about holes that have developed in the ozone layer, because the holes increase the ultraviolet radiation to which people are exposed. Chlorofluorocarbons (CFCs), which are used in aerosol cans and certain common refrigerants, destroy ozone in the upper atmosphere. Recent regulation of these chemicals has reduced but not eliminated their damaging effect on the ozone. Although the rising level of CO_2 in the atmosphere has many environmental effects, CO_2 does not harm the ozone layer.

In 1990, the U.S. Department of Agriculture (USDA) published a zone map of plant hardiness, which indicated where plants could be safely grown based on the coldest winter temperatures. In 2002, the USDA contracted the American Horticultural Society to revise the map, but the new map, reflecting a warming climate, was rejected. The American Arbor Day Foundation published a new hardiness zone map in 2006, and it reflected obvious changes in zone boundaries because of moderating winter temperatures.

These maps are important for commercial growers and landscapers, as well as for home gardeners. Changes in atmospheric temperature can affect the length of the growing season and sometimes alter the risk of frost damage when plants are vulnerable at the beginning and end of the season. Higher levels of CO_2 in the atmosphere may increase plant growth and productivity. Rainfall patterns are expected to change, as well, with some areas getting more annual rainfall and others getting less, especially in the southwestern states and California. In the future, it may be necessary to grow some crops at more northern latitudes because of changes in temperature and rainfall, but the quality of the soils may not be the same. Some wild species, especially trees, may find it difficult to move and adapt to these environmental changes.

As ice melts at the North and South Poles and in Greenland, rising sea levels will affect people living in coastal areas and on islands. Intrusions of sea water can cause salinity (salt) in soils to rise beyond what crops can tolerate, and they also can affect the salinity of irrigation and drinking water. A higher CO_2 level in the atmosphere also affects the pH of the oceans, as the level of carbonic acid increases. This change may have long-term effects on ocean species that have shells composed of calcium carbonate or on structures such as coral reefs. As temperatures rise, permafrost soils melt and release large amounts of methane (CH_4). Methane, also known as "natural gas," traps about 25–30 times more heat in the atmosphere than CO_2. The direct and indirect effects of the rising CO_2 level complicate predictions of the directions and effects of future climate changes.

The Impact of Agriculture on Soils

Although the carbon cycle is most appropriately applied to the whole world, we can examine a small portion of the cycle in an agricultural field. Before agriculture came to be practiced, organic matter was returned to the soil each year as resident plants and animals died and decayed. The organic matter in the soil is important for plant growth. It helps retain soil moisture and provides certain elements and nutrients essential for plant growth.

In agriculture, humans harvest organic matter produced through photosynthesis in the form of food and fiber crops, eliminating the contribution of this material to the organic portion of the soil. Many crops are grown in weed-free rows, leaving large portions of the soil uncovered by vegetation and reducing the amount of organic matter that may be contributed to the soil. Significant amounts of topsoil are simply blown or washed away in locations where vegetation is not present to hold it in place. At the same time, the organic matter already present in the soil continues to be degraded by soil microorganisms that use it as a source of nutrients and energy. Through cultivation and plowing, harvesting, wind and water erosion, and the continuous biological degradation of the remaining organic material, the organic matter in the soil slowly declines.

For many centuries, humans have tried to maintain the level of organic matter in soil by adding compost, manure, and other organic material. Recently, some farmers have adopted low-till or no-till methods, which leave the stubble from the previous crop on the surface to reduce erosion and decomposition of the organic matter. The new seeds are then planted directly into the stubble. The implications of these methods in terms of pathogen survival are discussed in Chapter 9. Many soils have not been successfully maintained, and some have even been abandoned. Productive agricultural soils are not a permanent feature of the landscape but rather a fragile resource. They were created over millions of years and have badly deteriorated in the brief time human beings have practiced intense cultivation.

Many soils have shallow organic layers or contain low levels of organic matter. Such soils deteriorate even more quickly and are much more vulnerable to destruction. These fragile soils are found in many tropical areas, particularly those in which rain forests have been cleared. The soils can be cultivated productively only for short periods.

The loss of arable soils is a critical issue for human survival. We return to the problem of conservation and maintenance of arable lands in Chapter 14.

Plant Leaves: To Capture Energy

Photosynthesis—the capturing of energy from the sun—takes place in plant parts that are green because of the presence of chlorophyll. The most abundant green tissue in higher plants is their leaves.

A cross section of the leaf of a typical broad-leaved plant reveals its anatomy (Figure 4.4). A single layer of epidermal cells lines each upper and lower leaf surface. In addition to the wall of the cell, a cuticle of water-repellent compounds (primarily cutin) protects its exposed surface. A distinct waxy layer usually exists on and blends into the cuticle. This waxy layer varies in thickness depending on the plant species. Usually, the wax and cuticle layers are thicker on the leaf's upper surface. The wax and cuticle also are thicker on the leaves of plants that are adapted for water conservation, such as evergreen and desert plants, and on the leaves of many plants in warm and humid tropical regions. The wax and cuticle layers also serve as barriers against invasion by many pathogens.

The epidermal layer is periodically interrupted by specialized guard cells around openings called stomata. The guard cells in the epidermis are capable of opening and

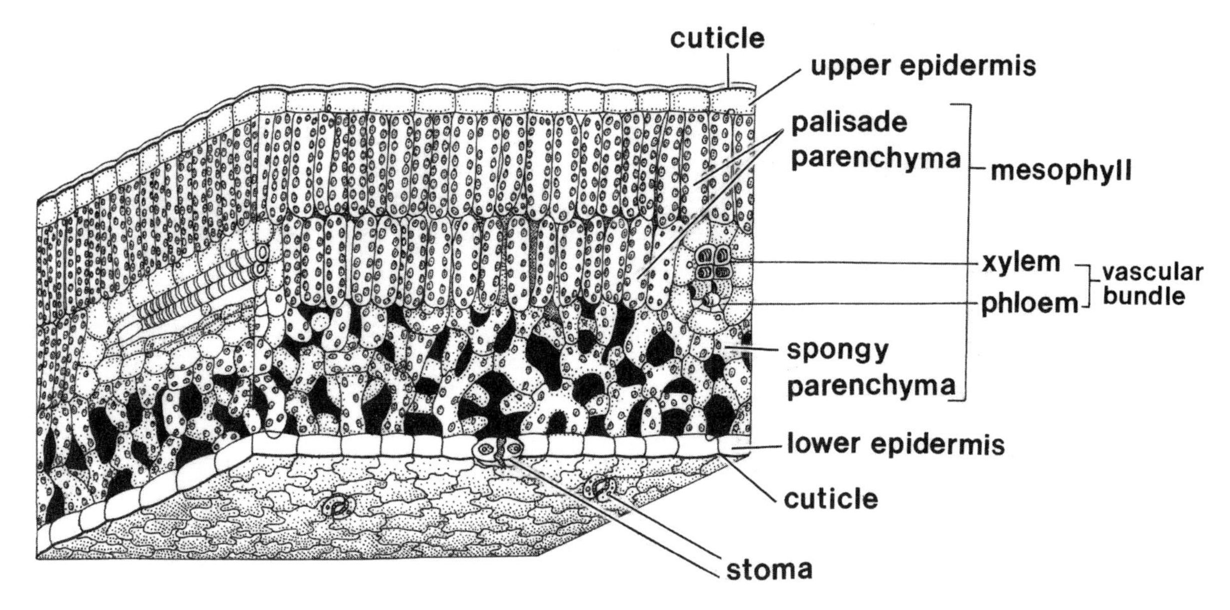

Figure 4.4. Cross section of a leaf, showing cells and tissues.

closing the stomata for control of gas exchange and transpiration (water loss). Stomata are usually more numerous on the leaf's lower surface. The stomata must be open for the absorption of CO_2 for photosynthesis, but when open, they allow water to escape through evaporation. When open, the stomata can serve as direct points of entry for plant pathogens.

Between the epidermal layers is the mesophyll, which is simply the middle of the leaf. It is filled primarily with thin-walled, living parenchyma cells of various shapes. In many leaves, one or several rows of column-shaped parenchyma cells (palisade parenchyma) are found just below the upper epidermis. Most of the leaf's photosynthesis takes place in these chloroplast-filled cells. Below the palisade layer lies the more loosely packed spongy parenchyma. These irregularly spaced cells are separated by air spaces.

Within the mesophyll are the vascular bundles, commonly called leaf veins. A vascular bundle consists of xylem tissue for water and mineral transport and phloem tissue for food transport (mostly sucrose from photosynthesis). Some pathogens use these transport tissues to move within their plant hosts. In addition, support tissues, such as collenchyma and sclerenchyma, are often associated with the vascular bundles and leaf margins. Familiar examples of these two types of tissues are the tough strings in celery (collenchyma) and the gritty stone cells in pear fruits (sclerenchyma). The thickened cell walls of these tissues lend strength to leaves and nonwoody stems.

Among flowering plants, leaf anatomy exhibits tremendous variation. Epidermal cells may produce various types and densities of leaf hairs, or trichomes. The internal pattern of parenchyma cells is variable. The number, pattern, location, and morphology of stomata vary, depending on the environment to which the plant is adapted and the botanical family to which it belongs. However, all leaves have a protective and relatively hydrophobic (water-repellent) surface, with natural openings at the stomata. The cuticle is least thick on the cells in the stomata, on newly expanding leaf tissue, in nectaries and on flower petals, and in hydathodes, which are openings along the sides of leaves for exudation of excess water. These less-protected areas can provide invasion sites for plant pathogens.

Even if a pathogen can penetrate the protective cuticle, it must face the multiple layers of the walls of the cells themselves. A cell wall consists primarily of cellulose, hemicellulose, and pectin, which occur in layers. All of these components are polymers, or long chains, of simpler molecules. Cellulose is the predominant component of cell walls and is a polymer of glucose, a simple sugar. Hemicellulose and pectin polymers are chemically similar. The intercellular (between-the-cells) layer that cements the cells together, called the middle lamella, consists primarily of pectin. Pectin is sold commercially as a thickener for jams and jellies.

How can a fungal spore penetrate such well-protected plant tissue? Most spores begin to germinate if a layer of water is present on the leaf surface for an hour or longer. After the water has been absorbed by the spore, germ tube growth begins. Water must typically be present for a total of 8–12 hours for infection to be successful, because the germ tube takes some time to grow and then remains sensitive to desiccation until penetration takes place. The time necessary for spore germination varies with fungal species, temperature, and other environmental variables and is one of the most important factors in a disease epidemic.

Many plant-pathogenic fungi have evolved elaborate mechanisms to penetrate host plants. Many produce a flattened structure called an appressorium at the infection site (Figure 4.5). The appressorium often produces a sticky substance that helps attach the germ tube to the leaf, preventing the spore from falling off before infection occurs. A narrow infection peg produced from the appressorium penetrates the cuticle with a combination of mechanical pressure and enzyme production, which softens and degrades the cuticle and epidermal cell walls. Many fungi produce various enzymes, such as cutinase, which degrade the cutin in the cuticle, plus pectinases and cellulases, which degrade pectin and cellulose, respectively, in the cell walls. (Note that the suffix -ase is used to designate an enzyme that degrades the named substance.) Such enzyme production not only expedites penetration, but it also provides important nutrients for fungal growth. Enzyme degradation of both pectin and cellulose releases glucose, an important energy source. As cells begin to die, the cell membranes become quite permeable, and other important nutrients begin to leak out of the cells.

Some fungi penetrate plants only via the stomata. They enter the spaces between the guard cells to reach the internal cells, where the waxy layer is absent and the cuticle is thin. Other fungi penetrate between epidermal cells, while still others penetrate epidermal cells directly. Many fungi take advantage of wounds caused by hail, insect feeding, or mechanical damage to facilitate penetration of the plant surface. Infection by some fungi results in lesions of dead cells that are visible only a few days later, whereas other fungi, such as rust fungi, absorb nutrients without triggering immediate cell death. The hyphae of such fungi grow between the parenchyma cells of the leaf and absorb nutrients through projections of the hyphae that push into the plant cell's cytoplasm without breaking through the cell membrane. These specialized projections, called haustoria (Figure 4.5), allow the host cells to remain alive but provide nutrients for the invading fungus.

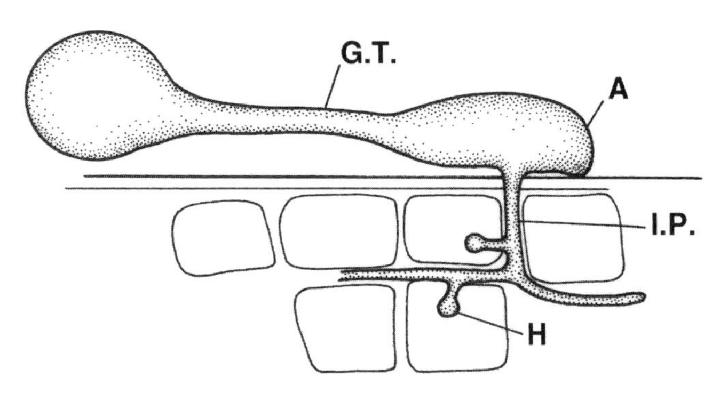

Figure 4.5. Spore germination. **G.T.** = germ tube; **A** = appressorium; **I.P.** = infection peg; **H** = haustorium.

Plant leaves are clearly adapted to optimize photosynthesis, minimize water loss, and protect against invasion by pathogens. Leaf function would be difficult for most plants, however, without the anchoring system of roots, which also deliver water to the stems and leaves. Even though H_2O and CO_2 provide most of the dry weight of plants, they also require other elements from the soil to complete many of the important molecules of life, such as chlorophyll, DNA, and proteins. Roots, which are belowground and out of sight, are often underappreciated when the major factors that affect plant health are considered.

Plant Roots: To Anchor and Absorb

Just as the aboveground parts of land plants are well adapted to withstand variations in the aerial environment, the belowground parts are generally well adapted to the soil environment. Environmental variations in soil moisture and temperature are usually more gradual and less extreme than the weather aboveground, but the range of variation is still considerable. In addition, plant parts that exist in the soil must withstand physical stresses from rocks and smaller mineral fragments. They also live in an intimate relationship with animals, plants, and microorganisms of the soil, with each organism influencing and being influenced by the others.

All vascular plants produce roots that provide anchorage and support for the aerial parts of the plant. Roots also are responsible for the absorption and conduction of minerals and water from the soil. They serve a storage function in many plants, in some cases resulting in large, starchy structures that we harvest and eat, such as carrots and sweet potatoes (but not potatoes, which are modified stems, rather than roots).

At seed germination, the primary root of a vascular plant must develop quickly to begin absorbing water for the growing seedling. Many seeds contain enough stored food to maintain the seedling until it can begin photosynthesis. However, moisture is needed immediately to keep the young tissues from drying out and to provide the water molecules used in converting complex stored foods (such as starch) into smaller molecules for respiration in the growing seedling.

Flowering plants, or angiosperms, are divided into two large groups based on, among other features, the morphology of their seedlings. If the seedling possesses one seedling leaf, or cotyledon—as do corn, grasses, lilies, and onions—it is called a monocotyledon or monocot (Figure 4.6). Monocots usually have leaves with parallel veins. Monocot seedlings have a short-lived primary root, which is quickly replaced with a fibrous mass of roots that arise from the base of the stem near the soil line. If the seedling possesses two seedling leaves, it belongs to the second group of angiosperms, the dicotyledons or dicots. Dicots usually have broad leaves with netlike veins and a taproot that continues to grow, producing many lateral branches. Familiar examples of dicots include carrots, chrysanthemums, dandelions, maple trees, and peanuts.

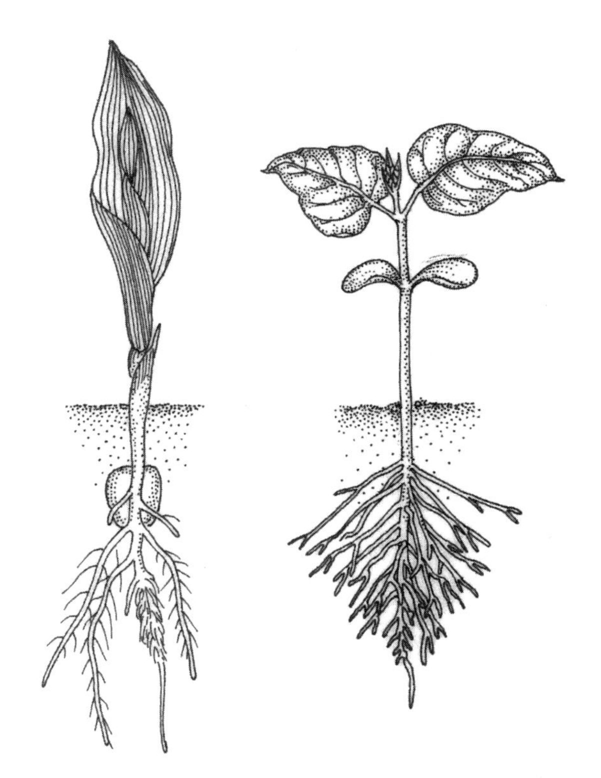

Figure 4.6. Left, germination of a monocotyledon (corn). The seedling possesses one cotyledon and adventitious roots. The leaves have parallel veins. **Right,** germination of a dicotyledon (bean). The seedling possesses two cotyledons (seed leaves) and a taproot that develops lateral branches. The leaves have netlike veins.

Although roots vary tremendously in structure and anatomy, those of vascular plants share several important features for efficient absorption and protection from the soil environment (Figure 4.7). Roots continually grow through the soil from the tip by means of an apical meristem, which is an area of undifferentiated cells. These cells show no specialization for a particular function but rather serve as the source of new cells for the growing root, similar to stem cells in animals. (Plants also have apical meristems in buds at the tips of the aerial parts.) The apical meristem of the root is protected by the root cap, a small group of cells that continually slough away as the root pushes past mineral fragments and other obstructions in the soil. Just behind the apical meristem, the cells continually divide to produce new cells. These new cells contribute to the zone of elongation, the area in which elongation of cells produces all increases in root length.

The new cells differentiate, becoming specialized for their various functions. Like a leaf, a root has an outermost layer composed of epidermal cells. Unlike the epidermal cells of a leaf, those of a root have protrusions called root hairs, which are quite small but greatly increase the root's absorptive surface area. The root cortex is composed of parenchyma cells just inside the epidermis, which function in storage rather than photosynthesis, as is the case with leaves. Water is easily absorbed through the root cortex, apparently moving mostly between cells. In contrast, absorbed minerals move through the cytoplasm of the epidermal cells and then of the cortical cells.

Once water and minerals have passed through the root cortex, they reach a barrier called the endodermis. Impervious material, called Casparian strips, connects the cells of this tissue layer, so that all the materials absorbed by the root must pass through the cytoplasm of the endodermal cells before entering the vascular cylinder in the center of the root. In a cross section of a carrot, it is possible to see the outer cortex and the inner vascular cylinder. Water passes through the endodermis by osmosis, a process in which water molecules move from areas of higher water concentration to areas of lower water concentration. Since soil water is usually more dilute than cell contents, water moves almost continuously into the vascular cylinder. When plants are exposed to very salty water, osmosis no longer moves water into the plants, which explains why most plants must receive fresh water to grow properly. The

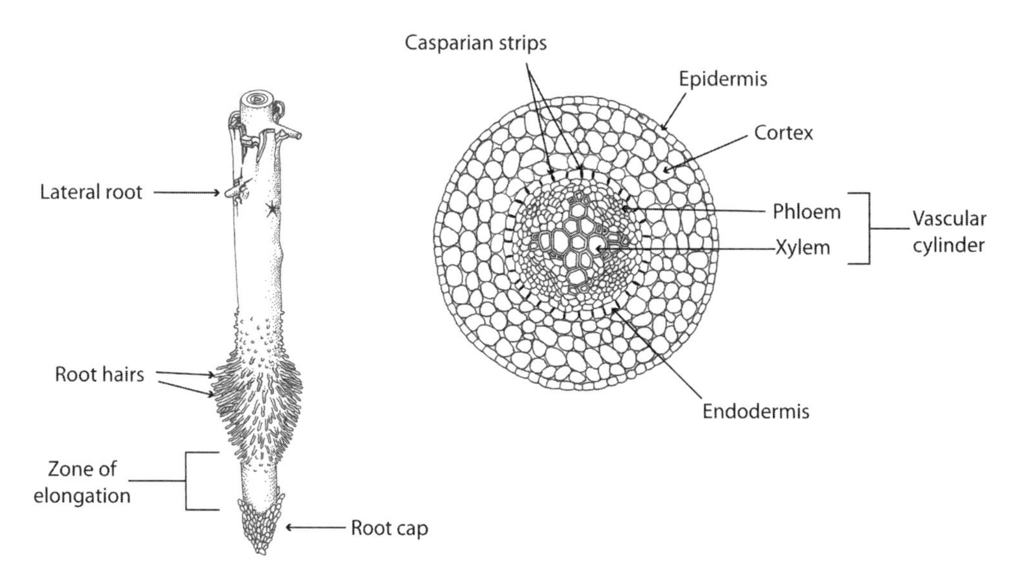

Figure 4.7. Left, a portion of a dicot root. Note that lateral roots emerge through broken cells of the cortex. The growing tip of the root is protected by a root cap. The surface area for water and mineral asorption is increased by root hairs. **Right,** a cross section of the main root.

movement of water into plant roots by osmosis requires no energy. In contrast, dissolved minerals in the soil water are transported to the xylem by active transport into the cells, which does require energy.

The vascular cylinder, inside the endodermis, is composed of several kinds of cells. Some of the cells play a role in secondary growth (see Chapter 12) and also serve as the origin of lateral, or secondary, roots. The xylem tissue, which conducts minerals and water upward, is in the center of the cylinder. Phloem tissues bring organic compounds produced during photosynthesis down from the plant shoot. These compounds are used in cellular respiration of the root and stored as excess food. As the root continues to grow, lateral roots arise in the vascular cylinder and exit by breaking through the cortex and epidermis.

All absorption of minerals and water takes place in the relatively short-lived and fragile root tip area, containing relatively young tissues. Root hairs are destroyed quite quickly, and soon secondary tissues begin to develop. For continued absorption, plant roots must continue to grow. Root development must be extensive below a large plant, because a balance is necessary between the photosynthetic area aboveground and the absorptive area belowground. Because the absorptive areas of plant roots are particularly fragile, they are easily lost during transplanting. Unless care is taken to minimize disturbance of the soil around the root system, plants may suffer transplant shock and grow slowly for some time after being moved. This is why trees are usually dug up with a ball of soil around the roots when they are moved from one site to another.

Science Sidebar

Why we need fresh water

Osmosis is the process by which both animals and plants absorb water at the cellular level. Water moves into cells only when the concentration of materials dissolved in the water is lower outside the cells than inside the cells. Fresh water usually has less than 1% dissolved salts and allows water to move into cells by osmosis. Salt water and heavily polluted water slow or stop the osmotic movement of water into cells. Thus, to survive, people stranded in the ocean must catch rainwater to drink. Reverse osmosis—a process used to create fresh water from salt water for human uses—requires enormous amounts of energy. Fresh water for drinking and irrigation is becoming an increasingly scarce resource in the world.

The Water Balance: A Lot but Not Too Much

Plants require much more water than animals, because they are continuously losing water through evaporation from their aboveground parts during transpiration. Moist cells must be exposed to the air to absorb CO_2 for photosynthesis, but this results in tremendous water loss for the plant, especially on a hot, windy day. The plant controls water loss via the protective waxy cuticle, releasing most water through the stomata, which open and close by means of guard cells. Although this gives the plant some control, the stomata must be open for extended periods to absorb enough CO_2 for photosynthesis. Thus, unlike animals, which need to drink only periodically, plants continuously absorb and transpire water whenever their stomata are open. To take in enough CO_2 for photosynthesis, plants lose more than 90% of the absorbed water to evaporation. A single corn plant can require 10–15 gallons (40–60 liters) of water on a sunny day.

Roots require many of the same factors for growth and life as aerial plant parts. Soil moisture is absorbed continuously. Even though this moisture is necessary, if water fills all of the spaces between soil particles, the root cells will be deprived of oxygen. Where the soil consists of clay, with small air spaces and poor drainage, crops are sometimes planted in raised beds to increase drainage and improve soil aeration. Plants grown in poorly drained soil deficient in O_2 grow slowly or wilt and die, because root development and function are reduced. The photosynthetic area aboveground must remain in balance with the absorptive capacity of the roots. Thus, plants with poor roots often have stunted growth and may exhibit symptoms of nutrient deficiency. Overwatered house plants may die if they are grown in pots without drainage holes, because the air spaces in the soil become filled with water, depriving the roots of oxygen and ultimately killing them. When soil is compacted by foot traffic or machinery, the pore space also is greatly reduced, leading to the decline and death of existing plants and preventing the establishment of new ones.

Minerals Are Essential

Just as certain minerals are necessary for good health in humans, certain minerals are necessary for plant growth. These chemical elements are released from the decay of organic matter and dissolved from eroding rock particles in soil. Although they are sometimes called "plant nutrients," they do not provide energy for the plant (and hence the mislabeling of products sold as "plant food"). Plants are autotrophs, producing their own food by trapping the energy of the sun through photosynthesis. Yet they still require additional elements to complete the molecules that form carbohydrates, proteins, and all the other compounds necessary for their growth and development.

The elements required by plants are divided into two categories, depending primarily on the amount necessary for adequate growth. Those elements needed in large quantities are called macronutrients, and those needed in small quantities are called micronutrients. Macronutrients include carbon (C), hydrogen (H), and oxygen (O), which are supplied by CO_2 and H_2O. C, H, and O comprise almost all of the dry weight of plants and other living organisms. Other macronutrients are dissolved in the soil water in various chemical forms and absorbed primarily through the roots. They include nitrogen (N), phosphorus (P), potassium (K), calcium (Ca), magnesium (Mg), and sulfur (S).

Important micronutrients, which also are called minor or trace elements, include iron (Fe), boron (B), manganese (Mn), zinc (Zn), copper (Cu), molybdenum (Mo), and chlorine (Cl). Although micronutrients are necessary elements, they can cause toxicity symptoms in plants when present in excessive quantities. Generally, a relatively narrow range separates the deficiency and toxicity levels of micronutrients. Even though toxicity resulting from the excessive supply of a macronutrient is relatively rare, both deficiencies and toxicities of micronutrients are commonly observed.

Table 4.1 lists some of the elements required by plants and their typical deficiency symptoms. Scientists usually study such symptoms by growing plants in a liquid or sand medium, into which each element is added in a measured quantity. This allows scientists to see the exact effects of varying the amount of a particular element or even eliminating it entirely. Such controlled conditions are not possible in nature, so the symptoms in the field or garden are not always as clear as the descriptions might suggest.

The diagnosis of a nutrient deficiency requires familiarity with the particular plant and its fertility history. For example, if a potted plant has been growing in the same soil for many years without the addition of fertilizer, some nutrient deficiencies should be expected. Likewise, fields that have recently received excessive rainfall are likely deficient in water-soluble nutrients, such as nitrogen.

Another Important Cycle: Nitrogen

Nitrogen is an important component of both proteins and nucleic acids. Because of this, it is commonly the limiting element in agricultural systems. Nitrogen is a mobile nutrient, which means it is transported from older tissues to growing tissues when adequate supplies are not available. This leaves the older tissues deficient in nitrogen and makes them yellow prematurely. Nitrogen-deficient plants typically look stunted and yellowed, with the lower leaves displaying the symptoms most severely (Figure 4.8).

Tremendous quantities of nitrogen-containing materials are added to agricultural soils each year. The apparent contradiction in this practice is that nitrogen in its gaseous form (N_2) is the predominant component of Earth's atmosphere (more than 78%), yet nitrogen is

Table 4.1. Some Important Elements for Plant Growth and Their Deficiency Symptoms in Plants

Element	Deficiency Symptoms
Nitrogen (N)	Stunted growth; light-green color; yellowing of leaves, beginning with oldest foliage
Phosphorus (P)	Stunted growth; dark-green, sometimes bluish to purplish coloration of foliage; thin and weak shoots
Potassium (K)	Dieback; poor shoot growth; yellow and necrotic spots on leaves; browning of tips and edges of leaves
Magnesium (Mg)	Symptoms appear on edges of younger tissue first; include yellowing and sometimes reddish color; leaves may appear cupped
Calcium (Ca)	Youngest foliage distorted and irregular; terminal buds may die; root growth poor
Sulfur (S)	Similar to nitrogen deficiency, except young leaves may be pale green first
Boron (B)	Poor growth of buds, new growth, and whole plant; surface and internal cracking on stems, fruits, and vegetables
Iron (Fe)	Young leaves light green to yellow, especially in interveinal areas; in severe cases, leaves become dry and may be shed
Zinc (Zn)	Interveinal yellowing; may become necrotic or purplish; leaves may be small; defoliation in some cases, beginning with lower leaves; reduced fruit production
Manganese (Mn)	Interveinal yellowing; necrotic spots on leaves common; whole leaves may become brown, in severe cases

Figure 4.8. Left, a petunia plant with severe nitrogen-deficiency symptoms, including stunting of growth and yellowing of lower leaves. **Center,** a plant with moderate nitrogen-deficiency symptoms. **Right,** a healthy plant.

Science Sidebar

How plants provide nitrogen for animals

Nitrogen is an important component of many complex organic compounds in living organisms. When plants are eaten by animals, amino acids from the plant proteins are used to create animal proteins. The so-called essential amino acids are those that we cannot synthesize through our own metabolic processes and that must be obtained by eating amino acids synthesized by other organisms. An easy way to get the full array of amino acids necessary for building human proteins is to eat meat and other animal products, which contain high concentrations of protein.

There is currently considerable interest in breeding plants, such as high-lysine corn, that contain high levels of the essential amino acids. If these plants were available, we could more easily obtain the necessary compounds directly from them. Producing plants with enhanced nutrient value would be a much more efficient way to feed the growing world population a diet with adequate protein, since considerable energy efficiency is lost when humans eat animals, rather than plants, to obtain these proteins.

an important limiting factor in agriculture. The reason behind this is that neither plants nor animals can use nitrogen in its N_2 form. Nitrogen fertilizers added to agricultural soils also pollute surface water and groundwater through runoff and percolation, respectively. Surface water pollution contributes to algal blooms, often resulting in premature aging, or eutrophication, of lakes and ponds. Excessive nitrates in groundwater also pose a health hazard, so water contaminated with nitrates must be treated before it can be consumed. Because nitrogen is a limiting factor in agricultural production, as well as an environmental pollutant, understanding the nitrogen cycle can lead to improved agricultural efficiency and to improved environmental protection.

Nitrogen is lost from agricultural lands when crops are harvested and animals graze on forage crops. It must be continually added if subsequent crops are to be grown. Nitrogen is not returned to the soil until plants and animals die and decay or the waste products of animals and humans are put into the soil. Farmers have long known that manure and plant debris are good sources of nitrogen and other important plant nutrients, and they have added them as soil amendments for use by subsequent crops.

A difficulty in modern agriculture is determining how to deliver sufficient manure and other sources of nitrogen to large fields to replace what is lost during harvest. Many people, including researchers and farmers, are interested in the forms of nitrogen available for plant uptake and how to maintain sufficient quantities for optimal plant growth. Because nitrogen is so important in biological processes, its chemical form constantly changes as a result of the activities of many kinds of living organisms. As in the carbon cycle, a balance in the nitrogen cycle is critical to the sustainability of life on Earth (Figure 4.9).

In natural ecosystems, organisms die and their tissues decay. Some of the nitrogen is converted into organic compounds used for the growth of the decay organisms, primarily bacteria and fungi. (This is the immobilization part of the cycle.) Some nitrogen is released from the decaying tissues in an inorganic form, ammonium (NH_4^+) (ammonification). Some of the ammonium is lost to the atmosphere after its conversion to ammonia, NH_3. This loss is noticeable in the odor of fresh manure, in which active microbial decay of organic matter is occurring. The urine of animals also releases significant quantities of NH_3 to the atmosphere and other parts of the environment.

Ammonium in soil is available for plant uptake for use in plant growth. It also has the advantage of being bound tightly by soil particles and is not subject to leaching into groundwater. NH_4^+ has a slight positive electrical charge, as indicated by the "plus" sign, which indicates that it will be held, as if by a magnet, by soil particles with a slightly negative electrical charge. Organic matter is especially effective at holding NH_4^+. Unfortunately, if NH_4^+ is not absorbed by plants, it does not remain in the soil for very long, because certain bacteria in the soil convert it to another inorganic form, nitrate (NO_3^-) (nitrification). Nitrate is more easily absorbed by plants than ammonium but has the major disadvantage of having a negative charge, which reduces its binding to soil particles. This explains why nitrate is highly vulnerable to runoff and leaching from agricultural soils and is a significant water pollutant.

Other bacteria convert nitrate back to the gaseous atmospheric form, N_2 (denitrification), under anaerobic conditions, which are common after heavy rains, in low spots where water collects, and in microenvironments, or small pockets of soil in which oxygen is absent. Significant amounts of nitrogen can be quickly lost from poorly drained soil because of denitrification.

In summary, nitrogen can be absorbed by plants in both the ammonium and the nitrate forms. However, some nitrogen is lost to the atmosphere in the form of ammonia, NH_3, during decay, and more is lost through denitrification, harvesting, leaching, and runoff.

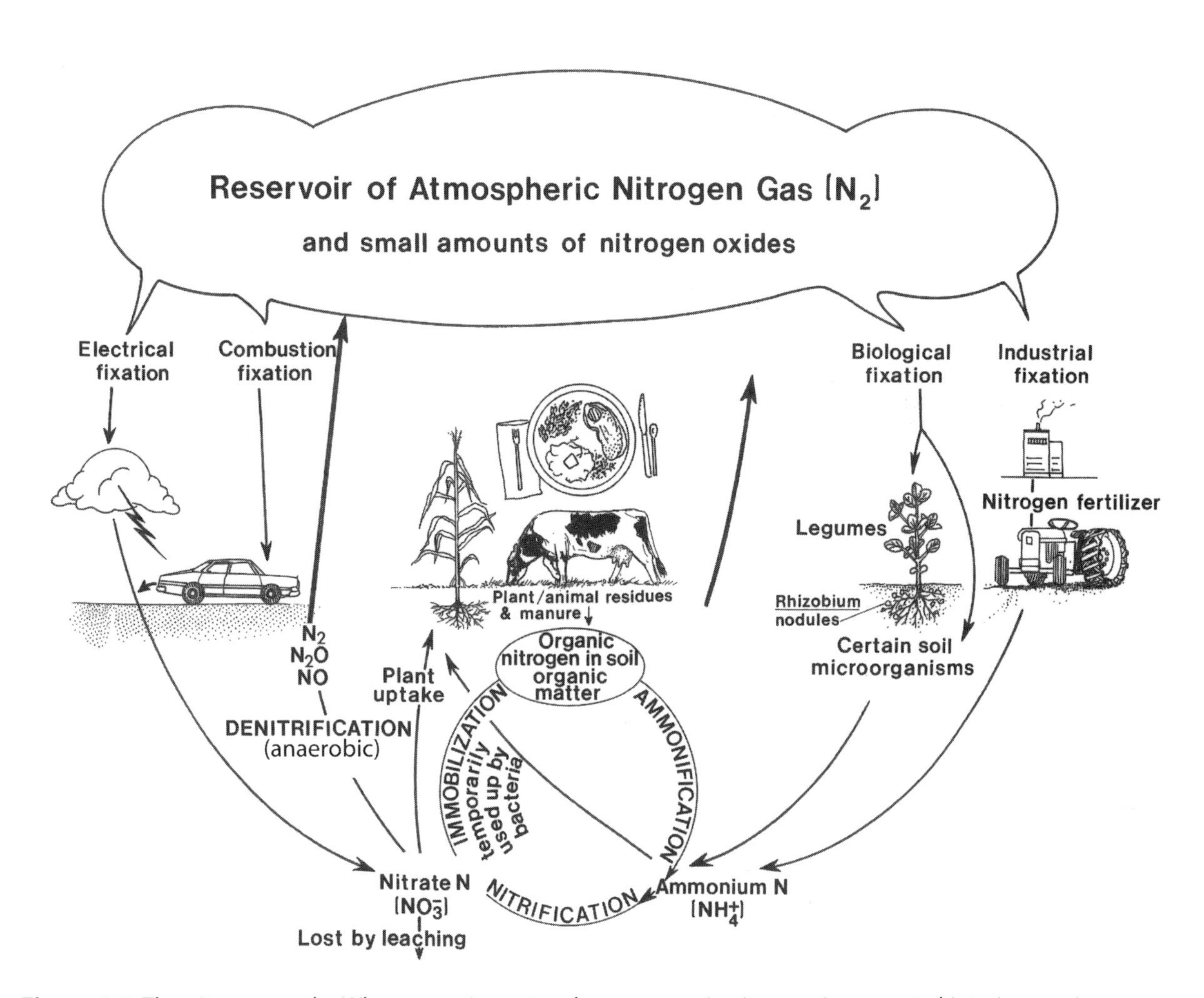

Figure 4.9. The nitrogen cycle. When organic matter decays, organic nitrogen is converted into inorganic ammonium (NH_4) for plant uptake. Ammonium may be converted to inorganic nitrate (NO_3), which also can be absorbed by plants; however, nitrate is subject to leaching from the soil. Under anaerobic conditions, nitrogen may return to the atmosphere. Other sources of nitrogen for plants include lightning, combustion, biological fixation, and fertilizers synthesized by industrial fixation.

Sources of Nitrogen for Plants: Lightning, Bacteria, and Fertilizer

Nitrogen can bc added to agricultural soils in several ways, besides through manure or plant debris. A measurable amount of nitrate is contributed to the soils each year by lightning, with the powerful electrical charge it sends through the atmosphere. Lightning converts atmospheric N_2 to nitrate, which is washed from the air during rain. A small amount of nitrogen also is added to both soils and the atmosphere in the form of nitrogen oxides from the exhaust of automobiles, trucks, buses, and airplanes.

Certain bacteria are capable of "fixing" atmospheric N_2 into ammonium, a form useful to plants. Some of these bacteria are free living in soil and water, and others are found in the root nodules of several kinds of plants (Figure 4.10). The best-known examples are found in the roots of plants in the legume family, including alfalfa, bean, clover, pea, peanut, and soybean.

Brazil and the United States are the world's biggest exporters of soybeans. The production of soybeans in the United States skyrocketed in the 20th century, resulting in several midwestern states being carpeted by alternating fields of corn and soybeans. Soy did not become an important part of the American diet, but soybean oil and the use of soybeans for animal feed created important markets for this crop, which originated in Asia.

Around 1900, David Fairchild, a USDA scientist, discovered that the higher yields of soybeans in Japan were caused by the presence of nitrogen-fixing bacteria in the soil. Based on this finding, the USDA recommended that American farmers mix some of the Japanese soil into their own fields. Although the presence of the bacteria greatly increased soybean yields, the imported soil carried with it a devastating pathogen—the soybean cyst nematode—which is now the most important soybean pest in the United States (see Chapter 9).

The nitrogen-fixing bacteria were originally classified in the genus *Rhizobium,* but more recent studies have found that they actually belong to a number of bacterial genera. Thus, they are often referred to by the informal name rhizobia. Those in the genera *Rhizobium* and *Bradyrhizobium* are closely related to *Agrobacterium tumefaciens,* the crown gall bacterium (see Chapter 5), and they exist in small galls on the root systems of their host plants, absorbing nutrients from them. The nutritional requirements of the bacteria are more than offset by the benefits they bring to the plant from their nitrogen-fixing activity. Particularly efficient nitrogen-fixing strains of rhizobia have been selected over the years and deliberately added to the seeds of legumes at planting time to encourage infection by these parasitic bacteria.

Figure 4.10. Root galls on peanut, a leguminous plant, caused by nitrogen-fixing bacteria.

A current focus of molecular biology is a better understanding of the genes responsible for nitrogen fixation. The goal is to increase the efficiency of microbial nitrogen fixation in agricultural soils and perhaps to enable crop plants that do not host nitrogen-fixing bacteria to fix their own nitrogen through genetic engineering. In the meantime, legume crops such as alfalfa and clover are commonly used in crop rotations. In fact, legume hay has a higher level of protein than grass hay for animals. When it is time to plant a nonlegume crop in a field, the remaining alfalfa or clover can be plowed into the soil to provide a source of nitrogen as the plant tissue decays.

Despite all the natural inputs of nitrogen, most crops still require the addition of this element. Most of it is supplied through synthetic fertilizers produced through industrial fixation processes. Industrial fixation, which is both energy intensive and costly, converts atmospheric nitrogen into ammonia. Ammonia can be combined with other chemicals to produce a wide variety of nitrogen fertilizers, including ammonium nitrate and ammonium sulfate. Millions of tons of nitrogen fertilizers are produced each year, contributing a large share of the energy cost of food production and accounting for modern agriculture's heavy dependence on fossil fuels. Organic materials—such as compost, manure, and sewage sludge—are not available in sufficient amounts to replace synthetic fertilizers. Improving our ability to provide optimal amounts of nitrogen to improved crop cultivars will increase not only the total food supply but also protein levels, which are insufficient in the diets of more than 1 billion people worldwide.

Recent concern over the problems of runoff and leaching of nitrate from soils has led to increased interest in slow-release fertilizers. Manures and other organic materials are natural slow-release fertilizers, because the nitrogen they deliver becomes available to plants gradually, as the organic materials decay and nitrogen is converted to inorganic forms. Organic materials such as compost and manure also provide other macro- and micronutrients to plants in the decay process and improve the soil structure. Nitrogen from synthetic fertilizers is identical to that from natural sources, except that the amount may be more concentrated. Thus, it is possible to cause nitrogen burn, or damage to plants, if too much synthetic fertilizer is added. Plant roots absorb nitrogen in the same way—in the form of ammonium or nitrate—whether the source is synthetic or natural. Slow-release forms of synthetic fertilizers also are widely available. One example is sulfur-coated pellets of urea. Soil bacteria metabolize the sulfur, and the nitrogen is released gradually. The goal is to provide a continuous supply of nitrogen to the plant without contributing to pollution.

An alternative approach is to apply synthetic fertilizers in small amounts and on a frequent basis ("spoon-feeding"), although doing so may increase application costs and soil compaction from additional passes of equipment through the field. Another method of applying synthetic fertilizers that is less likely to contribute to environmental problems involves the frequent addition of low concentrations of liquid fertilizers in irrigation systems ("fertigation").

Micronutrients: Small Amounts Are Critical

Most commercial fertilizers supply the three macronutrients required in greatest quantity for plant growth: nitrogen, phosphorus, and potassium. The chemical forms and relative amounts of these elements vary with the time of application, the soil type, and the crop. Micronutrients are not commonly added to agricultural soils, except in special circumstances. For instance, cole crops, such as cabbage and broccoli, are particularly susceptible to boron deficiency in areas where the soil is sandy.

In the majority of soils, most micronutrients are available at sufficient concentrations for good plant growth. When deficiency symptoms are observed, the micronutrient is frequently present but unavailable because the soil is very acidic (low pH) or very basic (high pH). pH is used to measure how acid or basic a soil is. The pH affects how soluble many chemicals are in water. As the pH of the soil water changes, some compounds

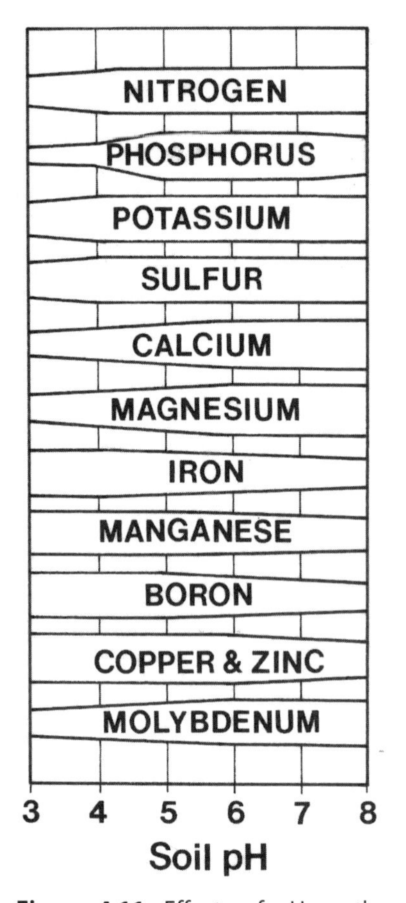

Figure 4.11. Effects of pH on the availability of elements in soil. At low pH, iron and manganese may reach concentrations that cause toxicity symptoms in plants. A number of important elements become less available for plant use in acidic (low pH) soils.

become more soluble, while others become insoluble and unavailable for plant uptake. Figure 4.11 illustrates some of these changes in solubility. Certain elements, such as iron and manganese, become especially soluble at low pH, to the point of causing plant toxicity.

Because the soil pH has such a significant effect on the availability of many nutrients, it should be adjusted to the recommended range of 6.0–7.0 (near neutral) for most plants before micronutrients are added in cases of suspected deficiency. This can be accomplished by using lime to increase pH or sulfur to reduce pH. If micronutrient amendments are still necessary after adjusting the soil pH, then care must be taken to avoid toxicity from adding excessive amounts.

Healthy Plants and a Changing Climate

To be healthy, plants require light, water, and minerals (Figure 4.12). Most food and fiber crops are grown in soils that supply both root support and minerals. As food and fiber demands increase, so will the need for the limited supplies of fresh water and arable soils. Many of the nitrogen needs of agricultural plants are already supplied by adding synthetic fertilizers, which require large inputs of energy for their manufacture. Global climate change will affect water supplies in some areas because of changes in rainfall patterns and incursions of salt water in coastal areas. Because of temperature changes, important crops may need to be produced in new areas in the future. The energy demands of agriculture may increase just when the goal is to reduce dependence on fossil fuels.

There is some evidence that plants growing in an atmosphere with a high concentration of CO_2 may need less water, because they do not have to open their stomata as long or as often to obtain enough CO_2 for photosynthesis. These plants may require less water, but they also may absorb less nitrogen and/or other nutrients, reducing the quality (protein content) of certain crops.

Plant pathologists often study the effects of CO_2 concentration on plant diseases, both in laboratories and under real-world conditions, by using environmental chambers, in which the atmosphere is artificially enriched with CO_2. So far, the studies have had variable results, depending on the pathogen involved and the plant species and its cultivars. In general, higher concentrations of CO_2 in the atmosphere benefit plant growth, increasing both plant size and density in agricultural fields. When plants have a dense leaf canopy, the leaves remain wet longer, and the lower leaves can become yellowed and necrotic. Both of these factors favor diseases caused by many fungal and bacterial pathogens. The reduced frequency of stomatal opening that occurs at higher levels of CO_2 has been shown to reduce infection rates by some pathogens that infect plants through the stomata.

Warmer temperatures can affect the survival of pathogens between crops and may increase the populations of some vectors—insects and other animals that transmit pathogens from one plant to another. Many fungi and bacteria grow more quickly at higher temperatures. Warm temperatures may increase the relative humidity of the atmosphere and slow the drying of foliage that is wet from rain or dew, which will enhance infection by almost all fungal and bacterial pathogens. Increased seasonal temperatures may allow diseases to begin to develop earlier in the growing season, thus permitting many of them to

Figure 4.12. A field of healthy plants (sunflowers)—the goal of plant pathologists.

develop to more significant levels over the course of a season. Drought conditions, which are expected to increase in some areas, could increase certain diseases, as well, by increasing plant stress and susceptibility, but those diseases that require moist conditions would decrease.

Very little is known about the effects of climate change on root growth and soilborne pathogens, but these effects are important. Higher temperatures can increase the rate of decay of organic matter in the soil, which releases various minerals for plant absorption. Environmental stresses, such as mineral deficiencies and toxicities, often affect the susceptibility of plants to various insect pests and pathogens. Although many climate change studies focus on atmospheric CO_2 concentration and water supplies, other environmental stresses, such as atmospheric ozone levels, interact with these factors to affect the final disease outcomes (see Chapter 14). There is no doubt that a changing climate, increasing energy costs, and reduced water supplies will impact the world's agroecosystems. The only questions that remain are how effectively and how quickly these challenges can be addressed.

CHAPTER 5

Single-Celled Pathogens: Bacteria

Healthy plants constantly interact with other organisms, including herbivores and microorganisms, and they have a variety of relationships with bacteria. As described in Chapter 4, many bacteria are important decay organisms, which release nutrients from organic matter for plant use. Some specific bacteria are responsible for fixing nitrogen from its gaseous form to a form that can be absorbed by plants. Some of these nitrogen-fixing bacteria live freely in the soil, and other special bacteria live in the root nodules of legumes. In recent years, fruits and vegetables have become carriers of human bacterial pathogens under certain circumstances.

Bacteria live on the surfaces of plants, both aboveground and belowground. Many of these bacteria are natural biological control organisms and have been studied as means of managing plant diseases. This subject is discussed in more detail in Chapter 9. Certain bacteria are plant pathogens that cause destructive diseases. Understanding the biology of bacteria will help us better understand these interactions.

What Are Bacteria?

Some very important plant diseases are caused by bacteria (singular: bacterium). As with fungi, the ability of bacteria to function as pathogens is linked to their biology. Even though infection by fungi or bacteria can cause similar symptoms of plant disease, bacteria are quite different from fungi.

One immediately obvious difference is that bacteria are much smaller than fungi (Figure 5.1). A typical bacterial cell is about 1 micron (micrometer) in diameter, whereas an average fungal spore might be 50–200 microns in diameter. Even a single fungal thread or hypha is commonly 5–10 microns wide and may be many millimeters long. Individual hyphae can actually be seen growing across a culture plate, whereas individual bacteria are visible only when viewed through a compound microscope at high magnification.

Stains can be used to make the bacteria more visible through a microscope. An important staining procedure for bacteria was discovered in 1884 by Danish microbiologist Hans Christian Gram. Gram was looking for a stain that would be absorbed by all bacterial cells, but the method he developed caused only some types of bacteria to be stained purple. His accidental discovery actually represented a means of separating bacteria into two biologically different groups. Those bacteria that stain purple, called Gram-positive, have a relatively thick, uniform cell wall. Those that do not retain the stain, called Gram-negative, have a thinner cell wall with an additional outer layer of polysaccharides and lipids. The layer containing the lipids protects the bacteria from retaining certain substances, including Gram's stain. Most of the bacteria that cause plant diseases are Gram-negative.

Bacterial cells are very different from the cells of plants, animals, and fungi. The cells comprising these multicellular forms of life are eukaryotic, whereas nearly all bacteria are single-celled prokaryotes (*pro* means "before," and *karyo* means "nucleus"). Prokaryotic cells differ from eukaryotic cells in that they do not contain nuclei, other organelles, or multiple highly organized chromosomes.

Fossil prokaryotes have been discovered in rocks approximately 2.5 billion years old, and prokaryotes remained the only form of life on Earth for more than 1 billion years. Another group of prokaryotes, the Archaea, were discovered in the 1970s. Archaea were originally thought to be found mostly in extreme environments, such as hot springs and vents on the ocean floor, but now they are known to exist in many ecosystems.

Some bacteria require only simple nutrients and CO_2 to synthesize their own organic carbon molecules and are called autotrophic. For example, cyanobacteria can photosynthesize. They were previously known as the blue-green algae, but the important cellular differences between prokaryotes and eukaryotes make them more closely related to bacteria than to algae—all of which are eukaryotes. Most bacteria are heterotrophs, like

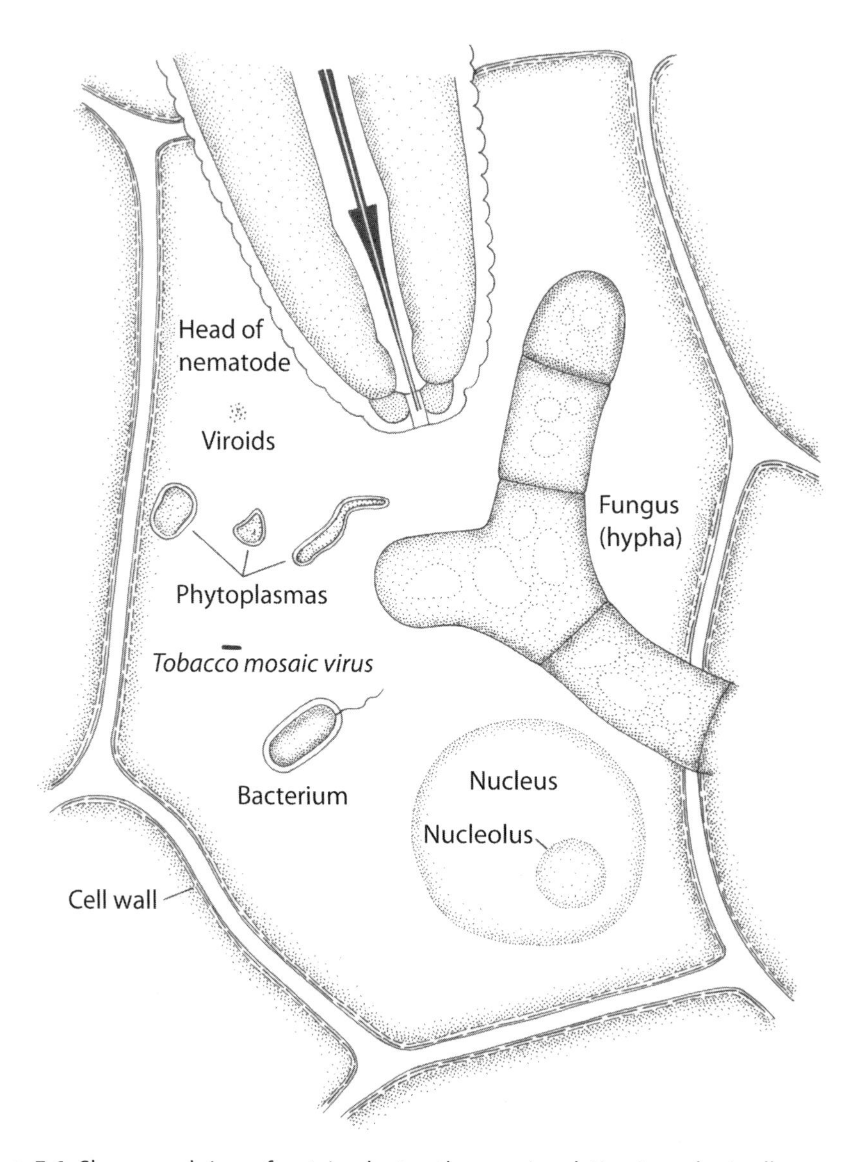

Figure 5.1. Shapes and sizes of certain plant pathogens in relation to a plant cell.

fungi, and require organic carbon from outside sources. Most bacteria also are saprophytes, obtaining nutrients from nonliving organic matter, and so they play an extremely important role as decomposers.

When bacterial cells divide, the cell membrane and cell wall grow inward to divide the cell in two—a reproductive process called binary fission. Most of the genetic information of a bacterium exists in a single circular chromosome that replicates, and each "daughter" cell receives a copy. Under optimal conditions, some bacterial cells divide in less than 20 minutes, so that a single cell can generate a population of billions of cells in less than a day. This enormous reproductive capacity accounts for the ability of bacteria to quickly colonize new food sources and to rapidly cause infection and disease. Unlike fungi, most bacteria do not produce specialized survival structures. Instead, they rely on their ability to multiply rapidly when conditions are favorable.

Bacteria commonly exist in three general shapes: bacilli, which are rod shaped; cocci, which are spherical; and spirilli, which are coiled. Common bacteria have cell walls that create the shapes just described. The bacteria with cell walls that are plant pathogens are all rod-shaped bacilli. Some bacteria, called mollicutes, have no cell walls. Among the plant pathogens, the mollicutes include phytoplasmas and spiroplasmas. Because they are enclosed only by cell membranes, they do not have the clearly defined shapes of walled bacteria. Finally, some bacteria produce long, branched filaments. These soil inhabitants, called actinomycetes, are known to produce inhibitory compounds (antibiotics) that affect the growth and development of other microorganisms. Some actinomycetes are responsible for the characteristic odor of moist soil.

Some bacteria move by means of flagella (Figure 5.2). The number and arrangement of flagella on a bacterial cell are important taxonomic characteristics. The flagella are so small and move so rapidly that special techniques must be used to make them visible under a microscope. Flagella do allow bacteria to move on their own, but bacteria are so small that they do not move long distances by self-propulsion.

Bacteria lack the sexual cycle of eukaryotes for genetic recombination, but they do have several means of increasing genetic variation. Deoxyribonucleic acid (DNA) can be moved from one bacterial cell to another by the transfer of genes from small, circular pieces of DNA called plasmids. Plasmids are not part of the cell's chromosome but rather exist independently. One bacterial plasmid found in a plant-pathogenic bacterium—the Ti plasmid—plays a critical role in genetic engineering (see Chapter 6). Pieces of DNA also can be transferred by viruses that infect bacterial cells, which are called bacteriophages.

Science Sidebar

Origin of the eukaryotic cell

The membrane-limited organelles of eukaryotes—such as the mitochondria and chloroplasts that function in respiration and photosynthesis, respectively—are believed to have originated from free-living prokaryotes that came to exist symbiotically within what was to become the eukaryotic cell. The mitochondria and chloroplasts divide independently of mitotic cell division in eukaryotic cells. They also contain their own genetic information in a form similar to that of bacteria.

Recent research has shown that the separate genetic information contained in organelles is a significant component of the total genetic makeup of a eukaryotic organism and can influence the interactions of plants and their pathogens. During sexual reproduction of eukaryotes, the organelles are all contributed by the female. This occurs because the male contributes only nuclear genes to the next generation, whereas the female contributes the cytoplasmic contents of its sex cell plus nuclear genes. The genetic information associated with these organelles is variously described as extrachromosomal (outside the nuclear chromosomes), maternal (contributed only by the female parent), or cytoplasmic (in the cytoplasm—that is, outside the nucleus) inheritance. We will return to the significance of these genes in plant disease in Chapter 7.

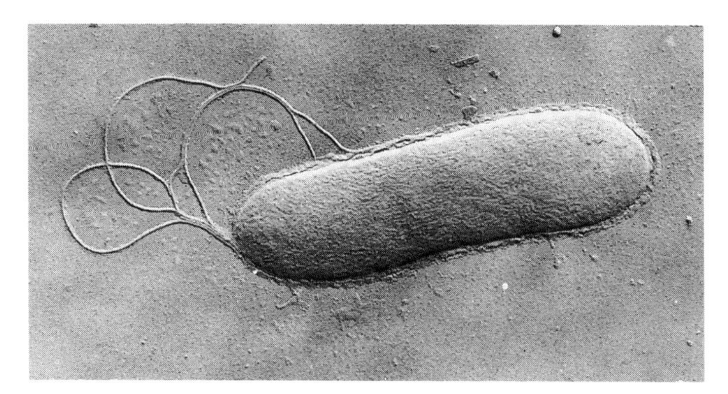

Figure 5.2. A bacterium, *Erwinia amylovora,* with flagella (electron micrograph).

The most common source of genetic change in bacteria is mutation: a mistake that occurs during replication of the DNA. Mutations occur only rarely but can be significant, depending on the generation time of an organism. Plants and animals reproduce only once or a few times a year—or even less often, as in the case of humans. The chance of a mutation occurring in bacteria is no greater than in these other species. However, some members of the population of bacteria that result from a single bacterium reproducing over a 24-hour period will contain mutations. Bacteria multiply so rapidly and exist in such large populations that they are able to adapt rapidly to changes in their environment through the genetic variation provided by mutation.

Of the approximately 9,000 classified bacterial species, only a relatively small number are actually pathogens that cause disease in animals, people, or plants. Except for a few species that may be able to infect humans under extremely unusual conditions, the bacterial species that cause plant diseases threaten only plants.

Science Sidebar

Biofilms

Biofilms are aggregates of bacterial cells that stick to each other or to a surface. Familiar examples include the plaque on teeth and the slippery layer on rocks in river beds and ponds. Biofilms can be hazardous when they clog the filters and medical tubing used in hospitals, but they can be beneficial when they form on particle surfaces in wastewater treatment plants and enhance the degradation of pollutants in dirty water. Scientists have discovered bacteria that can help clean up oil spills and other kinds of environmental contamination.

Bacteriologists are studying biofilms to better understand how these bacterial populations communicate with each other. This chemical communication, called quorum sensing, is important in the production of toxins and other chemicals necessary for successful infection by pathogens. Disrupting biofilm formation may be possible in cases where it is a precursor to health problems, such as gum disease, infections in hospital patients, and bacterial invasion of plants. Finding new approaches to managing bacteria is increasingly important as antibiotic resistance becomes more prevalent.

The Impact of Bacteria on Our Health and Our Environment

Bacteria are all around us. Most of them are not pathogens and may even be beneficial to our health. It has been estimated that approximately 1,000 species of bacteria are found on the skin of a healthy person! In fact, the number of bacteria in and on our bodies is greater than the number of cells in our bodies. Sometimes, however, bacteria can invade our bodies and cause disease—just as they can invade and cause disease in plants.

Most people are familiar with the use of antibiotics to fight bacterial infections. Many antibiotics are produced by soil microorganisms to inhibit the growth of other soil microorganisms. Antibiotics may provide an advantage in the highly competitive soil environment. Humans grow these microorganisms in laboratories and harvest the antibiotics for use against bacterial infections.

When a single type of antibiotic is used repeatedly against a bacterial infection in humans, animals, or plants, resistant strains of bacteria may develop. A classic example is the resistance of some strains of gonorrhea bacteria to penicillin. The ability of bacteria to develop resistance very quickly is one reason that medicinal antibiotics are available by prescription only. Even though scientists continually search for new antibiotics, the safe, well-known types should be used with care to prolong their effectiveness. A life-threatening situation could develop and a cure might be delayed when one antibiotic is found ineffective and others must be tried. In fact, many people question the use of antibiotics in hand soaps because of the risk of selecting more bacteria with resistance. Your hands can be cleaned effectively with thorough washing with soap and water or an alcohol-based sanitizer.

Of the antibiotics currently produced each year, approximately 50% are used in medicines to treat people, 40% are used in animal care, and less than 1% are used to treat plant diseases. Nearly all of the antibiotics used on plants are used to manage a single disease: fire blight of apples and pears, which is discussed later in this chapter.

Antibiotics are not commonly used for plant diseases because of their high cost, their ineffectiveness against many bacterial diseases, and the rapid selection of resistant bacteria that develop with repeated applications. In addition, there are concerns about the potential impact of these antibiotics on human health. If resistant plant-pathogenic bacteria become more common, the resistance genes may be transferred to other bacteria in the environment and eventually to human pathogens. Unlike eukaryotes, bacteria commonly transfer genes between species. A similar concern exists with the routine use of antibiotics in animal feed. The more resistant bacteria there are, the greater the chance that human pathogens will pick up resistance genes from them.

Although a single bacterial cell can do little in the environment, its ability to multiply rapidly when favorable conditions exist provides this tiny organism with a major advantage. In some ways, bacteria can work together and act something like a multicellular organism. Some bacterial populations form biofilms, which help them attach to surfaces and protect them from certain antibiotics and other environmental threats. Although most bacteria rely on rapid multiplication for survival, some Gram-positive bacteria produce thick-walled internal structures called endospores, which protect the bacterial DNA from heat, desiccation, and other environmental stresses. Fortunately, the bacteria that cause plant diseases do not produce endospores, so they are less able to survive adverse environmental conditions.

Science Sidebar

Endospores

When viewed with a microscope, an endospore looks like a small, white egg inside a stained bacterium. Anthrax of humans and animals is caused by *Bacillus anthracis,* a bacterium that produces endospores. When diseased animals die and decay, the endospores are released into the soil. Later, the endospores may find their way into a cut or scrape on another animal and infect this new animal host. The anthrax bacterium was used in recent attempts at bioterrorism, because the endospores could be dried into a powder and dispersed for inhalation by unsuspecting victims.

A related bacterium, *B. thuringiensis,* also produces endospores. When certain insects ingest these endospores, a toxin is produced, killing them. These endospores make it possible to mix a dried product in water for use in sprays to control many insect pests, such as the gypsy moth and corn ear worm. This bacterium is not harmful to people and other organisms, so it is commonly used as a biological control agent and can be used in producing organic food.

Two important human pathogens are anaerobic bacteria that produce endospores. The tetanus bacterium (*Clostridium tetani*) survives in soil as endospores until a deep puncture wound introduces it into another animal host. There, the growth of the bacterium produces a potent neurotoxin. *C. botulinum* is another anaerobic organism that produces endospores that are found in soil and can contaminate fruits and vegetables. If home-canned food is not processed at the correct temperature and pressure to destroy the endospores, the bacteria will multiply easily in the closed container and poison the person who eventually eats the contents. These bacteria produce a deadly nerve poison called botulinum, which can paralyze the muscles necessary for breathing and thus cause respiratory failure. Botulinum toxin is the active ingredient in BOTOX, a product that causes a temporary, relaxed paralysis of the muscles at injection sites for people seeking more youthful-looking faces.

Science Sidebar

Bacterial contamination of food

The ability of bacteria to colonize fruits and vegetables can have health implications for people. Today, many stores offer a variety of bagged salad greens and precut vegetables and fruits. The fruits and vegetables provide nutrients and the containers retain moisture, so contaminating bacteria can multiply, especially if refrigeration is not consistent.

A number of cases of food poisoning have recently been reported from animal pathogens, such as *Salmonella enteritidis* and *Escherichia coli* strain O157:H7. The source of these bacteria can be contaminated water used in irrigation of imported produce or contamination by farm workers who do not have access to sanitary facilities in agricultural fields. Previously, these bacteria were most commonly found in eggs and hamburger, respectively. Most people cook eggs and meat products, which can kill many bacterial contaminants. However, consumers eat many fruits and vegetables raw, increasing the danger of food poisoning. In Asia, where human waste has been used for many years to fertilize fields, food is commonly prepared by stir frying at high temperatures, which helps to kill bacterial contaminants.

Figure 5.3. Thomas J. Burrill in a laboratory with students (1882).

How Bacteria Are Isolated and Identified

Bacteria are ubiquitous and found on most plant surfaces. Diseased plant tissues are rapidly colonized by saprophytic bacteria as soon as dead cells are present, so the accurate diagnosis of a plant disease is often confounded by the presence of such secondary invaders. It is difficult to distinguish pathogenic bacteria from the saprophytic species that also are present.

In 1866, shortly after fungi were shown to cause disease in plants, French scientist Louis Pasteur first proved, in experiments with wine, that bacteria are not produced by spontaneous generation. Over the next few years, the role of bacteria in human and animal diseases became clear, and Koch's postulates allowed pathogenicity to be proven (see Chapter 1). Bacteria were generally accepted as animal pathogens, and fungi were considered the agents of plant disease.

In 1878, studies by American plant pathologist Thomas Burrill implicated a bacterium as the disease agent of the fire blight in North America that was causing the death of apple and, especially, pear trees (Figure 5.3). At first, Burrill thought that fire blight was caused by a fungal pathogen, but then he noticed that fluid from diseased tissue was "alive with moving atoms known in a general way as bacteria." He was able to transmit the disease from infected to healthy plants, which subsequently became diseased, but he did not isolate the bacterium in pure culture. That feat was accomplished by another American scientist, J. C. Arthur, who completed Koch's postulates with the fire blight bacterium in 1885 and thus proved that a bacterium could cause a plant disease.

The fire blight bacterium is now called *Erwinia amylovora.* The genus *Erwinia,* which includes many species, honors Erwin F. Smith, an important American plant bacteriologist who contributed greatly to our understanding of plant diseases caused by bacteria, beginning in 1895. Since Smith's last name was so common, scientists used his first name for the genus.

In the late 1800s, laboratory techniques were too primitive for scientists to accurately identify bacteria. One important advance involved sterilization, usually by heat, of the nutrient media in which bacteria could grow. Early experiments were performed using various broths, wine, urine, and other liquids. Most bacteria will grow in a variety of liquid nutrients, including raw milk. After the liquids were heated to kill the bacteria already present, samples containing the bacteria to be studied were introduced and then the bacteria would multiply. Most of the samples were mixed populations of bacterial species. One important limitation of using liquid nutrient media is that in a mixture, all of the bacteria grow together, so that contaminants are not always perceived.

The sterile cut surfaces of vegetables such as carrots and potatoes also were used to culture bacteria. This approach worked because most bacteria can grow using the rich nutrients provided by the wounded plant cells. Today, bacteria are isolated on sterile nutrient media in laboratories. The addition of a gelling substance to a liquid nutrient medium provides a moist, semisolid surface on which individual colonies of bacteria can be grown. Nutrient media today often include agar, derived from seaweed, which gels after it is boiled and allowed to cool. When bacteria are streaked across the sterile, gelled surface of the medium, individual bacterial cells multiply to form small circular, individual colonies, usually in a day or two (Figure 5.4). This technique was an important advance in the study of bacteria in pure cultures of a single species.

Even after a pathogenic bacterium has been isolated, it is often difficult to identify. The commonly used bacterial characters—shape and Gram stain—are of limited utility for plant-pathogenic bacteria, since all are rod shaped and most are Gram-negative. Plant-pathogenic bacteria are commonly identified to genus based on their ability to grow on so-called selective media. These media contain various antibiotics and other nutrient mixtures known to allow the growth of bacteria of some genera but not others.

Bacteria also can be identified by other characteristics. Some bacteria can grow in the absence of oxygen. For instance, these anaerobes are able to grow when mineral oil or another oxygen-blocking agent is layered over a liquid medium inoculated with bacteria, whereas most other plant-pathogenic genera are not.

Bacteria in the genus *Pectobacterium* cause soft rot when they produce enzymes that destroy pectin, the intercellular substance that cements plant cells together. Anyone who has kept vegetables in the refrigerator too long knows the mushy texture and foul smell of

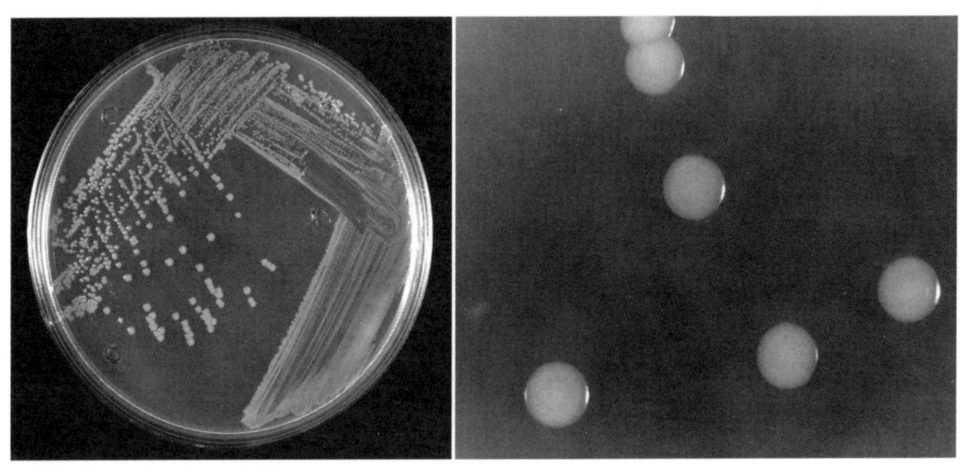

Figure 5.4. Bacterial colonies on a nutrient agar plate. **Left,** dilution streaking to obtain individual colonies; **Right,** close-up of colonies.

Science Sidebar

Isolating a bacterial pathogen

Bacteria found in infected plant tissue are often a mixture of the pathogen and saprophytic secondary invaders. One simple isolation technique relies on the assumption that pathogenic bacteria are probably present in a higher concentration in the infected plant tissue than are the contaminating saprophytes.

First, infected tissue is surface disinfested with alcohol or a dilute bleach solution to kill contaminating microorganisms. Then, the tissue is macerated in sterile water to release bacteria from it into the liquid. The sample of liquid containing the bacteria is diluted with more sterile water in a series of clean sample bottles. A small amount of liquid from each bottle is then streaked across the surface of a petri plate of nutrient agar medium, using a sterile tool such as a bent glass rod. Petri plates are the covered, circular dishes found in laboratories everywhere. They were invented by R. J. Petri, a 19th-century microbiologist and student of Robert Koch, to prevent contamination from airborne microbes.

In 24–48 hours, individual bacterial cells usually produce circular colonies that are visible on the medium. Plates containing samples from the lower dilutions often have many colonies of variable colors and morphologies. In plates containing more dilute samples, fewer colonies should be present. These colonies are most likely to be the pathogenic species, because the contaminating bacteria have been reduced and perhaps eliminated by dilution.

Each tiny, circular colony grows from a single cell and should therefore be a pure culture of the causal agent. Of course, it is only presumptive evidence that the bacterial colonies that appear after extensive dilution from the original sample are actually the pathogen. Using Koch's postulates, a pathogenicity test must be done on a healthy plant to prove that the isolated bacterium is the causal agent of the disease.

lettuce, potatoes, and other vegetables infected with soft rot bacteria. These bacteria can be identified by growing them on a medium containing pectin and observing pits that develop in the surface as the bacteria degrade the pectin.

Other identifying characteristics include the ability of some bacterial species to fluoresce, or glow, under ultraviolet light when grown on certain media or to take on a gummy or yellow appearance in culture. *Xanthomonas* species are named for their distinctive yellow color, compared to the cream color of most bacteria. The abundant sticky gum produced from the cell walls of *Xanthomonas* bacteria is harvested and used as a thickener called xanthan gum in many foods, such as creamy salad dressings.

Most diagnostic laboratories routinely complete a number of tests to arrive at the correct genus name for a bacterial isolate. Although a genus name for a bacterium is usually accepted by most scientists, identifying bacteria to species is quite another matter. Today, bacterial genus and species designations are based on similarities in DNA.

Bacteria as Plant Pathogens

To cause a plant disease, a bacterium must survive for periods of time when actively growing host plants are not present. Then, the bacterium must be dispersed to and finally invade a healthy plant. It should not come as a surprise that bacterial plant diseases are much more common in warm and humid climates. When soils do not freeze, some bacteria survive as saprophytes in plant debris and soil. Many plant-pathogenic bacteria die when plant tissue decays, especially in the dry, cold conditions of winter soils. In temperate climates, bacteria survive in protected locations, such as cankers (sunken diseased areas) on perennial plants, or in seeds or other propagative parts, such as bulbs, corms, rootstocks, and tubers. Bacterial plant diseases are more common in crops that are vegetatively propagated, because the handling and wounding of these vegetative parts by humans is an effective means of spreading bacteria between plants.

Some very important plant diseases are caused by bacteria that are seedborne. For example, snap beans are susceptible to three seedborne bacterial diseases: common blight, brown blight, and halo blight. In the past, seed was commonly treated with hot water to kill the bacteria, but the process often had adverse effects on the seed's viability. Many seed-production farms now exist in the dry western areas of the United States, where bacterial diseases are much less common than in other areas. European countries that lack arid areas obtain vegetable seed grown in dry areas of Africa or the U.S. West. Under dry conditions, bacteria-free seed can be harvested, tested, and certified as meeting acceptable standards of freedom from pathogens for use in more humid areas. Many bacterial diseases of annual flowers and vegetables have been greatly reduced by the use of certified seed.

As mentioned previously, no plant-pathogenic bacteria form endospores, so they remain quite vulnerable to environmental extremes. They are particularly susceptible to desiccation. The vulnerability of plant-pathogenic bacteria to destruction limits not only their ability to survive in the absence of a host plant but also their ability to disperse rapidly and to distant places. Because no endospores are produced and the bacterial colonies comprise sticky masses of cells, plant-pathogenic bacteria are unable to disperse a long distance through the air, as many fungal spores do.

For bacteria to move from one plant to another, they must rely on passive dispersal by water, wind-blown water, insects and birds, and movement of soil on tools, machinery, and agricultural workers. When dew and rain are frequent, bacteria can be splashed and spread by tools and machinery. In the United States, such growing conditions are found primarily in the southeastern states and in greenhouses, so bacterial diseases are more common in these locations.

Agricultural implements, such as pruning shears, not only disperse bacteria but also create wounds for entry. Not disinfesting a tool between cuts is something like performing surgery on one patient after another without cleaning the scalpel. New infections can occur with every cut of a susceptible plant. Humans are responsible for most long-distance movement of plant-pathogenic bacteria. Bulbs, fruits, plants, seeds, and soil clinging to tools and machinery have carried bacteria to all parts of the world during trade and commerce (Figure 5.5).

As discussed in Chapter 4, plant tissues present a formidable barrier to invasion by bacterial cells. Many pathogenic bacteria exist in a saprophytic state on leaf surfaces until conditions are conducive for infection. In most cases, the bacteria must increase in number for the enzymes and toxins they produce to be present in sufficient quantities to affect the nearby plant cells. Bacteria can enter healthy plant tissues only through wounds or natural openings—points of vulnerability where the cuticle is thin or nonexistent, such as stomata, newly expanding leaf tissue, nectaries, hydathodes, and lenticels (the lens-shaped air-exchange openings on bark and stems)

Science Sidebar

Pathovars

For many years, a newly discovered plant-pathogenic bacterium was identified to genus and then often given a species name that reflected the host plant from which it had been isolated. However, many bacteriologists believed the number of species was getting too large, because the discovery of each new disease often led to identification of a new bacterial species. These scientists did not believe that differences in host range alone necessarily indicated sufficient biological differences to justify a separate species designation.

In 1980, a new taxonomic category was created to try to solve some of the problems of naming new species, subspecies, strains, and races of bacteria. When two strains of bacteria appear to be identical with respect to morphology and biochemical and physiological tests—and vary only in host range—they are differentiated by the term pathovar (abbreviated pv.). For example, halo blight of bean (*Phaseolis vulgaris*) is caused by *Pseudomonas savastanoi* pv. *phaseolicola*.

Adopting this new taxonomic category greatly reduced the number of plant-pathogenic species and more accurately reflected their known biologies, as currently understood. Today, bacterial taxonomy is in a state of flux because of the tremendous amount of new information being discovered through molecular biology. Many names are changing, and the term pathovar may be discarded in the future.

Figure 5.5. Some means of pathogen dispersal.

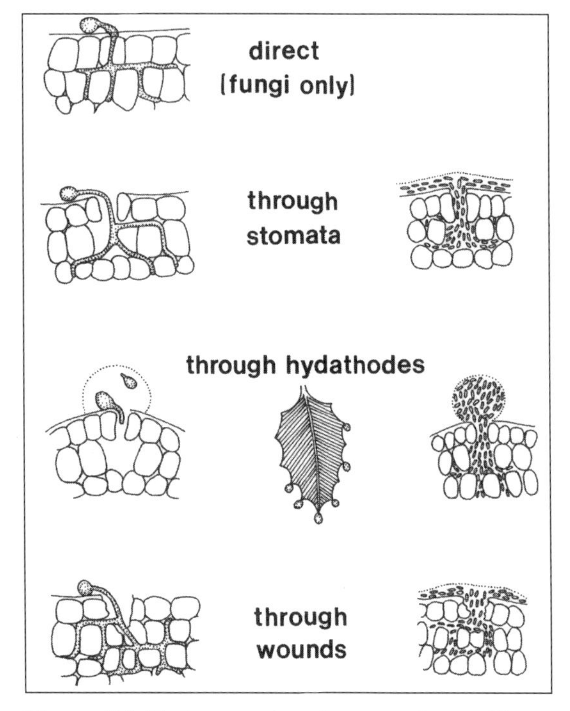

Figure 5.6. Various modes of penetration and invasion of a leaf by fungi (**Left**) and bacteria (**Right**).

(Figure 5.6). Wounds not only expose vulnerable plant cells to bacteria but also contain torn cells, which provide nutrients for the rapid multiplication of the bacteria and the subsequent increases in the enzymes and toxins they produce.

Clearly, it is quite a challenge for plant-pathogenic bacteria to survive, spread, and invade new hosts. Some bacteria, however, are highly successful and therefore highly destructive plant invaders. Their stories are told in the sections that follow.

Fire Blight: Why Most Pears Are Grown in the Far West

Many important plant diseases are caused by bacteria, including the previously mentioned fire blight of pome fruits. Pome fruits include apple, crabapple, and pear.

E. amylovora is a common bacterial pathogen of many North American native plants in the rose family, the Rosaceae. Disease development is usually quite limited on native hawthorns, mountain ash, serviceberry or shadbush, and other closely related plants. *E. amylovora* became an economically important pathogen as a result of the intercontinental movement of plants and agricultural practices. When European colonists arrived in North America, they brought with them apple and pear trees, which were quite susceptible to *E. amylovora*. Unlike many situations in which a pathogen causes an epidemic after being inadvertently introduced to a native plant population, this situation involved the introduction of highly susceptible foreign plant species to a native bacterium.

Pear trees are especially susceptible to fire blight. Whole trees can die during a single season, leaving them with blackened foliage and twigs that demonstrate the origin of the name fire blight. The first recorded epidemic was in New York State in 1780, long before

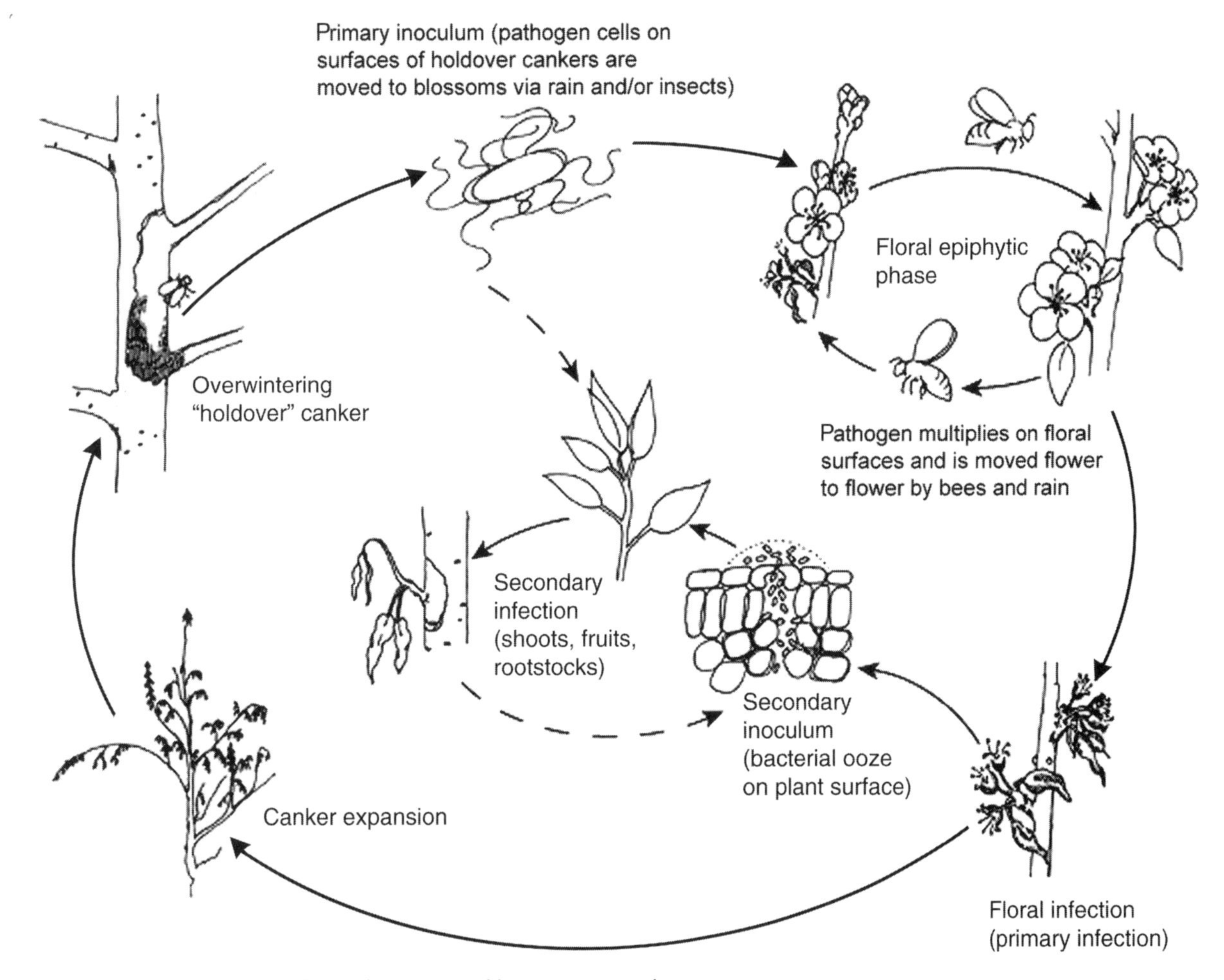

Figure 5.7. Disease cycle of fire blight caused by *Erwinia amylovora.*

the role of bacteria in plant disease was discovered. Epidemics of fire blight were limited in the beginning. Orchards were small and often planted from seeds, so there was more genetic diversity among the trees. As orchard size increased and trees were grown as clones created by grafting, fire blight became a devastating disease.

Fire blight bacteria survive winter in the margins of cankers, which are sunken, infected areas in woody branches and trunks. In the spring, the bacteria begin to multiply and ooze out of the cankers through cracks in the bark. The sticky bacteria become attached to insects, which are attracted to the sweet ooze. When the insects visit the blossoms of fruit trees and the wild relatives of those trees, they deposit the bacteria. Bacteria in ooze also are distributed in splashing and wind-blown rain (Figure 5.7). They are able to survive on the surfaces of flowers and in the nectaries until environmental conditions are optimal for infection.

Most infections occur in the flowers and on young, expanding leaves, where insects, humans, and storms cause wounds. After beginning an infection in the flower, the bacteria multiply quickly and produce toxins and enzymes that cause the tissues to blacken and die. The infection spreads down the stem of the flower spur, and unless pruned off, the infection may eventually reach the trunk of the tree. Extensive trunk cankers can kill a tree.

Science Sidebar

Molecular interactions between hosts and pathogens

The molecular interactions of *E. amylovora* and its host plants have been studied in great detail in recent years. One surprising discovery has been that some genes that make plant pathogens able to attack their hosts are similar to those of animal pathogens. Scientists are now using plant pathogens to study these genes, because it is easier to study genetic mutants in a plant system than in an animal one. However, the results might be useful in protecting both plants and animals from their bacterial pathogens.

One important area of study in plant pathology is investigating how a plant knows when it is being invaded by bacteria. A specific protein produced by *E. amylovora,* called harpin, is recognized by plants as the signal to turn on their defense responses. Harpin is named for the hrp (pronounced "harp") bacterial genes necessary for pathogenicity. This protein is available as a commercial plant-protection product called EMPLOY.

The early fire blight epidemics that devastated eastern pear orchards caused farmers to abandon most of them. Pear orchards were established in the western states instead, as agriculture developed there. *E. amylovora* also was carried west in the infected tissues of the pear and apple trees brought to establish the new orchards.

Between 1900 and 1910—before management strategies had been designed—fire blight epidemics destroyed 95% of the pear trees in the San Joaquin Valley of California and other areas throughout the U.S. West. Production in the United States is still centered in the dry western states, because pear orchards are nearly impossible to maintain in humid areas. Protecting the trees from fire blight requires the careful pruning of cankers and the use of chemical sprays. Fire blight also is an important disease of apple trees, but they are less susceptible than pear trees and can be maintained in humid regions with appropriate management.

Copper sprays are toxic to bacteria but also can damage plant tissue and developing pears. Such damage is called phytotoxicity. An alternative spray is an antibiotic such as streptomycin. Antibiotic therapy for trees is relatively expensive, and if used frequently, the bacterial population may become resistant to the antibiotic and cause disease despite its application. As discussed previously, some scientists favor banning agricultural antibiotic sprays and restricting the use of antibiotics for plant disease to cases with less environmental exposure—for example, the injection of trees infected with phytoplasmas, as discussed later in this chapter.

Biological control of fire blight is sometimes successful with the use of other bacteria isolated from the surfaces of host plants. Commercial formulations of these bacteria are applied to flowers, where they colonize the areas where fire blight bacteria usually multiply and begin infection. These antagonistic bacteria multiply so rapidly that the pathogen cannot establish itself and cause disease.

Careful pruning of the overwintering cankers remains the most effective management practice for fire blight. To be certain that all infected tissue is removed, it is important to cut well below any apparent branch cankers. Pruning shears should be disinfested with a 10% bleach solution or alcohol between cuts, so bacteria are not spread to other pruning cuts on the same tree or other trees.

A number of plants common in home landscapes and gardens are susceptible to fire blight, including cherry, crabapple, pear, quince, raspberry, and rose. Thus, it is important not to spread the bacteria via pruning if fire blight is found on a property. Infected prunings should be burned or placed in the trash to avoid the spread of bacterial ooze to healthy plants.

E. amylovora was eventually transported to the Eastern Hemisphere. In 1919, fire blight was first reported outside North America in New Zealand. In 1957, it was discovered in England, perhaps after being carried on infected fruits or shipping crates from New Zealand. The disease then spread to continental Europe, initially appearing in 1966 on wild hawthorns on islands off the coast of the Netherlands. The appearance of the disease in this location suggests the bacteria were carried by birds that found hawthorn fruit attractive. These early areas of infection were burned in an attempt to eradicate the disease, but it is now firmly established in the Netherlands. Although pear orchards can be carefully pruned, the common hedgerows of hawthorn remain a significant source of fire blight bacteria, which are carried by insects to the orchards.

Fire blight has continued to spread and is now present throughout the British Isles, most European and Mediterranean countries, and as far as the Middle East, where it threatens not only pome fruits but also many rosaceous ornamentals. In 1995, fire blight was observed in the Po River Valley of northern Italy, the largest area of pear production in the world. The Italian government has destroyed hundreds of thousands of trees in an attempt to eradicate *E. amylovora.* Australia is fighting to prevent the introduction of fire blight from New Zealand, given the high cost of managing the disease and its impact on trade to other countries in which it has not yet been found, such as Japan. Fire blight also has spread in the Western Hemisphere into Canada and some countries in Latin America. Certainly, fire blight caused by *E. amylovora* will continue to have important economic impacts for years to come.

Citrus Canker: The Recent Bane of the Florida Citrus Industry

In Florida, the citrus industry has traditionally worried most about the danger of frost, but in the fall of 1984, a serious bacterial disease problem was reported in citrus nurseries. The new disease caused lesions on citrus leaves and fruit, making the fruit unmarketable (Figure 5.8). The new disease—citrus canker—was not really new to the United States. It had been introduced previously in the early 1900s and was reported to have been eradicated from Florida in 1933, following the burning of millions of citrus trees at a cost of about $6 million in that state alone.

The 1933 eradication of citrus canker from the citrus-growing states of the United States is the only successful eradication of a well-established pathogen ever reported. No citrus canker was reported again in any citrus-growing state until 1984. To prevent reintroduction of the disease, stringent border inspections of plants and fruits were maintained. Between 1973 and 1978, citrus canker on infected fruits and plants carried by travelers was intercepted at borders 2,603 times. Even with these precautions, however, many scientists feared the quarantine would have only temporary success (see Chapter 3).

Surprising everyone, the 1984 report of bacterial citrus canker came not from a commercial orchard but from a nursery, where workers take particular care to protect plants from pathogens. The symptoms of the disease were atypical, but the bacterium was identified using standard methods as *Xanthomonas axonopodis* pv. *citri.* State and federal regulatory agents were required by law to initiate eradication programs, and more than 20 million citrus trees were subsequently destroyed.

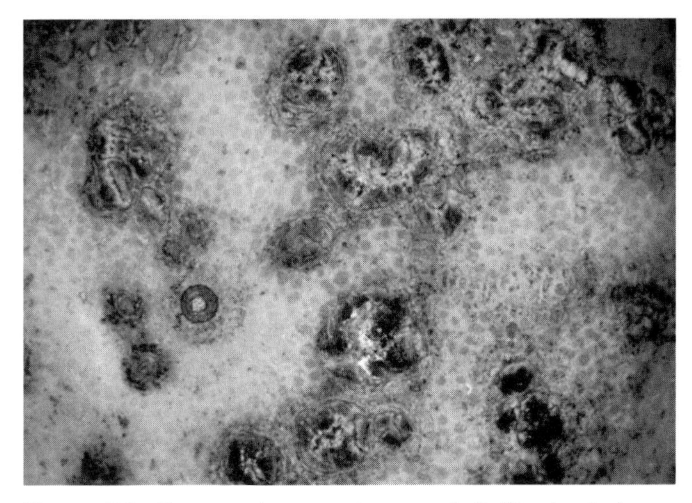

Figure 5.8. Citrus canker symptoms on fruit. The bacteria are not harmful to people.

Compared with many parts of the United States, Florida—with its warm and humid climate—is quite conducive to the survival and spread of bacteria. In addition, citrus is a vegetatively propagated crop, produced primarily by bud grafting. The propagation techniques require taking specific precautions to prevent the spread of bacteria to new plants. There also is considerable concern that the genetic uniformity of citrus plants makes them particularly vulnerable to major epidemics. Nearly all of the U.S. citrus cultivars are susceptible to infection by the citrus canker bacterium. Grapefruit is particularly susceptible.

In 1984, quarantine restrictions were established in Florida. They required the removal and destruction of not only infected trees but all citrus trees within 1,900 feet (579 meters) on both commercial and residential properties. Citrus workers wore clothing that could be removed at the site, and all machinery, tools, and vehicles were thoroughly cleaned after being used in a citrus grove. Shipments of plant material and fruit were restricted to non-citrus-growing states.

Despite the joint state and federal restrictions, citrus canker continued to spread throughout many areas of Florida. Many homeowners were angered as personnel from the Division of Plant Industry inspected and removed citrus trees from their yards. Often, the trees were not even diseased. Regardless, the goal was to produce citrus-free barriers around areas of outbreaks and thus prevent movement of the bacteria. Unfortunately, a series of major hurricanes cut across Florida in 2004 and 2005, resulting in more widespread outbreaks of the disease.

After an expenditure of nearly $1 billion, the eradication campaign was declared a failure, and citrus canker is now considered endemic in Florida. Disinfested fruit has been determined safe for shipment to other citrus-growing states, but quarantine restrictions remain on the movement of citrus plant material to Arizona, California, and Texas. Like citrus growers in Asia, Brazil, and many other areas of the world, those in Florida now must rely on cultural practices to manage the disease, such as the erection of windbreaks, careful pruning, and application of copper-containing sprays. Florida's citrus industry will continue to face the significant problem of reintroduction of citrus canker to the vegetatively propagated, genetically uniform citrus crop grown in a warm and humid climate.

In 2006, the state's citrus growers were dealt another blow with the discovery of a new bacterial disease, Huanglongbing (HLB), or citrus greening disease. This disease is unusual in that the pathogen, *Candidatus* Liberibacter asiaticus, is restricted to the phloem of the plant. It affects yield and quality and can eventually kill a tree. There is no known cure. The bacteria are transmitted by a tiny sucking insect vector, the Asian citrus psyllid. Quarantines are now in place for both the bacterial pathogen and its psyllid vector in all citrus-growing areas.

The value of the 2007–2008 Florida citrus crop was estimated at $1.2 billion, down 19% from the previous season. Nationwide, the citrus industry is worth more than $9 billion. In 2010, the U.S. National Academy of Sciences released a report suggesting that the only likely solution for this new plague is genetic engineering, because so little genetic resistance is available in existing citrus plants. This technology is discussed in detail in Chapter 6.

Fastidious Bacteria: The "Fussy" Pathogens

The HLB pathogen just described belongs to a group of bacteria that are difficult or impossible to grow on culture plates and are restricted to phloem or xylem tissues of plants. Such bacteria are called fastidious, because they require specific growing conditions.

For decades, many of the diseases caused by these bacteria were considered virus diseases. Like diseases caused by viruses, these diseases have symptoms that usually include stunting, yellowing, and various growth distortions. As with virus diseases, the pathogens cannot be easily cultured from diseased tissue, and transmission usually requires insect vectors, a living bridge (such as grafting), or the parasitic plant dodder. Today, we know that these diseases are not caused by viruses but by fastidious bacteria, such as phytoplasmas and spiroplasmas.

Phytoplasmas are fastidious bacteria that cause more than 200 plant diseases. They were originally called mycoplasma-like organisms (MLOs). They received this somewhat clumsy designation because mycoplasmas were first discovered as animal pathogens in the late 1800s, causing diseases such as pleuropneumonia. When organisms that resembled animal mycoplasmas were seen in plant cells in 1967, they were designated MLOs until more could be learned about them. Now that biochemical and genetic information has been obtained, the genus *Phytoplasma* has been proposed but not yet officially approved.

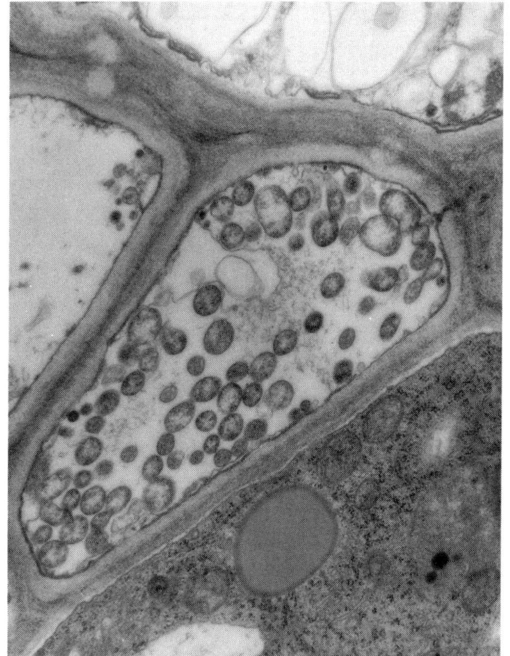

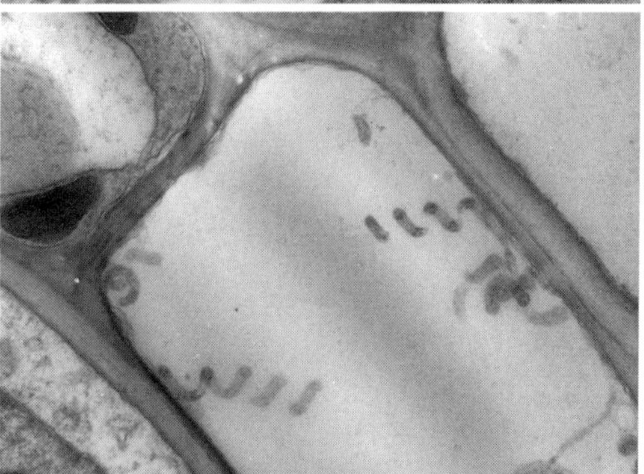

Figure 5.9. Phloem cells of a plant infected with phytoplasmas (**Top**) and spiroplasmas (**Bottom**) (electron micrographs).

Phytoplasmas are obligate pathogens that have not, so far, been grown in culture. They are bacteria that lack cell walls and are restricted to the phloem of their host plants (Figure 5.9, Top). Many infected plants are induced to form a proliferation of branches in certain areas, a symptom described as a witches' broom. Although this is a common symptom of phytoplasma infection, a witches' broom can be stimulated by other pathogens, including mistletoes and rust fungi, and by certain arthropod pests. Phytoplasmas are spread between plants by insect vectors with piercing-sucking mouthparts called leafhoppers. In cool regions, phytoplasmas are carried by leafhoppers that are introduced annually in air masses from warmer areas. Phytoplasmas can even reproduce in their vectors. Some scientists have speculated that phytoplasmas originated as insect pathogens that were inadvertently injected into the nutrient-rich and protected environment of the phloem by leafhopper feeding, becoming plant pathogens as well.

When phytoplasmas were discovered in 1967, an entire group of "virus" diseases characterized by a symptom called yellows had to be redefined. Another common symptom of phytoplasma infection is an off taste in the fruits and vegetables from infected plants. (Eating the produce is not harmful, however.) Two economically important phytoplasma diseases are aster yellows, which causes problems in many annual flowers and vegetables, and X-disease of peach and cherry trees, which was so named because no one could determine its cause for many years.

Management of a phytoplasma disease depends on the specific host plant that is threatened. For an annual plant, vector management with insecticides may reduce the leafhopper population sufficiently to prevent economic damage. Phytoplasmas are sensitive to tetracycline antibiotics. Trees may be injected with tetracycline to prolong their lives, although the treatment must be repeated, because it causes only a remission in symptoms. This may be a short-term, cost-effective practice in a high-value cherry or peach orchard. Another effective practice involves removing wild cherry and plum trees near orchards, because the wild trees may serve as sources of phytoplasmas, which can be transmitted by leafhopper vectors.

In the Western Hemisphere, lethal yellowing of coconut palms is a severe phytoplasma disease for which few management strategies exist (Figure 5.10). High-value trees may receive tetracycline injections, but such an expensive treatment is not practical for the numerous trees in the poor areas of some Caribbean islands and coastal Mexico.

In Miami, Florida, more than 90% of the coconut palms have been eradicated. For tourists, these palms are merely part of the atmosphere of a semitropical vacation, but for many poor people, coconut palms are significant to their survival. Called the "trees of life," the coconut palms provide shade, food, and shelter. They also provide a source of income from the sale of several products derived from them: copra, which is the flesh or meat of the coconut, and the oil that is extracted from it, as well as charcoal, which is produced from the trees' shells.

Many trees have already been lost and replaced by resistant types, such as the Malayan dwarf coconut palm, but they take several years to mature. Unfortunately, some of the resistant cultivars are becoming diseased. Many alternative palm species not susceptible to lethal yellowing are available for use in landscaping, but both coconut and date palms are susceptible to the disease. This fact will continue to cause hardships for the people who depend on these important agricultural species.

Despite the negative effects of phytoplasmas on plants, one phytoplasma revolutionized the poinsettia industry with its beneficial effect. In Mexico, poinsettias grow wild and reach the sizes of shrubs and small trees. In the flower industry, poinsettias are vege-

Figure 5.10. Diseased coconut palms. The trees are dying of lethal yellowing, a disease caused by a phytoplasma pathogen in the Western Hemisphere.

tatively propagated from cuttings. At some point, a phytoplasma was introduced accidentally. It did not have a negative effect on the leaves or colorful bracts, which are often called the "flowers." Rather, the witches' broom symptom created the compact form of the poinsettia commonly seen today in Christmas displays.

Spiroplasmas are another important group of fastidious bacteria that cause several plant diseases (Figure 5.9, Bottom). As their name suggests, these organisms exist in helical forms. Some spiroplasmas have been cultured on nutrient media and been designated with scientific names. For example, corn stunt, caused by *Spiroplasma kunkelii,* is a relatively minor spiroplasma disease in the United States, but it has modern and perhaps historical significance in Central America.

By the time Hernán Cortés and the Spanish conquistadors arrived in Mexico around 1520, more than 50 large cities on the country's Yucatán Peninsula had already been abandoned. The disappearance of the region's Mayan inhabitants has been the subject of much speculation. In 1980, J. L. Brewbaker suggested that a virus disease called maize mosaic led to the Mayan people's starvation and the collapse of their empire before the Spanish invasion. L. R. Nault has provided evidence that the virus and its vector adapted to maize in Central America in the post-Columbian period. He has further suggested that corn stunt is a much more severe disease in the Yucatán Peninsula, where a leafhopper, *Dalbulus maidis,* thrives and transmits the spiroplasma so efficiently that corn still cannot be grown in the region. Corn stunt disease, along with maize mosaic, may have contributed to the destruction of the staple food of the Maya and thereby made Mexico vulnerable to Spanish conquest.

The fastidious bacteria just described are all restricted to phloem tissue. *Xylella fastidiosa* is an important member of the group of xylem-limited bacteria (XLB), which are clearly identified by their name. The XLB cause symptoms associated with the inability of the infected xylem to distribute water and minerals, such as stunting, wilting, and necrosis, as well as discoloration of the leaf edge called leaf scorch. *X. fastidiosa* is widespread in the Western Hemisphere and infects a variety of plants, causing important diseases such as bacterial scorch of shade trees and Pierce's disease of grapes (Figure 5.11).

Xylem is not a particularly attractive food source for most insects. It is quite dilute and does not contain the high concentrations of sugars found in phloem, but it does contain amino acids and some other organic molecules. Special leafhoppers, called sharpshooters, and spittle bugs routinely feed on the nutrient-poor xylem fluid. If *X. fastidiosa* is acquired by the insect from a plant, the bacteria develop into a biofilm along the esophagus of the insect vector and obtain nutrients as the xylem fluid passes through it.

X. fastidiosa causes citrus variegated chlorosis in citrus crops in Brazil, where it is especially severe in oranges. Brazil produces more oranges by far than any other country, and almost all of the nation's orange juice is exported. The state of Florida is the world's second-biggest producer of orange juice, but almost all of the juice produced there is consumed in the United States. Brazilian orange juice has become an important source of juice for U.S. consumption. In 2000, Brazilian scientists succeeded in sequencing the genome of *X. fastidiosa.* By understanding which genes are necessary for transmission of the pathogen, it may be possible to interfere with its spread by insect vectors.

Figure 5.11. Scorch symptom on a leaf from a sweet-gum tree infected by *Xylella fastidiosa.*

Although *X. fastidiosa* has been present in California for years, a new vector—the glassy-winged sharpshooter—became established there in the 1990s. Unfortunately, this vector multiplies quickly and develops large populations on citrus, avocado, and other important California crops. It can move easily to vineyards growing near citrus trees. The arrival of this new vector will likely increase the spread of the bacterium and threaten the California wine industry, along with a number of other important crops.

Crown Gall: The Natural Genetic Engineer

A final example of an important bacterial plant disease is crown gall disease, caused by *Agrobacterium tumefaciens.* Unlike the previously described diseases, crown gall is not characterized by symptoms such as cankers and dying tissues but rather by galls, or tumors, which were referred to in the past as "plant cancer" (Figure 5.12). Since these masses of undifferentiated plant cells are often seen at or near the soil line on the crown of the plant, the disease became known as crown gall. Although much smaller than crown galls, the galls produced on legume roots are caused by closely related bacteria that fix nitrogen.

A. tumefaciens lives in the intercellular spaces of the gall tissue. It survives for years in soil in association with the roots of many plant species and enters susceptible plants through wounds when they are planted in infested soil. The host range of this pathogen is very wide. Susceptible plants include apples, euonymus, grapes, raspberries, roses, and many other species of woody and nonwoody ornamentals. After infection occurs, galls appear and continue to grow, eventually crushing the vascular tissue and distorting the growth of the plant. The galls can be removed to make the plant more aesthetically pleasing, but that will not eradicate the infection. Special care must be taken to prevent the spread of crown gall bacteria between plants during pruning or grafting.

Figure 5.12. Crown gall on euonymus caused by *Agrobacterium tumefaciens.*

A commercially successful biological control exists for this disease. A closely related strain of the bacterium, *A. radiobacter* strain 84, can be applied to root systems of many susceptible plant species before they are planted. This bacterium produces an antibiotic that is specifically antagonistic to the crown gall bacterium. In addition, because this nonpathogenic bacterium becomes established on the root system or graft union first, it effectively competes with the crown gall bacterium in the soil to help prevent infection.

Some years ago, *A. tumefaciens* attracted considerable research attention from scientists who were trying to understand animal cancer. Although that research did not elucidate any new understanding of animal cancer, it did provide key information about crown gall disease. For example, studies showed that infected plant cells were transformed into uncontrollably dividing tumor cells, which continued to divide even when the bacteria were removed from the plant tissue. Thus, researchers established that *A. tumefaciens* is a naturally occurring genetic engineer. This bacterium contains a tumor-inducing (Ti) plasmid, which inserts some of its genetic information into the nucleus of the plant cell, causing it to divide repeatedly and form a gall. At first, the phenomenon seemed nothing more than a biological curiosity. Today, the crown gall bacterium is the workhorse of scientists who wish to insert foreign genes into plants.

To understand this discovery and the fundamentals of genetic resistance, it is necessary to understand the structure of DNA and how it functions in cells. Genes and genetic engineering are the topics of the next chapter.

CHAPTER 6

People Improving Plants: Genes and Genetic Engineering

People have long recognized that certain traits are passed on from parents to children, from animals to their young, and from plants to seedlings. Selective breeding has resulted in larger farm animals, faster racehorses, and higher-yielding crops. The means by which traits are inherited by successive generations, although recognized and exploited for centuries, has only been understood in recent years. New discoveries in this field are being made so rapidly that it is difficult to predict the possible innovations that may occur in even the next few years. Because the genetic information in an organism's deoxyribonucleic acid (DNA) contains the "blueprint" for all cellular control, some basic understanding of genes and their structure is necessary to appreciate modern biology.

Before we examine the molecular structure of DNA, let us consider a gene simply as a locus, or site, on a chromosome. The description of meiosis in Chapter 2 indicated that offspring receive one set of chromosomes from each parent. Thus, for any particular genetic trait, the offspring receive a gene on one chromosome from one parent and another gene for that trait on the homologous, or similar, chromosome from the other parent. When genes exist in at least two alternative forms, the forms are called alleles. Offspring may receive different alleles of the same gene on the homologous chromosomes contributed by both parents. An individual that receives two identical alleles of a gene is homozygous for that trait, and an individual that receives two different alleles of a gene is heterozygous.

A familiar example of a trait produced by gene alleles is attached versus nonattached earlobes in humans. The trait of earlobes that hang down freely from the sides of the head is dominant over earlobes that are attached—a recessive trait. The actual genetic constitution of an individual is called the genotype, but the resulting appearance or function of an individual is called the phenotype. Individuals with free earlobes have the same phenotype but may have different genotypes. They may have two dominant alleles for free earlobes, or they may have one free earlobe allele and one attached earlobe allele and be heterozygous. If both parents are homozygous for either the dominant or recessive trait, they will always have children with the same trait. Two heterozygous parents have a 25% probability of producing a child with the recessive trait—attached earlobes (Figure 6.1).

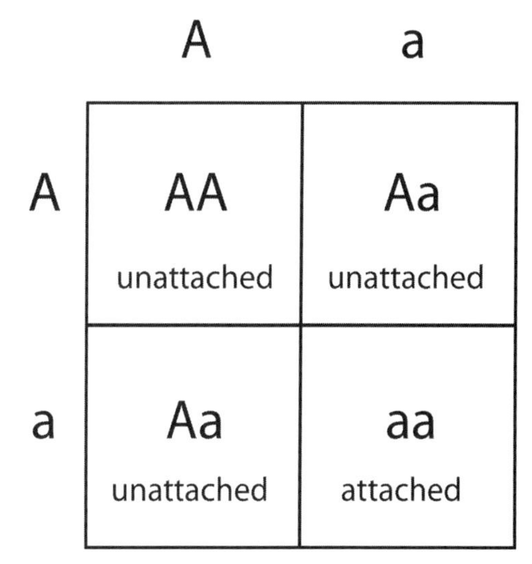

Figure 6.1. The allele for unattached earlobes is dominant over the recessive attached earlobe allele in humans. Individuals with the unattached earlobe phenotype may have homozygous (*AA*) or heterozygous (*Aa*) genotypes. Two parents with unattached earlobes may have a child with attached earlobes (*aa*), if each is heterozygous (*Aa*) for unattached earlobes.

Mendelian Genetics: Predicting the Results

The beginning of modern genetics can be traced to 1865, when the famous studies of Gregor Mendel were first published. As a friar in Austria, Mendel conducted many scientific experiments involving weather and physics. He is best known, however, for his observation that genes exist as units of genetic information.

Mendel recognized that certain traits are inherited in a predictable manner. He made specific crosses between garden peas with different traits, such as flower color, seed color, and smooth or wrinkled seed coat. These traits exist as alleles of genes, as just described for earlobe attachment in humans. Mendel chose pea plants and carefully controlled the pollination of flowers so he would know the phenotype of each parent. He then observed the phenotypes of the resulting progeny and determined that certain traits exist as distinct units that are inherited independently. His experiments explained how factors that we now recognize as genes are passed on from parents to offspring to maintain distinct traits.

Mendel conducted his pea experiments in his free time. He worked as a teacher—a role for which he apparently had little talent, since he never obtained his teaching credentials. He did, however, have a good supply of dried peas to throw at students who dozed off in his classes.

People sometimes comment that certain traits "skip a generation." This is a somewhat inaccurate explanation of a common genetic occurrence. Mendel observed the phenomenon of dominant and recessive traits, which explained for the first time how traits might disappear in one generation and reappear in the next. If two parents are homozygous for the same allele, they will produce only offspring with their phenotype. For example, in peas, red flower color is dominant over white flower color. Homozygous red-flowered parents will produce only red-flowered offspring, and homozygous white-flowered parents will produce only white-flowered offspring. A cross between a homozygous red-flowered parent and a homozygous white-flowered parent will produce only heterozygous red-flowered offspring. The white flower trait will seem to have disappeared in that generation (Figure 6.2).

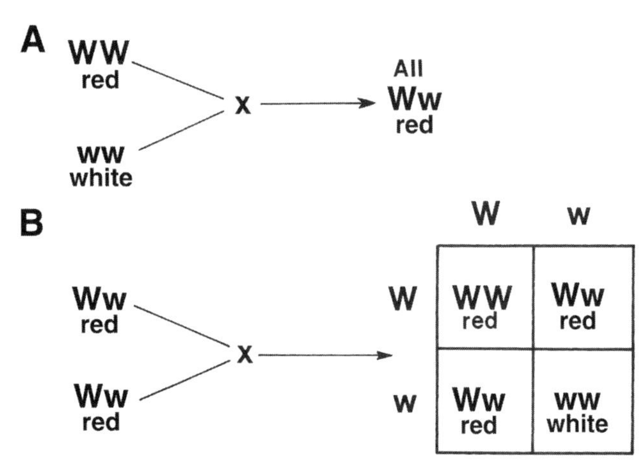

Figure 6.2. The allele for red flowers is dominant over the allele for white flowers. White flower color is a recessive trait that appears only in the homozygous condition (*ww)*. **A,** if homozygous red-flowered peas (*WW*) are crossed with homozygous white-flowered peas (*ww*), the next generation will be all heterozygous and red flowered (*Ww*). **B,** if these heterozygous progeny are crossed, the next generation will appear in a 3:1 ratio of red-flowered plants to white-flowered plants. One-fourth of the plants will be homozygous for red flowers, and half of the plants will be heterozygous for red flowers.

If those heterozygous offspring are crossed, however, one-fourth of the next generation will be homozygous for the recessive allele and thus have the phenotype of the recessive allele. That is, one-fourth of the offspring will have white flowers. The other three-fourths will have a red-flowered phenotype consisting of two different genotypes. One-fourth of the offspring will be homozygous and dominant, and half of the offspring will be heterozygous. Mendel observed identical patterns for all the other pea traits he studied.

Mendel recognized that genetic traits can be discrete factors that are contributed randomly by the parents. Because the genes for the traits he studied happened to be on different chromosomes, he also observed that inheritance of one genetic trait may be independent of the inheritance of other traits. Thus, the alleles for seed color are inherited independently from alleles for flower color and the other traits he studied.

In these simple cases, Mendel was able to identify the important principles of inheritance long before the roles of chromosomes and DNA were discovered. However, his work remained virtually ignored until the early 20th century, when cell biology had progressed to the point that his studies could begin to be explained.

Of course, not all inheritance phenomena are as simple as the ones described by Mendel. For example, many alleles are not strictly dominant or recessive. In some plant species, a cross between red-flowered plants and white-flowered plants produces pink flowers in the heterozygous offspring, a phenomenon referred to as incomplete dominance (Figure 6.3). Also, genes may be very close to each other on a chromosome and thus be inherited as a unit. Such genes are termed "linked," because they are inherited identically as a result of their close proximity on the chromosome. In addition, many phenotypic traits are not governed by a single gene but by several or many genes. Such polygenic traits govern many important and complex characteristics of plants and animals.

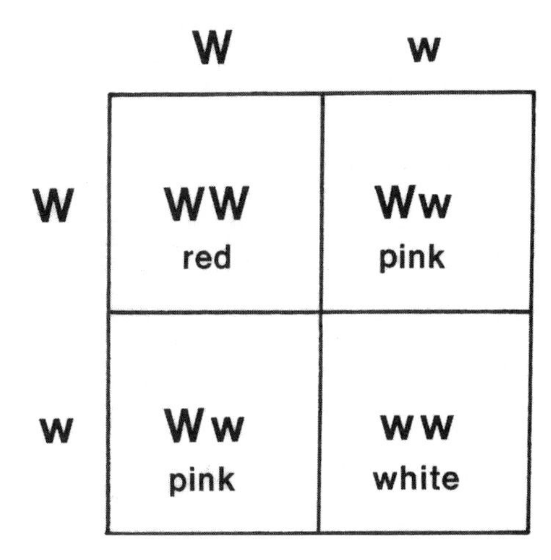

Figure 6.3. In snapdragons, the allele for red flower color is incompletely dominant over the allele for white flower color. Heterozygous individuals have pink flower color.

The discrete nature of genes and the process by which genetic traits are passed to the next generation, as first recognized by Mendel, are fundamental to our understanding of genetics: the science of inheritance and evolutionary change. Mendel's selection of a plant with distinct characteristics (although perhaps resulting from luck), his diligence in making thousands of crosses among plants of many successive generations, and his genius in recognizing the patterns of inheritance gave us the elementary principles of genetics that govern even the most complex phenomena.

DNA and the Genetic Code: The Secret of Life

While important genetic studies continued throughout the first half of the 20th century, the concept of the gene itself remained somewhat obscure. Chromosomes, composed of protein and nucleic acid, had already been recognized as the source of genetic information. The processes of mitosis and meiosis explained how each cell receives its genetic information and how chromosome numbers remain stable across generations. Many detailed genetic studies revealed inheritance patterns of various traits, both independent and linked, and the occurrence of mutations, or mistakes, had been observed. How so much complex genetic information could be stored, retrieved, and replicated was totally unknown.

Many clues had suggested that the molecule DNA was probably the key element in the mystery. For instance, haploid gametes (sex cells) contain half as much DNA as the diploid cells of the parent. In addition, chemical analyses had revealed that different species contain varied amounts of purines and pyrimidines, the nitrogenous bases that are components of DNA. In 1953, Francis Crick and James Watson published their model of DNA, based on the experimental evidence compiled by many other scientists. The discovery of the structure of DNA is one of the most important in all of biology, because it suggested how genetic information is coded and replicated and how mutations occur.

DNA is a long, double-stranded, helical molecule composed of many nucleotide units. Each unit consists of a sugar (deoxyribose), a phosphate group, and a nitrogen-containing, or nitrogenous, base. Four nitrogenous bases are found in DNA: the purines, adenine (A) and guanine (G), and the pyrimidines, cytosine (C) and thymine (T). In the double strand of DNA, the nitrogenous bases pair in a distinct pattern: A always pairs with T and G always pairs with C to form the "rungs" of the DNA helical "ladder" (Figure 6.4).

When the DNA of an organism replicates, the double-stranded molecule opens, and each strand then serves as a template (or mold) to form a new, complementary strand. This process is quite complex and requires a number of different enzymes to open the double strand and complete the replication, but the means of replication was deduced almost immediately from the Watson-Crick model of the structure of DNA. Since chromosomes contain enough DNA to account for millions of base pairs, the discovery of the structure of

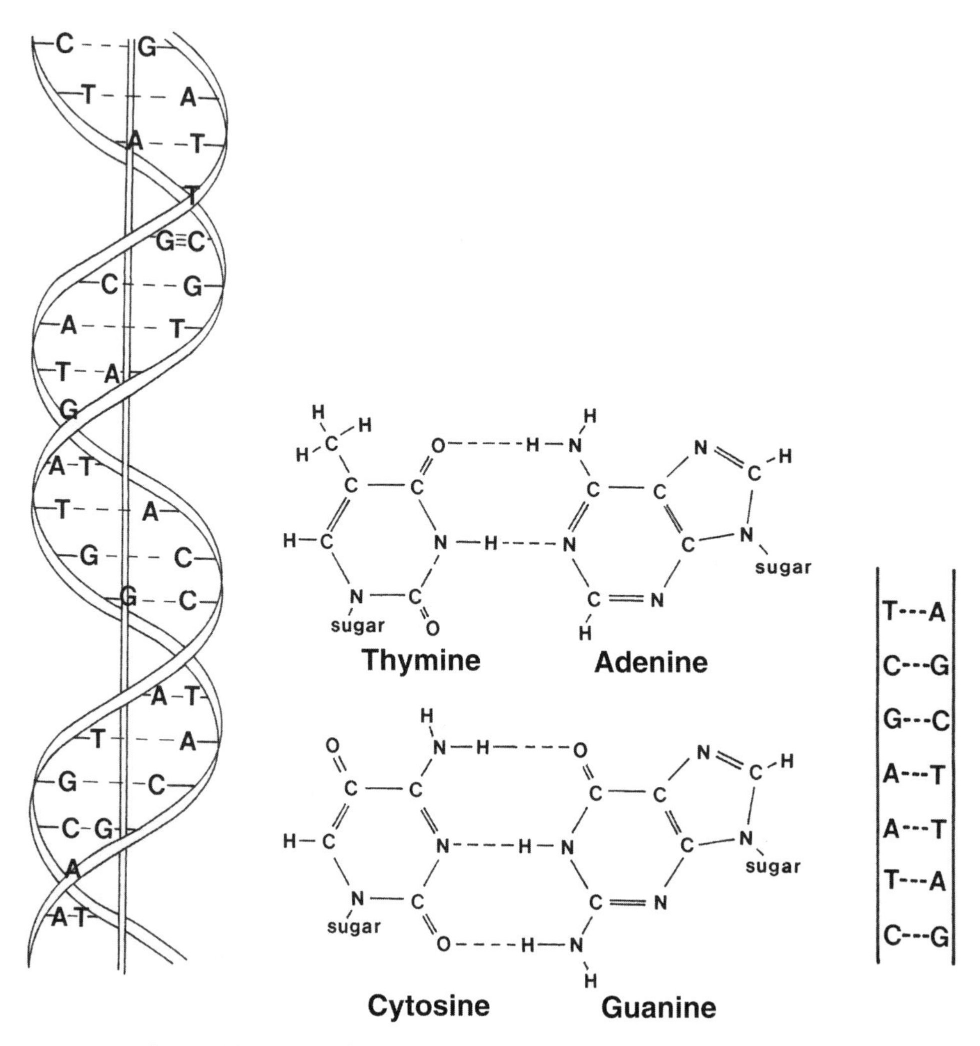

Figure 6.4. The double helix. **Left,** deoxyribonucleic acid (DNA) is a double-stranded, helical molecule composed of nucleotide units. Replication of DNA occurs through the pairing of nucleotide bases to form new strands. **Center,** diagram of the molecular structure of the base pairs of DNA. Sugar indicates the bond to deoxyribose. Phosphates connect the sugars to make the "backbone" of DNA. A nitrogen base plus deoxyribose and a phosphate group constitutes a nucleotide. Adenine (A) always pairs with thymine (T), and cytosine (C) always pairs with guanine (G). **Right,** pairing of the nucleotide bases in a short DNA segment.

this molecule also revealed a means for storing tremendous amounts of genetic information in a code based on the pattern of base pairs.

Further studies demonstrated how the genetic information stored in base-pair patterns of the DNA molecule can be transformed into cellular products. Scientists already knew that proteins play important roles in cellular control, particularly the specialized proteins called enzymes. Enzymes are complex molecules that catalyze chemical reactions in living organisms. They are responsible for controlling the synthesis and degradation of the numerous compounds necessary for life. Scientists also knew that proteins are made from chains of "building blocks" called amino acids. It seemed reasonable that the chains of nucleotides in DNA somehow coded for the chains of amino acids of proteins, many of which were the enzymes that governed the chemical reactions in living organisms.

Production of Proteins: Translating the Message

How do the chains of A, T, C, and G bases in DNA code for the patterns of amino acids found in proteins? Since there are 20 different common amino acids and only four different DNA bases, groupings of at least three bases are necessary to provide enough different combinations to code for the various amino acids. If a single base codes for a single amino acid, then codes for only four amino acids are available. If combinations of two bases are used for the code, then only 16 combinations are available, which is still not enough to code for 20 different amino acids. Sixty-four combinations can be made using various groups of three bases. As shown in Table 6.1, the extra combinations are all used, because

Table 6.1. The Genetic Code: Triplet Codons in Ribonucleic Acid (RNA) and Their Corresponding Amino Acids

Codon	Amino Acid	Codon	Amino Acid	Codon	Amino Acid	Codon	Amino Acid
UUU	Phenylalanine	CUU	Leucine	GUU	Valine	AUU	Isoleucine
UUC	Phenylalanine	CUC	Leucine	GUC	Valine	AUC	Isoleucine
UUG	Leucine	CUG	Leucine	GUG	Valine	AUG	Methionine
UUA	Leucine	CUA	Leucine	GUA	Valine	AUA	Isoleucine
UCU	Serine	CCU	Proline	GCU	Alanine	ACU	Threonine
UCC	Serine	CCC	Proline	GCC	Alanine	ACC	Threonine
UCG	Serine	CCG	Proline	GCG	Alanine	ACG	Threonine
UCA	Serine	CCA	Proline	GCA	Alanine	ACA	Threonine
UGU	Cysteine	CGU	Arginine	GGU	Glycine	AGU	Serine
UGC	Cysteine	CGC	Arginine	GGC	Glycine	AGC	Serine
UGG	Tryptophan	CGG	Arginine	GGG	Glycine	AGG	Arginine
UGA	None (stop signal)	CGA	Arginine	GGA	Glycine	AGA	Arginine
UAU	Tyrosine	CAU	Histidine	GAU	Aspartic	AAU	Asparagine
UAC	Tyrosine	CAC	Histidine	GAC	Aspartic	AAC	Asparagine
UAG	None (stop signal)	CAG	Glutamine	GAG	Glutamic	AAG	Lysine
UAA	None (stop signal)	CAA	Glutamine	GAA	Glutamic	AAA	Lysine

Notes: U = uracil, C = cytosine, G = guanine, A = adenine. The codons in deoxyribonucleic acid (DNA) are complementary to the RNA codons. C and G are complementary in both RNA and DNA. U in RNA is complementary to A in DNA.

each amino acid is coded for by more than one three-base group, and three of the combinations are stop signals. The three bases that code for an amino acid are called a triplet codon. This pattern of nucleic acid codons that match particular amino acids is the genetic code, and it is nearly identical in all living organisms.

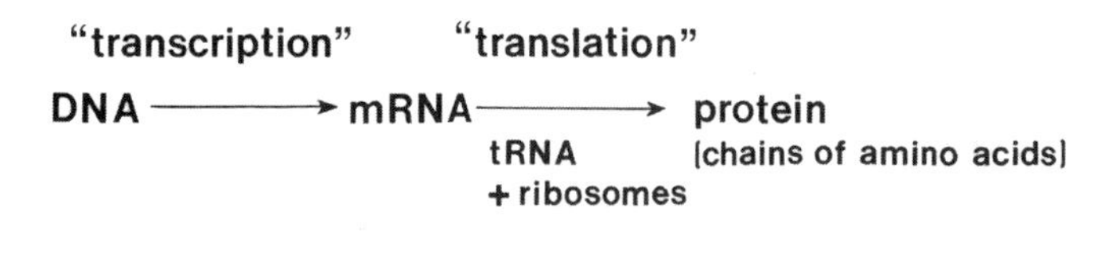

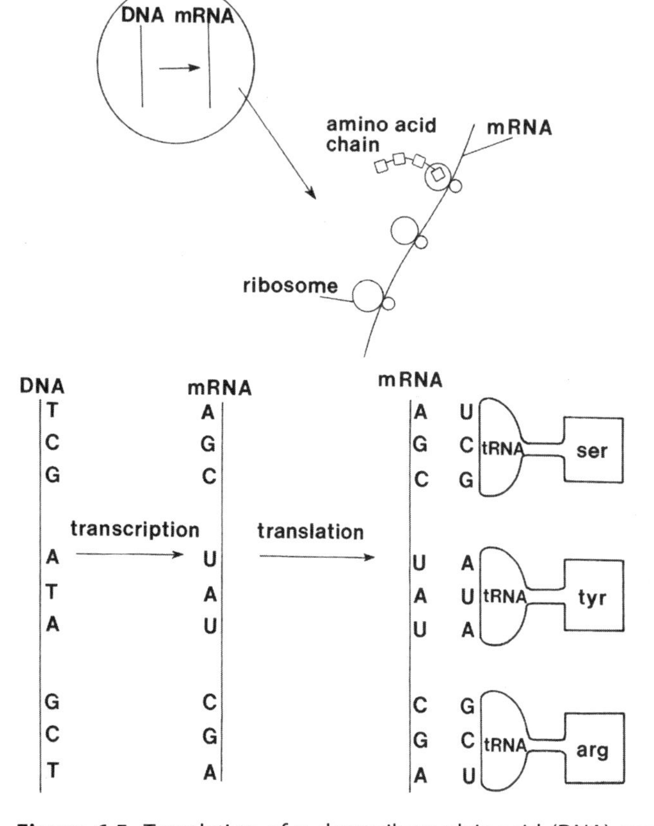

Figure 6.5. Translation of a deoxyribonucleic acid (DNA) genetic message into a protein. Ribonucleic acid (RNA) differs from DNA in having ribose sugar units, rather than deoxyribose, and in having uracil (U), rather than thymine (T), as a nucleotide base. **Top,** overview of transcription of DNA to messenger RNA (mRNA) and translation to a chain of amino acids (protein) with the help of transfer RNAs (tRNAs). **Center,** after transcription from DNA, mRNA leaves the nucleus and moves to the cytoplasm. tRNAs carry amino acids and match with a triplet codon on the mRNA to create a chain of amino acids in the order coded for in the DNA. **Bottom,** transcription and translation. Nucleotide bases: C = cytosine, G = guanine, A = adenine. Amino acids: ser = serine, tyr = tyrosine, arg = arginine.

DNA contains all of the information necessary for the functioning of an organism. It is the "blueprint" that never leaves the cell's nucleus. When a gene becomes active, only part of the DNA message is used. How can the information coded in the base pairs of a piece of DNA in the nucleus become translated into the amino acid chains of cellular proteins often found in the cytoplasm? The key is an important intermediary molecule called ribonucleic acid (RNA).

RNA is closely related to DNA but differs in two important ways: the sugar component is ribose, not deoxyribose, and uracil replaces thymine as one of the four nitrogenous bases. In RNA, the base adenine (A) pairs with uracil (U) rather than thymine (T), as it does in DNA. Many types of RNA exist in cells, and one of their functions is to mediate the transfer of information in the DNA nucleotide code to the amino acid chains of proteins.

A type of RNA called messenger RNA (mRNA) carries the DNA code for each specific gene into the cytoplasm. A section of the DNA serves as a template for a process called transcription, in which the DNA code is paired with a mirror-image chain of nucleotides in an mRNA (Figure 6.5). Just as the double strands of DNA match each other with base pairings, mRNA matches the pattern of a specific segment of DNA. After transcription, the mRNA leaves the nucleus and becomes associated with components of cells called ribosomes in the cytoplasm.

In the ribosomes, two other kinds of RNA participate in the production of proteins. Ribosomal RNA (rRNA) functions to bring together mRNA and the third type of RNA: transfer RNA (tRNA). tRNAs are small molecules that contain two important regions. One region binds to a specific amino acid, and the second region is composed of a triplet codon. The tRNAs carry their amino acids to the mRNA, matching their triplet codon region to the correct codon on the mRNA. This causes the amino acids to line up in a chain in the order originally specified in the nuclear DNA. The resulting chain of amino acids becomes a protein.

All of these processes are carefully controlled by a complex series of enzymes that determine the type and amount of amino acid chains created. The resulting protein, usually an enzyme, is the translation of the DNA message into a protein molecule that can do its work in the cell.

The basic genetic code, translated into proteins via RNA, is essentially the same for all living organisms. It has therefore been called the universal genetic code. The entire DNA code, or genome, has been sequenced for more than 5,000 different organisms. In the coming years, new technologies will increase this number by orders of magnitude. The human genome sequence, completed in 2003, consists of approximately 3.9 billion base pairs and is approximately 6 feet long. It is amazing to think that this huge amount of information is tucked into the nucleus of every human cell and must be replicated every time a cell divides. Even more amazing is the genome of wheat, which can be a hexaploid with six sets of chromosomes and more than 17 billion base pairs of DNA.

New genetic engineering techniques, which are described later in this chapter, have shown that DNA sequences from one kind of organism can function in a totally unrelated organism. This remarkable phenomenon illustrates that the genetic code of DNA is the common thread that connects all forms of life.

Science Sidebar

RNA interference

Not all of the bases of a DNA molecule are transcribed into mRNA and then translated into proteins. Much of the DNA that does not code for proteins appears to code for a variety of small RNAs. Scientists have wondered what happens to all of the mRNA that is produced by various genes as they are activated over time.

In 2006, a Nobel Prize was awarded to Andrew Z. Fire and Craig C. Mello, two scientists who discovered a natural process in the cell called RNA interference (RNAi). A double-stranded RNA with one chain that is complementary to a messenger RNA can cause the mRNA to be destroyed by the cell. This appears to be a natural process that allows cells to control the amount of protein that is produced after a gene in the nucleus is activated and begins to produce mRNA.

Scientists are now using RNAi to block specific mRNAs, a process that has been called gene silencing. Experimentally, RNAi can be used to block the mRNA of a gene that has been identified in DNA to determine the gene's function. RNAi also may be used therapeutically. Although it is not yet possible to repair DNA mutations in the nucleus, RNAi can block the functioning of aberrant DNA segments by targeting their mRNAs for destruction.

Mutation and Evolutionary Change: Survival in a Changing World

The discovery of the structure of DNA led to a clearer understanding of how genetic information is stored, replicated, and translated into proteins for cellular control. It also helped explain how mistakes known as mutations might occur.

The concept of mutation was first suggested in 1901 by Dutch scientist Hugo de Vries. It was used to describe new genetic characteristics that arose suddenly and had not been previously observed in the parental lines. Studies conducted since 1951 have shown that mutations occur in many different ways. Some of the simplest can be explained by the structure of DNA. Certain chemicals and physical damage may cause insertions, deletions, or mispairings during replication that affect one or a few nucleotides. Such small changes may cause shifts in the reading of the genetic code, so that a very different protein (often nonfunctional) is translated.

Science Sidebar

Genes in eukaryotes

In prokaryotes (bacteria), a sequence of DNA is simply transcribed into mRNA and then translated into a chain of amino acids. Thus, in prokaryotes, a single gene can be identified as the section of DNA bases that codes for a particular protein.

In eukaryotes, the sequence of DNA that codes for a particular protein does not usually exist as a single linear sequence. Rather, it consists of a number of shorter segments (exons) along the DNA that are interrupted by sections of bases that do not function in the coding of the amino acid sequence (introns). Long segments of RNA are transcribed from the DNA and then processed by enzymatic action into the actual mRNA, which will move into the cytoplasm for translation into an amino acid chain.

Thus, the identification of a gene as a section of DNA bases that codes for a particular protein is more complex and more difficult in eukaryotes than in prokaryotes. Eukaryotic DNA contains large sections of nontranslated bases and many identical copies of certain base sequences, the functions of which are not yet completely understood. Recent research has shown that the same gene can code for a variety of proteins, depending on how the exons are spliced together after transcription into RNA.

Larger changes in the structure of the DNA are possible as well. Barbara McClintock received a Nobel Prize in 1983 for her identification of small, mobile portions of chromosomes called transposons, which can randomly move to different sites on the chromosome, causing changes in the genetic information. In addition, chromosomes are known to break and reattach in ways that also change the genetic information. For instance, a piece of a chromosome may break off and reenter the same chromosome with its gene sequence backward from its previous orientation. A chromosome segment also may attach to a different chromosome.

All these various "mistake" mechanisms have been discovered in investigations of sudden phenotype changes in offspring that could not be explained by examining the previous genetic histories of the parents. Mutations are an important source of genetic variation that allows organisms to adapt to their changing environments. Mutations also may have deleterious effects, such as increasing susceptibility to diseases, including cancer. Many human cancers have been linked to specific gene mutations or to series of mutations that accumulate over time. Such mutations may reduce the production of an important enzyme or cause a different enzyme to be produced, which may affect control of the growth and development of cells and tissues.

People are warned to avoid exposure to X-rays and ultraviolet light, because such radiation is capable of causing mutations. The ozone layer in the upper atmosphere helps shield Earth from ultraviolet light. Its vulnerability to destruction by industrial activities may result in increased mutations among living organisms. Chemicals to which people are exposed in their work and diet also may act as mutagens. Many compounds have been banned or restricted to protect the genes of human beings.

The Gene-for-Gene Hypothesis: How Pathogens "Track" Their Plant Victims

Let us now consider how this improved understanding of genetics and gene structure is useful in agriculture and, specifically, in plant pathology. For years, farmers have been selecting crop plants for their various desirable traits, including yield, quality, and pest and pathogen resistance. The rediscovery of Mendel's work at the beginning of the 20th century led plant breeders and pathologists to study specific genes for desirable traits. In the area of disease resistance, single dominant genes were found to confer resistance to specific pathogens in various crops.

As mentioned in Chapter 1, *Solanum demissum* served as a source of resistance genes that were crossed into cultivars of *S. tuberosum,* conferring resistance to *Phytophthora infestans.* The effect of these resistance genes was specific to *P. infestans* and affected no other potato pathogen. The initial success of the resulting resistance was so complete that many scientists believed late blight had been eliminated for all time. Unfortunately, the continued use of single-gene resistance often selects for new races of the pathogen population, resulting in the failure of the resistance.

We have already referred to the genetic interactions between hosts and pathogens. Although the actual interaction between organisms that allows pathogenicity to occur is complex, some specific factors have been discovered. For example, some pathogens produce enzymes or toxins that contribute to their ability to attack the host plant. A resistant plant may produce a chemical that blocks the function of one of these pathogen enzymes, or the toxin-binding site may be different in the cells of a resistant plant, so that a pathogen toxin is no longer effective. With time, genetic change in the pathogen—through sexual reproduction or mutation—may result in an enzyme of a slightly different shape that cannot be blocked or a toxin with improved binding ability, so that host resistance is overcome. This description (although greatly oversimplied) suggests the kind of stepwise evolutionary interactions that occur between hosts and pathogens in natural communities.

H. H. Flor, working at North Dakota State University in the mid-1900s, was the first plant pathologist to simultaneously study the genetics of a plant—flax (*Linum usitatissimum*), from which linen fibers are derived—and its pathogen, *Melampsora lini,* a rust fungus. Based on Flor's studies, the gene-for-gene hypothesis was proposed, in which a successful infection is governed by genetic factors in both the host and the pathogen.

In the example of potatoes and *P. infestans,* genetic studies have shown that potato plants with a resistance gene can be infected only by certain races of the pathogen. When a whole field of potatoes with a particular resistance gene is planted, the initial effect is a tremendous reduction in the amount of disease, because most genetic races of the pathogen have been eliminated. However, the surviving races capable of infecting the resistant plants infect and multiply on the plants. The resistant potato cultivars eventually become as severely diseased as the original susceptible ones. Farmers often say that the resistance has "broken down," when in fact, the pathogen has overcome the resistance with genetic change.

Genetically uniform, monoculture agriculture is particularly vulnerable, because it puts selection pressure on the pathogen to adapt to the genetic makeup of the host. The pathogen must "adapt or starve." Flor's work explained why the use of single-gene resistance in an agricultural crop is often effective only temporarily. Gene-for-gene systems have been identified in plant interactions with many kinds of plant pests, including bacteria, nematodes, viruses, and even insects.

Science Sidebar

The Ames test

Although we know that some chemicals can cause mutations, we do not want to wait years to see whether they do. In the 1970s, Bruce Ames, at the University of California, Berkeley, developed a test that can identify mutagens quickly. The Ames test detects the ability of a chemical to cause mutations in bacteria. The test can be performed more quickly and inexpensively than animal tests, because bacteria multiply so rapidly.

Ames test results are used as preliminary evidence that a particular chemical can cause mutations—and perhaps increased cancer—in humans. The Ames test does not prove that a chemical causes cancer, but it can provide a warning that the chemical should be investigated further for its potential detrimental effects on human health.

One of the most difficult aspects of these genetic interactions is that pathogens generally have the advantages of shorter life cycles and much greater reproductive capacities than the plants we want to protect. The pathogen may be able to quickly overcome genetic resistance in the host plant. This is particularly true of pathogens that reproduce many times during a growing season and are dispersed via the air, such as leaf spot fungi, rust fungi, and downy mildew pathogens.

Certain kinds of crops are especially vulnerable to resistance failure. For example, vegetatively propagated crops—such as potatoes—are genetically uniform, which places strong selection pressure on the pathogen to overcome the resistance. Another example is perennial crops—such as apples, asparagus, grapes, and turfgrass—which remain in the same soil for many years and thus provide pathogens sufficient time to overcome resistance. In contrast, with annual crops—such as corn and soybeans—farmers can alternate crop species or cultivars of the same crop from year to year. The longer the normal life of a plant, the more time the pathogen has to overcome the resistance. For instance, it is likely that the rapidly reproducing, airborne coffee rust fungus, *Hemileia vastatrix,* might overcome single-gene resistance in a coffee tree before the tree could even mature enough to produce a harvestable crop.

Genetic Resistance to Disease: Defending the Plants

The popularity of a new cultivar is considered both the reward and the doom of a plant breeder. The more widely a single genetic line of a crop is planted, the more likely a devastating epidemic will arise from a new pathogen race. This has been dubbed the "boom-and-bust" cycle of plant breeding (Figure 6.6).

Although some spectacular resistance failures have plagued plant breeders, some single-gene resistance factors have been quite effective for many years. For example, resistance genes to various forms of the fungus that causes Fusarium wilt (*Fusarium oxysporum*) have been very durable, because the pathogen produces only one generation of spores at the end of the growing season and the spores remain soilborne. The reproductive capacity and dispersal ability of this fungus are obviously much more limited than those of a pathogen such as *P. infestans.* The same genetic interactions exist, but the probability of selecting a new fungal race capable of overcoming the resistance is much lower because of the biology of the pathogen.

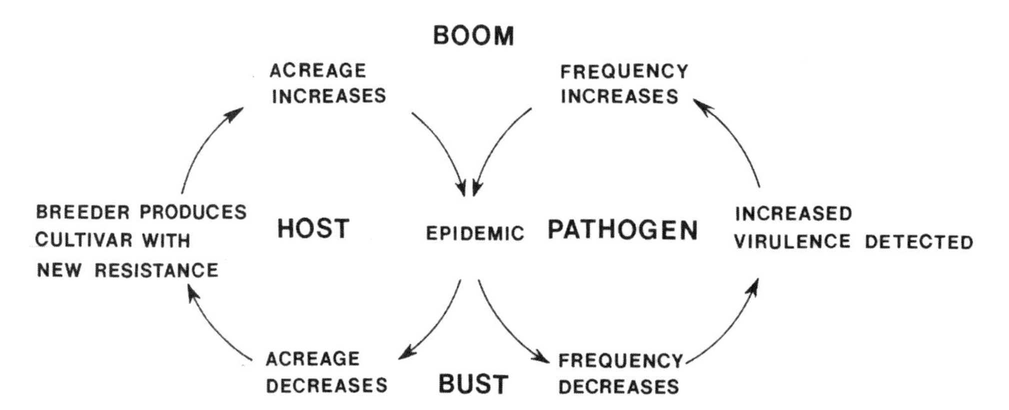

Figure 6.6. Boom-and-bust cycle. As a resistant cultivar becomes popular, the selection pressure for the pathogen to overcome the resistance becomes greater. As the new pathogen race increases, the cultivar can no longer successfully resist infection.

Tomato plants with single-gene resistance to Fusarium wilt have remained resistant for more than 30 years. Resistance first failed in places growers planted tomatoes in the same land repeatedly for many years. Eventually, a new race appeared in the fungus population that was capable of infecting the tomatoes, despite the resistance gene. Fortunately, the dispersal of the new race was slow, because the pathogen was soilborne. In most areas, where tomatoes are rotated with other crops and planted in the same land only once every three or four years, resistance has remained more durable. Even if some spores of the new race are produced, most germinate and starve or are destroyed in the intervening years without a host plant.

Single-gene resistance can be useful for rapidly reproducing airborne pathogens when it is deployed appropriately. It remains the most common means of combating rust diseases in grain crops, but its use requires plant breeders to continually assess the races that exist in the rust population. Breeders can then change the resistance genes in the grain cultivars from year to year, according to the genetic status of the pathogen population. This important subject is discussed in more detail in Chapter 11.

Even though single-gene resistance can be useful and effective, many plant breeders have pursued a form of genetic resistance effective against all possible races of a pathogen. Their goal is to produce a resistance that cannot be overcome by genetic changes in the pathogen. This type of resistance has been described using various terms, including durable, general, nonspecific, field, horizontal, and polygenic. (Despite use of the term general, this resistance is effective against all races of the pathogen for which it has been selected, but it does not necessarily confer resistance to any other pests or pathogens.) The theoretical basis for such resistance is that it involves many genes, each of which contributes to the resistance in some small way. Because many genes are involved, a pathogen probably would not be able to simultaneously make all the genetic changes necessary to overcome the resistance.

General resistance is identified in the host plant by a reduced amount of disease, even when the plant is exposed to a variety of races of the pathogen. Disease might be reduced because lesions develop more slowly, spores take longer to appear on lesions, or fewer spores are produced on a lesion compared to these factors in a plant lacking general resistance. Plant breeders hope that with time, they will be able to increase the levels of general resistance in various crops. Some potato cultivars are now available with levels of general resistance to *P. infestans* that reduce the amount of fungicide needed to protect the foliage from infection (Figure 6.7). General resistance to rust diseases in certain grain

Figure 6.7. Potato leaf from a susceptible plant (far right) and leaves from plants with varying levels of general resistance. General resistance functions against all races of the late blight pathogen, *Phytophthora infestans.* In resistant plants, lesions are smaller, develop more slowly, and produce fewer sporangia.

crops is referred to as "slow rusting." It can be combined with single-gene resistance for added protection against rust diseases.

Genetic resistance has always been a popular means of plant disease management, because it requires no special inputs by growers during the growing season and is compatible with other agricultural activities. Research in the area of genetic resistance to plant pathogens has received renewed interest in recent years, for several reasons. In many parts of the world, subsistence farmers do not have the equipment or money needed for frequent pesticide applications. It also has become clear that intensive pesticide use can have detrimental environmental effects.

Advances in modern genetics and cell biology have greatly increased understanding of the genetic interactions among fungi, bacteria, nematode, viruses, and their host plants. This improved understanding has allowed plant breeders and plant pathologists to develop breeding programs based on scientific genetic principles, rather than the random selection of naturally occurring genetic variations. In addition, researchers are beginning to understand how plants turn on defense reactions in the early stages of invasion by pathogens. This has led to the discovery of means to induce these defenses to help protect plants proactively. This approach is discussed further in Chapter 8.

Tissue Culture and Biotechnology: Modern Tools

One of the major limitations faced by plant breeders is the reproductive isolation between plant species. Because of pollen incompatibilities, it is not usually possible to successfully cross plants unless they are very closely related. For instance, if pollen from an oak tree is transferred to the flower of a potato plant, no fertilization will occur and no seeds will be produced. However, after the structure of DNA and gene sequences were discovered, scientists began to speculate about the possibilities of transferring specific genes between species. This technique, which is called genetic engineering, is particularly applicable to plants because of the way they develop.

Each somatic (nonreproductive) cell of a living organism contains the full complement of DNA in its nucleus, yet as an organism grows and develops, each cell becomes specialized. A particular cell uses only some of the total genetic information for its functions. For example, leaf epidermal cells contain the same DNA information as root cortex cells, but the appearances and functions of these two cell types are very different. Various growth hormones and other complex controlling mechanisms determine the genetic information that is used in a specific cell. Unlike animal cells, which become predestined for particular functions early in their development, many plant cells retain the ability to change functions and take on new roles at different times.

Cuttings from many plants develop roots when placed in water. As early as 1902, Austrian botanist Gottlieb Haberlandt studied these developmental changes and described plant cells as totipotent. That is, any plant cell should be capable of developing into a whole, mature plant. This concept has always been popular with science fiction writers, who have written stories about clones of people created from adult cells, since all cells have the complete DNA blueprint.

Haberlandt's speculation about plant cells remained theoretical until 50 years later, when American plant physiologist F. C. Steward grew cells from mature carrot root phloem in coconut milk, which provided nutrients and growth hormones. He succeeded in growing new, whole carrot plants from cells that had come from specialized tissues of a mature plant. Steward's work opened an important door for genetic engineering. If a specific gene could be transferred to the nuclear DNA of a single plant cell, then that cell could be regenerated into an entire plant, in which each cell would contain the new gene in its nucleus.

Besides specific gene transfer, other nontraditional breeding methods have resulted from the culture of individual plant cells in nutrient media. For instance, entire haploid plants have been grown from haploid cells cultured from pollen and ovules. One advantage of haploid plant culture is that the full genotype is apparent, because recessive factors are not masked by dominant genes on homologous chromosomes. Since many agriculturally important plants are polyploids (having more than two sets of chromosomes), they are difficult to work with. Haploid plant culture from pollen and ovules reduces the chromosome numbers by half and thus simplifies genetic studies. Because agriculturally important potato cultivars are tetraploids (having four sets of chromosomes in each nucleus), haploid potato culture provides a means of reducing *S. tuberosum* to a diploid state, so that it can more easily be crossed with wild diploid *Solanum* species.

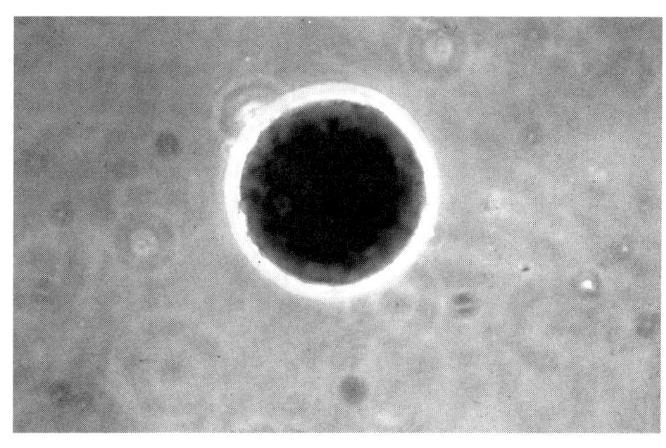

Figure 6.8. Cucumber protoplast stained with neutral red. The cell wall has been removed enzymatically, leaving only the cell membrane (light micrograph).

Through careful degradation with enzymes, the cell wall can be removed from a plant cell, leaving the cell contents surrounded only by a cell membrane (Figure 6.8). This protoplast can be grown like a single-celled microorganism in liquid nutrient media. Some protoplasts have been induced to fuse, and successful nuclear fusions have even occurred. New plants have been regenerated from the fused protoplasts. Protoplast fusion presents a possible means of combining plants that cannot be crossed by traditional breeding methods. For example, cultivated bananas are sterile triploids. They produce no seeds. Protoplast fusion between banana cultivars can be used to increase genetic diversity in these plants, which no longer undergo sexual reproduction. Large numbers of protoplasts can be exposed to chemicals such as pathogen toxins, and the survivors can be regenerated into whole plants that may then be resistant to the pathogen that produces the toxin.

The ability to manipulate plant cells and tissues in culture and subsequently regenerate whole plants has become an extremely important area of plant biotechnology and may result in important plant improvements in the coming years. Commercial success in plant biotechnology has already occurred in the rapid propagation of pathogen-free planting material from tiny pieces of plant tissue, which is discussed in Chapter 13.

Recombinant DNA and Genetic Engineering: Precise Genetic Change

The ability to regenerate whole plants from single cells has stimulated great interest in the genetic engineering of plants. Plant breeders using traditional breeding methods must often perform crosses between plants for many generations to create a new plant cultivar with a desirable genetic trait. Although breeders have been successful with this means of genetic modification, they also must be on guard for new gene combinations that confer undesirable traits. For example, potato breeders must test tubers from breeding lines for the poisonous alkaloids that tend to accumulate, mostly in the leaves. Every time plants cross, new gene combinations are made—including the various mutations that have occurred.

At this time, classical breeding remains the most efficient method to introduce new genetic traits into plants that are sufficiently closely related for hybridization. However, genetic engineering offers the potential to transfer any genetic sequence to the nucleus of any plant cell. It provides a more precise means of plant modification, and it allows genes from other organisms to be used to improve plants.

There are three important steps in such a procedure: (1) identification of the specific gene sequence that controls the desired trait, (2) isolation of a functional gene sequence from the donor organism, and (3) insertion of the gene sequence into the nucleus of the receptor organism, followed by expression of the desired trait. Because we know that DNA codes for proteins, it is possible to determine the function of a DNA sequence by figuring out what protein it codes for using the triplet codons in Table 6.1. Conversely, it is possible to identify a protein, determine its amino acid sequence, and work backward in Table 6.1 to determine the DNA sequence that codes for that protein.

Identification of a gene sequence is simplest in prokaryotes, in which genes are coded in a linear sequence. For eukaryotic genes, it is often necessary to remove and further process large DNA sequences to ensure that an entire message is transferred. To begin genetic engineering, the DNA from the donor organism is purified from other cellular components. DNA is not cut with a sharp-edged tool. Rather, isolation of a gene sequence and its insertion are accomplished with special enzymes called restriction enzymes, which were discovered in the 1970s. Restriction enzymes cut DNA at particular nucleotide sequences, allowing a section containing the desired sequence to be isolated. Ligases are enzymes that seal the cut ends of DNA, allowing the transferred sequence to be inserted into another piece of DNA. These enzymes have been referred to as the "scissors" and the "glue" of genetic engineering.

The earliest genetic engineering experiments were accomplished with bacteria. The first step was to determine the actual base sequence of DNA molecules in bacteriophages (viruses that infect bacteria) and in bacterial plasmids. Plasmids are relatively short, circular pieces of DNA in bacterial cells that are independent of the main chromosome. After repeated enzymatic cutting of the DNA and analysis of the nucleotides on the ends of the DNA pieces, scientists were able to make genetic maps of the base sequences of these relatively short DNA pieces.

Once scientists knew the base sequences, they could use enzymes to insert new gene sequences into viruses or plasmids. The new DNA, containing the introduced sequence, is known as recombinant DNA (rDNA). When a population of normal bacteria is then exposed to the genetically engineered viruses or plasmids, the gene can be transferred to a few cells by natural means of genetic recombination. When the recipient cell absorbs the recombinant DNA and begins to express the gene by producing the protein it codes for, the cell is said to be transformed. The genetic engineering process is then complete. The transformed cell can be isolated and grown in pure culture.

Many such experiments have been performed since 1973, and the results have been quite interesting. It is remarkable that a gene from a eukaryotic organism can function in a bacterium. For instance, the genetic information necessary for the production of insulin in humans has been transferred to a bacterium, so that inexpensive insulin can now be produced by bacteria, reducing the need for its isolation from butchered pigs and cows. Because the insulin produced by bacteria is identical to that produced by human cells, problems with allergies have been eliminated for diabetic individuals who are sensitive to insulin derived from animals. Bacteria can serve as tiny "factories" to produce a variety of useful compounds through genetic engineering.

Crown Gall Bacteria: The Natural Genetic Engineers

After establishing the techniques for the genetic engineering of bacteria, scientists turned to the problem of inserting a gene into a plant cell. Chapter 5 described the crown gall bacterium, *Agrobacterium tumefaciens,* as a natural genetic engineer. During the same years that researchers developed techniques for bacterial genetic engineering, they discovered that *A. tumefaciens* possesses a unique plasmid that transfers a segment of its DNA into the nuclear DNA of the infected plant host cell. This plasmid was subsequently named the tumor-inducing (Ti) plasmid, because the presence of a segment of it in the nucleus of a plant cell causes the uncontrolled cell division and tumor growth characteristic of crown gall disease. Scientists observed that after the plasmid gene transfer, gall formation would continue, even if the bacteria were removed. They began to realize that crown gall bacteria had great potential for gene transfer to plants. Today, *A. tumefaciens* is the workhorse of genetic engineering for plants.

Scientists discovered that only a small section of the Ti plasmid is transferred into the plant cell nucleus. In addition, only the two ends of the gene sequence are needed to control gene transfer. The nucleotide sequence between the ends can be removed to prevent the bacteria from causing crown gall disease in the plant without losing the genetic engineering ability of the bacteria. So, the tumor-inducing sequences were replaced by a smaller, well-known plasmid from the common bacterium *Escherichia coli.* To transfer a gene to a plant cell, the DNA sequence of the gene is first inserted into the *E. coli* plasmid. Recombination between the *E. coli* plasmid and the Ti plasmid results in the transfer of the desired gene into the Ti plasmid. When plant cells are exposed to the Ti plasmid, the gene is inserted into a chromosome in the nucleus of a plant cell. Using tissue culture techniques, the plant cell can then be regenerated into a whole plant, in which each cell contains the new gene (Figure 6.9).

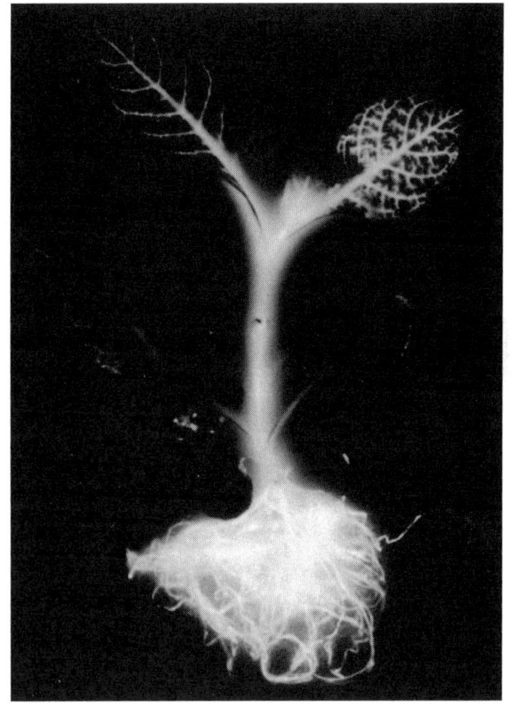

Figure 6.9. Genetically transformed tobacco plant that received the luciferase gene responsible for the luminescence of fireflies. The photograph was taken over a 24-hour period.

The Ti plasmid provides an efficient and genetically predictable means of gene transfer in plants. However, it cannot be used in all situations. The most common alternative technology is biolistic transformation using the so-called gene gun. This device propels tiny metallic particles coated with the desired DNA sequences into plant cells, with the goal of permanently incorporating the foreign DNA into the nuclei of some of the cells.

Whether using the Ti plasmid or a gene gun to insert the new gene (or genes), entire plants must be regenerated from a single transformed cell, so that each cell will contain the desired gene. In addition, it is usually necessary to grow the plant and to produce seeds to be certain that the next generation expresses the genetic change in the desired manner. Organisms that have been changed by genetic engineering are called transgenic (Figure 6.10).

The potential for improving plants using genetic engineering is vast. Possible improvements include enhancing the amount and quality of protein in various food crops and perhaps even inserting genes that control the fixation of atmospheric nitrogen to a form usable by plants, reducing the need for fertilizers. As our understanding of resistance mechanisms in plants improves, the transfer of genes to make plants more resistant to diseases and pests will become possible. Of course, this will not eliminate the need for traditional breeding methods to improve and maintain more complex characteristics. Still, genetic engineering provides an important new tool for rapid transfer of specific gene sequences.

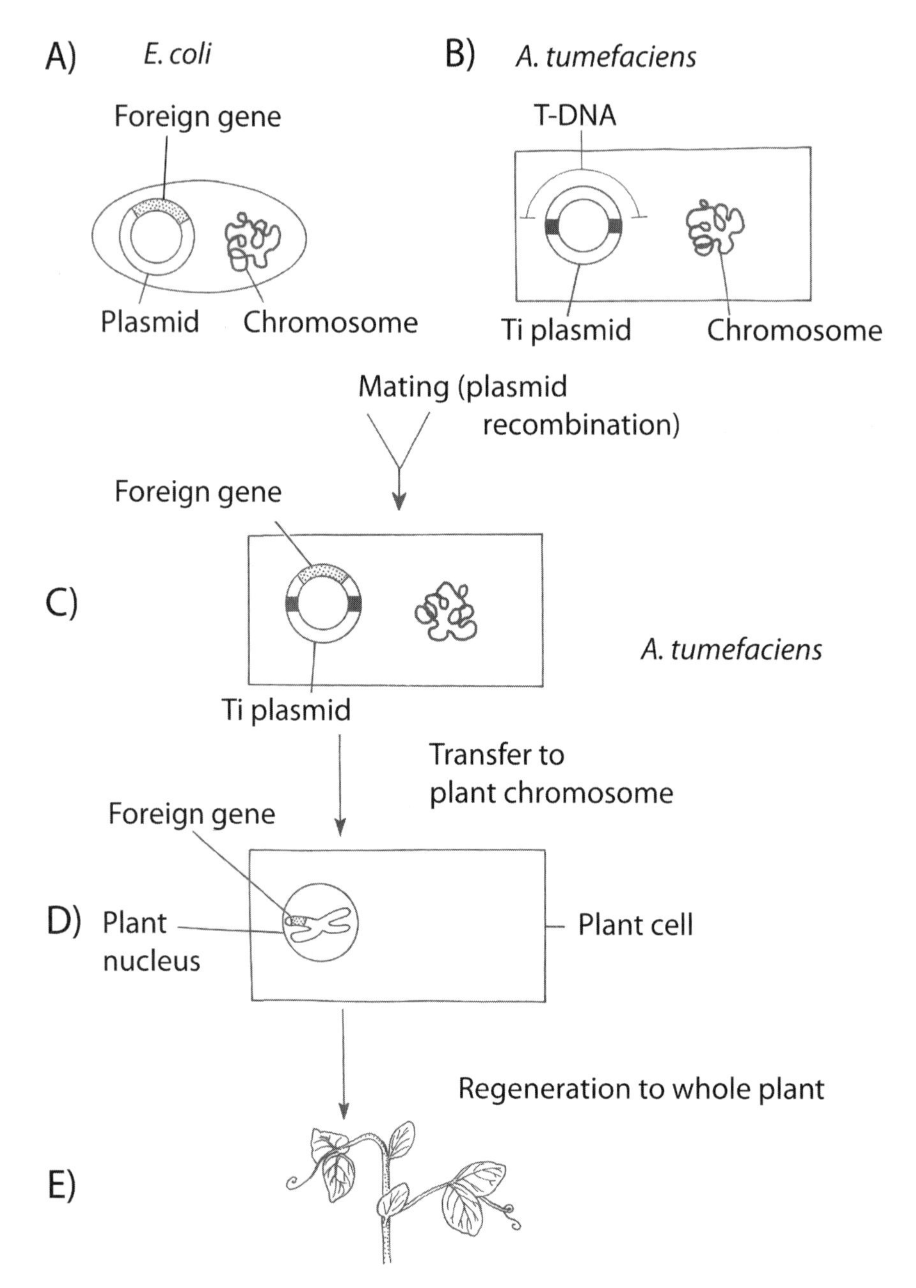

Figure 6.10. Plant genetic engineering using the tumor-inducing (Ti) plasmid of *Agrobacterium tumefaciens,* the crown gall bacterium (**B**). Using recombinant DNA techniques, a foreign gene is inserted into a plasmid of *Escherichia coli* (**A**). Through plasmid recombination, the foreign gene is inserted into the T-DNA section of the Ti plasmid (**C**). *A. tumefaciens* transfers the foreign gene to a chromosome in the nucleus of a plant cell (**D**). Each cell of a plant regenerated from the transformed cell contains the foreign gene (**E**).

Genetically engineered crops were introduced in 1996. Since then, the acreage planted with these crops has increased at least 10% each year, reaching more than 250 million acres (100 million hectares) in 2008. Initially, most of the acres of genetically engineered crops were planted in developed countries, but between 2003 and 2008, the acreage planted in developing countries increased rapidly. By 2008, 14 countries had significant acreages of genetically engineered crops, including countries in North America (United States, Canada), South America (Brazil, Argentina), Europe (Spain), Africa (South Africa), Asia (China, India), and Australia. Eleven other countries across five continents had begun to plant these crops as well.

The four major genetically engineered crops are soybean, maize (corn), cotton, and canola. In 2009, 70% of the soybeans, 46% of the cotton, 24% of the maize, and 20% of the canola grown worldwide were genetically engineered. In the United States, more than 80% of the 2009 cotton, maize, and soybean crops were genetically engineered. Note that most of these early applications are in field crops. Many people mistakenly believe that all of the changes they notice in fresh fruits and vegetables at the grocery store are due to genetic engineering. Many of these changes—for example, the availability of orange cauliflower and seedless watermelon—were produced by traditional breeding.

To date, genetic engineering has produced three outcomes with significance for the commercial market. The first is the development of herbicide-resistant crops. Glyphosate is a nonselective herbicide that kills many species of weeds. Plants have been produced that are glyphosate resistant, so that a single herbicide (Roundup) can be used after the crop has emerged without causing harm. This technology has allowed farmers to adopt no-till methods, in which seeds are planted directly into the stubble from a previous crop to help reduce soil erosion. Use of this technology also has helped reduce groundwater contamination, by decreasing the need to apply herbicides to bare soil before planting. Unfortunately, repeated use of glyphosate over a number of years has selected for glyphosate-resistant weed species in some areas. There is some concern that the glyphosate resistance genes could cross into weeds via pollen, but this could happen only with weeds closely related to the crop species.

Insect resistance is the second commercial success resulting from genetic engineering. *Bacillus thuringiensis* is a soil bacterium that produces a chemical that is converted into a toxin when ingested by certain insects, killing them. Endospores of *B. thuringiensis* (Bt) have been used as an insect spray by organic growers for years because this bacterium is safe for people and the environment. Genetic engineers have transferred the Bt toxin genes into plants, so that the plants now produce the chemical. Insects that feed on these plants, such as the corn root worm, are poisoned. This method has proven quite effective for insect management, but farmers must grow non-Bt crops nearby to reduce the development of resistance in the pest populations. Although this practice is required by government regulators,

Figure 6.11. Transgenic tomato plants are resistant to *Tobacco mosaic virus.* The plants on the left have been transformed using the Ti plasmid. The coat protein gene of the virus was transferred to the DNA in the nucleus of the plant cell. Virus-resistant plants were regenerated from the transformed cells. Both sets of plants were inoculated with the virus, but only the plants on the right became diseased.

it has not been strictly enforced. This lack of enforcement worries those who feel that better stewardship of these valuable genes will lengthen the time they remain useful.

A third common application of genetic engineering of plants has been to produce resistance to virus diseases. It is very difficult to find virus-resistance genes in plant populations for traditional breeding. Tobacco plants were first genetically engineered to resist infection by *Tobacco mosaic virus* through the transfer of the gene for the virus coat protein. The same technology has been used to produce virus-resistant papaya, squash, tomato, and other crops (Figure 6.11) (see Chapter 13).

After the first decade of commercial genetic engineering, several groups have analyzed the results. Based on those analyses, we can consider what the future might bring.

Regulation of Genetic Engineering: Keeping It Safe

The various types of genetic engineering and cell manipulation that have been described can be grouped under the broad heading "biotechnology." As happens when rapid change of any kind occurs, people are divided over its value. Some people feel that biotechnology is unnatural and dangerous, whereas others find the potential improvements it offers too great to ignore.

Even finding a term for the products of genetic engineering has been difficult. The terms transgenic organism and genetically engineered (GE) are widely used in the scientific community, and in the media, the term genetically modified organism (GMO) is popular. Describing these crops as genetically engineered is probably most accurate, because the term genetically modified suggests that only genetically engineered plants have had their genetic material modified. However, examination of the wild relatives of the plants grown as crops makes it obvious that farmers have been selecting plants with desirable genetic differences since agriculture began. For the past century, agronomists have been creating new plants through plant-breeding activities. For example, the familiar vegetables broccoli, brussels sprouts, cabbage, cauliflower, kohlrabi, and kale are all genetically different plants, derived through traditional breeding and selection from the same plant species, *Brassica oleracea.* Another familiar example is the many breeds of dogs, all of which belong to the same species. Clearly, all domesticated plants and animals have been genetically modified, even though they have not all been genetically engineered.

Recent studies have shown that natural genetic exchanges between various species via bacterial plasmids and viruses are much more common than most people recognize. Few people understand that the crown gall bacterium naturally transfers bacterial genes to plants—an example of transkingdom genetic engineering. Gene exchange in nature occurs among plants, fungi, and animals. We are exposed to a wide variety of DNA, its mutations, and the proteins it codes for in everything we eat, inhale, and touch. We maintain trillions of genetically changing microorganisms in most of our body parts, especially our intestines and skin. Although genetic change is a natural process, genetic engineering conducted by scientists in laboratories poses potential dangers that must be evaluated and respected.

In the United States, genetic engineering is highly regulated by the federal government. Before any genetic engineering experiment can be performed, permission must be obtained. To date, more than 10,000 research applications have been made for genetic engineering involving plants. Experiments using recombinant DNA are permitted only in special laboratories. Genetically engineered plants and microbes are grown in special containment facilities, where all waste is carefully sterilized before disposal. The eventual outdoor release of the products of such experiments is more controversial.

Some people believe the organisms produced in recombinant DNA experiments represent a special danger—one greater than that posed by genetic recombination through the so-called natural means of selection and plant breeding discussed previously. However, a ruling by the U.S. Department of Agriculture (USDA), published in the Federal Register on June 16, 1987, concluded that organisms produced through recombinant DNA techniques pose no greater danger than organisms produced through naturally occurring processes. In 1989, a distinguished group of ecologists published ecological considerations and recommendations for the planned release of genetically engineered organisms. They emphasized that risk is associated with an organism's phenotype (its actual activity and survival in the environment), rather than its genotype (the source and combination of its genes).

In the United States, genetically engineered plants do not have to be labeled as such unless they produce significantly different nutritional levels, allergens, or other factors. Of the millions of proteins in our diet and environment, only a small number are known allergens. The changes in plants caused by genetic engineering are actually easier to monitor than those that result from traditional breeding, which also can produce new gene products. There is even a genetic engineering project designed to eliminate the allergens in peanuts to make this high-protein food safer. Just as no label is required on plants improved through traditional breeding, no label is warranted for most genetically engineered plants, in the opinion of most scientists.

In 2010, the National Research Council of the National Academy of Science published a report on the impact of genetically engineered crops on farm sustainability. It found that these crops can help reduce pesticide use and protect soil and groundwater, but plants with herbicide resistance and insect resistance must be used appropriately to prevent loss of function from overuse. The danger of overusing one or a few genes is just as risky in genetic engineering as in traditional methods of crossing genes into plants. The boom-and-bust cycle of plant breeding was observed long before genetic engineering even was developed. This same cycle can certainly occur with plants created through this new technology.

In addition to the herbicide resistance, insect resistance, and virus resistance just described, other features of plants have been developed or are being developed through genetic engineering. These features include improved product quality (flavor, appearance, and nutrition); resistance to fungal diseases; resistance to cold, drought, frost, and salinity; increased content and quality of gluten in wheat; natural decaffeination; more efficient use of nitrogen; production of vaccines; and lower costs of paper production from trees with reduced lignin. Without genetic engineering, it is impossible to use certain genes in wild relatives of important crop plants because of pollen incompatibilities with domesticated versions of the same species. For sterile crops, such as bananas, genetic engineering opens the door to improving disease resistance and nutrition.

Science Sidebar

Regulation of genetic engineering

Regulatory oversight of genetically engineered plants is the responsibility of three agencies of the U.S. government. The USDA's Animal and Plant Health Inspection Service (APHIS) regulates field testing to ensure that the release of a genetically engineered plant does not pose a risk to agriculture or the environment. The same organization protects U.S. borders from foreign pathogenic invaders that might threaten the nation's crops. Products of genetic engineering must be evaluated in the same way as imported plants at quarantine facilities.

The Environmental Protection Agency (EPA) regulates genetically engineered plants if they produce substances that control pests. The EPA regulates these plants in the same way it regulates other kinds of pesticides. Finally, if a genetically engineered plant is a food crop, then the Food and Drug Administration (FDA) determines if it is safe to eat. The FDA tests for changes in nutritional levels and the production of potential allergens.

In developing countries, where fertilizer is prohibitively expensive and soils are poor, genetic engineering holds great potential for improving human nutrition. Future nutritional improvements are likely in the staple crops of many of the world's poor people. For example, genetically engineered "golden rice" produces the precursors of vitamin A, helping to prevent blindness in countries where much of the daily diet is rice. Genetically modified cassava—a starchy staple root crop common in Africa—may be engineered to contain enough protein and minerals to provide the equivalent of a day's nutrients.

Both natural processes and human technology have been genetically modifying plants and animals through the centuries. In addition to the traditional techniques of crossing and selecting plants, we now have the tools to move specific genes. Genetic engineering is a new tool for use in supplementing existing standard plant-breeding methods to improve our food and fiber plants in a world where the climate is changing, the population is increasing, and the amount of agricultural land is shrinking.

CHAPTER 7

Roles People Play: Epidemics and Their Management

"The history of mankind is the story of a hungry creature in search of food," wrote historian Hendrik Van Loon. The need for food has always governed the movement and activities of people. Early humans were foragers, sometimes following seasonal food supply patterns but always moving on as supplies became depleted. Groups with small populations could support themselves in this way without too much interference from their neighbors.

With the advent of agriculture, people were able to live in one place. As food production became more efficient, some people produced more food than they needed. Making this food available to others provided them with the time to build cities, write books, create art, and pursue scientific studies. However, the potential for famine was always there, along with the threat of illnesses and the fear of invading armies.

Today, most of us live in a time and a place in which having food surpluses seems a greater problem than having food shortages. We have access to diverse food crops from many different regions of the world. For us, localized epidemics and crop losses pose only a momentary inconvenience, even though they may ruin individual farmers. Only a small segment of the U.S. population (2–3%) is employed in agriculture, so most Americans are removed from direct experience with food production. Even those with backyard gardens perceive crop failure as a small risk, since they can go to the grocery store and buy what they need if their tomatoes or carrots fail. However, highly mechanized agriculture and the rapid transit of crops in refrigerated trucks, railroad cars, and airplanes across continents are very recent phenomena. When the store shelves are well stocked with inexpensive foods, it is easy to lose sight of the precarious future of tiny seeds planted in a field and exposed to the unpredictable elements of weather, insects, diseases, and soil fertility. These elements determine the outcome of the harvest several months later.

Early farmers were almost totally dependent on their own food production, so significant losses meant going hungry or even starving. Because all agricultural tasks were performed by human or animal power, production was limited. Since cultivation was done by hand and herbicides were not used, crops faced stiff competition from weeds for water and nutrients.

Today, a large variety of food crops is available. However, as explained in Chapter 3, until approximately 500 years ago, each continent had access to only some of these important plant species. Before refrigeration became commonly available, food storage was generally unsuccessful, so supplies dwindled quickly and could be restocked only after the next growing season. In many developing countries today, people's diets are severely limited by the lack of refrigeration and storage facilities that can protect food from insects and decay.

All agricultural crops are, by nature, vulnerable to losses caused by plant pathogens. The planting of many individuals of the same plant cultivar, variety, or species over many thousands of acres (or hectares) invites epidemics. Plant disease epidemics have plagued humans since agriculture began, just as human diseases have decimated the populations of large cities. How frightening crop epidemics must have been to those early farmers, and how frustrated modern farmers still are to watch their crops wither or wilt.

Plant Disease in Traditional Agriculture: Facts Versus Myths

In the early days of agriculture, were plant diseases more of a problem or less of a problem than they are today? A good argument could probably be made for either case.

Ancient writings and biblical passages describe plant disease epidemics—especially outbreaks of diseases that were particularly recognizable, such as mildews and rusts. Losses caused by soilborne pathogens, such as damping-off and root-rot organisms, were probably considered fertility problems ("tired land"), rather than disease problems. Farmers do not always distinguish among the causes of poor harvests, which can result from flooding, drought, temperature extremes, insect pests, soil fertility problems, poor seed, or a variety of disease problems. All of these factors can play a role in the final harvest, so ancient crop protection probably involved a general approach to crop health. If so, ancient farmers were already practicing what modern scientists recommend today in their complex programs of integrated pest management (Figure 7.1).

Because early farmers' understanding of agricultural biology was primitive, their attempts at crop protection were based on religion and superstition. Ancient agricultural and historical records from throughout the world contain references to the use of sacrifices, prayers, and ceremonies to ensure a good harvest. Many of these rites are described in *The Golden Bough,* by Sir James Frazer. When animals were sacrificed, when blood was applied to the fields, or when human sexual activities took place where crops were to be planted, the overall objective was to produce a good harvest, not to manage a specific problem.

Figure 7.1. Mixed cropping in the highlands of Ecuador.

Religious figures and ceremonies for specific plant pathogens did exist, however. For example, the Roman rust god received sacrifices to protect the grain crops, and St. Anthony and other saints received requests for protection from the "holy fire" of the Middle Ages, caused by the rye ergot fungus. Many ceremonies were performed by ancient tree worshipers who believed in the magical powers of the tree pathogen mistletoe. These ceremonies are discussed further in Chapters 10, 11, and 12.

Agricultural methods from past centuries—some of which are still practiced successfully by subsistence farmers in many parts of the world—may seem primitive, upon initial consideration. In fact, they reflect careful observation and decision making by intelligent human beings over many generations. For instance, the role of certain environmental factors in plant disease epidemics has been known for many centuries. The ancient Greeks and Romans recognized that wheat rust was most severe in wet years and in low-lying areas of fields, where moisture accumulated. Of course, they did not understand the role of water in spore germination and its subsequent effects on the development of an epidemic caused by a foliar fungal pathogen. They could not control the weather any more than we can today, but their early observations provided important clues to future scientists, who had access to better tools for the study of pathogenic microbes.

In some ways, ancient agriculture was probably better adapted than modern agriculture to the general prevention of plant diseases. If we think about what agriculture was like many centuries ago, several important factors should come to mind. For example, ancient farmers worked in relative isolation, both from their immediate neighbors and from farmers on other continents. Farming was done in small patches surrounded by wild vegetation, which helped block the movement of spores and insects. Even though new kinds of plants (and inadvertently, the pathogens they carried) were popular cargo for ancient explorers, relatively little plant material typically survived the explorers' long ocean voyages. For the new plants that did survive, their dispersal on land via horse-drawn carts or packs of walking travelers was quite slow. Compare this slow dispersal against the tremendous quantity of plant material confiscated today at any international airport in a single day. Certainly, much of the plant material carried by modern travelers goes undetected and finds its way into new habitats.

Another factor worth considering is the genetic makeup of the crops planted in ancient times. They were not the seeds of careful genetic crosses and many generations of inbreeding and back-crossing to obtain maximum yields. Rather, they were the result of many years of survivor selection by farmers. These were seeds that would grow into plants that could produce some yield, despite competition from weeds, insects, pathogens, and poor soil. In plants selected under such circumstances, the levels of general resistance to a multitude of problems ensured at least some harvest, even if not the maximum possible. In addition, early crop species were grown in their centers of origin, so that natural crossbreeding with closely related wild species probably continued. Genetic diversity also was much greater in the past, because most crops were planted from seeds rather than vegetatively propagated.

Genetic diversity was substantial, both within a crop species and across the many species planted. Cautious farmers planted many kinds of plants when possible, in the hope that some food would be harvested even if certain crops did poorly. Genetic diversity between farms was probably substantial, as well, since farmers selected seed from their own plants each year, rather than purchasing it from a major seed company, as modern farmers do.

Even centuries ago, problems of soil fertility and buildup of soilborne pathogens were observed. When the human population was small and land was plentiful, early farmers simply moved to new locations when crop yields declined because of pathogens, pests, or reduced soil fertility. These nomadic farmers used slash-and-burn agriculture, a practice still used in many developing countries. The heat from burning plant material reduces many insect and pathogen populations and releases nutrients that can be used by future plants. As the human population grew in many areas, all suitable land was eventually cultivated and communities became more settled. The reuse of agricultural land became more important, and crop rotation came into practice.

The Inca of Peru developed detailed rotation laws for the production of potatoes and other crops, and the laws were enforced with the threat of death for those who did not comply. According to one of these laws, potatoes were to be planted in the same soil only every seventh year. The biological significance of this law was determined only recently, when the golden nematode—a parasitic, root-feeding worm native to South America—was discovered in Europe and the United States. Experiments demonstrated that an acceptable yield of potatoes could be produced if the crop was not grown repeatedly in one field. In the intervening years, the population of this obligate parasite declined to below a damaging level, so planting potatoes became safe once again. This rotation plan is biologically sound, even if it is not commercially acceptable to most modern farmers.

Although traditional farmers did not understand the intricacies of soil microbiology, their careful observations led them to effective practices for reducing soilborne pathogens. In the United States in the early twentieth century, before prices of farm land became prohibitive and population pressure increased, farmers grew a wide diversity of crops in rotation patterns that traditionally included leaving land fallow, or uncultivated, for a year or two.

Over the years, observant farmers have used trial and error to develop many cultural practices that help reduce disease. Even though the farmers knew little, if anything, about the microorganisms that cause plant diseases, in some cases, their methods turned out to be accurately connected with the life cycles of pathogens in ways that help protect crops. Even now, subsistence farmers in many parts of the world cling to their traditional methods in the face of modern agriculture, and these methods are often appropriate for the climates and environments in which the farmers live.

An example of such a method is *frijol tapado,* or "covered beans," which is the practice of planting bean seeds in carefully selected weed patches and then cutting and chopping the weeds until the seeds are covered by the resulting mulch (Figure 7.2). This method was described as an established means of bean culture in Central and South America by the early Spanish explorers. Today, we know that it protects the growing plants from splashed soil carrying the spores of the fungus *Thanatephorus cucumeris,* which causes a severe bean disease called "web blight." In a warm, moist tropical climate, beans planted the higher-yielding "modern way" in weed-free rows, succumb quickly to this disease.

Figure 7.2. A, bean plants grown through *frijol tapado* mulch. Mulch prevents weed growth and conserves moisture. **B,** web blight (*Thanatephorus cucumeris*) destroys beans grown without mulch.

Another example is the folk wisdom that led to laws in Europe and the British colonies in New England banning the planting of common barberry near wheat fields. These laws were put in place more than 200 years before Anton deBary, the famous 19th-century German mycologist, discovered the important role that barberries play in the life cycle of the wheat stem rust fungus.

Despite these examples, traditional methods are not always more appropriate or more suitable than modern methods. The Romans sacrificed red animals, such as cows and dogs, to their rust god to protect their wheat fields, but there is no evidence that rust disease was subsequently reduced. We must admit that the multitude of potential pathogens and pests that threaten our crops present an array of complex problems. When we concentrate on solving one or a few specific problems, we often neglect other important factors and find ourselves ill prepared to protect our crops.

Ecological Agriculture: Striking a Balance

Recent intensive study of microbial ecology and the genetic interactions of pathogens and plants has greatly increased our understanding of the complexities of the agricultural ecosystem. An ecosystem includes all the living organisms in a natural community and their various interactions. Agricultural ecosystems, or agroecosystems, are different from natural ecosystems, because the ecological balance is being manipulated by humans, who are trying to tip it in favor of the crop to be harvested. Such an unbalanced system is difficult to maintain and often requires continuous inputs, such as fertilizer, cultivation, and irrigation.

The concept of a modern agroecosystem forces us to accept the fact that all agricultural activities have multiple effects on the interactions among the various members of the community. Plant pathologists often spend years studying the details necessary to understand even one aspect of a single host/pathogen relationship. At the same time, entomologists are studying similar problems with insect pests, and soil scientists and agronomists are trying to understand the complexities of soil structure and fertility. Eventually, it is farmers who must attempt to apply all this detailed information in a way that results in optimal crop production.

In this chapter, we look at some of the various methods used to protect plants from disease. We discuss these methods in a general way, even though some important types of plant pathogens have not yet been introduced. Diseases mentioned in previous chapters are used as examples of the relative successes and failures of various management methods in relation to the types of epidemics that occur. The rationale behind each choice of method can then be applied to the diseases presented later in this book. As those diseases and their pathogens are introduced in the following chapters, imagine the approaches to management that might be effective. This will help you better understand how plant pathologists determine what they will recommend to growers.

Several terms are used to describe activities that reduce plant disease. The word control is often used to denote such activities, but in recent years, the word management has become more popular.

This small difference in word choice represents a great philosophical change on the part of plant pathologists. In the early years of plant pathology, disease-free plants were the ultimate goal of control programs, and scientists even hoped to eliminate certain pathogens completely. As the science of ecology became more sophisticated, however, it became clear that eliminating most pathogens from an agroecosystem is neither realistic nor necessary. In addition, the economic and environmental costs of trying to achieve complete control of a particular disease usually cannot be justified.

The idea of disease management emerged from a better understanding of agroecosystems. When a disease is properly managed, rather than controlled, a low level of the disease is acceptable, because eliminating it would cost more than the value of any additional yield that might result. A disease management program is founded on detailed economic and biological studies and is thus much more complex than a control program, but disease management represents a more ecologically sound approach to crop protection. In the following discussion, the word control is used to identify a specific method of disease reduction that is generally incorporated into a disease management recommendation comprised of a number of measures.

The list of possible management measures could fill an entire page, but reading it would not be very interesting or instructive. A rational means must be used to choose methods that are appropriate for a given crop and a given disease. When certain key aspects of the life cycles of both the host plant and the pathogen have been determined, a reasonable strategy for disease management is feasible. These choices are closely tied to understanding how a plant disease epidemic develops in a particular crop and the economics of crop production.

Epidemiology: Disease Increase over Time

The word epidemic is often used to indicate the rapid and widespread development of a disease. In fact, the word more accurately refers to the increase of disease in a population over time, which can occur slowly or rapidly. Technically, the word epidemic refers to the increase of disease among humans. The increase of disease among animals is called an epizootic, and the increase among plants is an epiphytotic. Although some scientists prefer to use the specific terminology, many use epidemic and the related word epidemiology to describe the study of disease increase in any kind of population.

To appreciate the importance of epidemiology in modern plant pathology, imagine a field of wheat in which some plants have suddenly appeared to be diseased. Then consider these questions:

What can you expect to see happen over the next few weeks? Will all the plants die, leaving no grain to harvest? Will only the currently infected plants die or produce less grain at harvest time? Are the plants that appear healthy already infected but not yet symptomatic?

Is the pathogen air dispersed, or is it carried in water droplets or by sticking to passing insects? Perhaps the pathogen is soilborne. If so, will you see it spread slowly, causing patches of diseased plants surrounded by healthy plants?

If the pathogen is air dispersed, is it already producing inoculum that will cause disease in farms nearby, or has inoculum from a neighbor's farm caused the disease? Can this same crop be planted in the same field next year, or might the epidemic begin even earlier and develop with greater intensity?

Epidemiologists attempt to answer such questions by describing disease development patterns during a single season and from year to year. In some cases, epidemiological studies involve the creation of complex mathematical equations that model increases in disease or changes in the pathogen population. Such studies can lead to more general models that have broad application to a variety of disease situations.

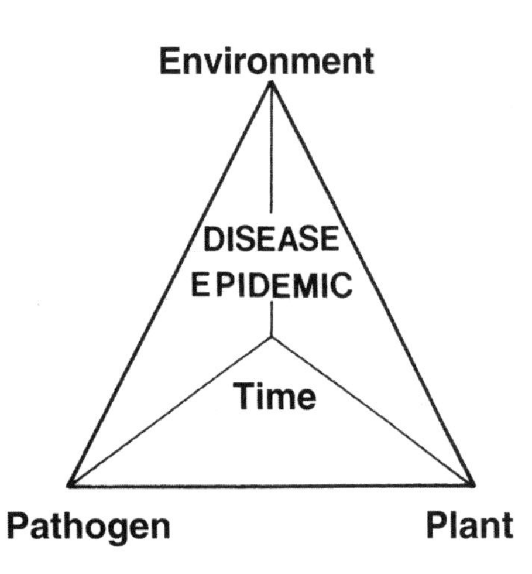

Figure 7.3. The epidemic pyramid. Time is added as the fourth factor in the development of an epidemic, acting in combination with the three factors of the disease triangle: a susceptible plant, a virulent pathogen, and a favorable environment.

Several general approaches can be taken to plant disease management. It may seem quite hopeless even to think that plants can be protected from the variety of rapidly reproducing microorganisms that are potentially or actually present. Remember from Chapter 1, however, that although pathogens may always be present, disease development occurs only when three major factors interact in a suitable fashion. As illustrated by the disease triangle, the three components necessary for disease are a susceptible plant, a virulent pathogen, and a favorable environment.

When we consider a plant disease epidemic, we can expand the two-dimensional disease triangle to form a three-dimensional disease pyramid which includes time as the fourth factor (Figure 7.3). For an epidemic to occur, the three original factors—plant, pathogen, and environment—must interact over a period of time. Then, the reduction of plant disease can be approached from two perspectives: managing the pathogen or protecting the plant.

Exclusion: Keeping the Pathogen Away

If the pathogen has not reached the plant, its arrival can be prevented by several methods. These management methods are grouped under the heading "exclusion." Exclusion is an excellent management strategy when it works, because without the pathogen, there can be no disease. The general goal of exclusion is to manage the distribution of the pathogen. If economics was not a factor in plant production, plants could be grown as sterile organisms in pathogen-free soil enclosed in sealed containers receiving only filtered air. This extreme form of pathogen exclusion is employed in some special cases, as discussed in Chapter 13, but obviously, it is not practical for most crops.

Although total pathogen exclusion is not usually possible, planting pathogen-free seeds and propagative parts is a useful means of disease management. To prevent the establishment of the pathogen in or on the seed, many seeds are tested and certified as meeting standards of freedom from pathogens. Since large propagative parts are particularly likely to carry pathogens that will cause severe losses (namely, systemic bacteria, fungi, and viruses), the production of pathogen-free propagative parts is an important agricultural industry. Examples of these products include potato tubers free of mycelium of *Phytophthora infestans,* citrus nursery stock free of citrus canker bacteria, and raspberries and many woody ornamentals free of crown gall bacteria. Quarantines and agricultural embargoes between countries, regions, and states also are important means of pathogen exclusion (see Chapter 3).

Eradication: Reducing the Pathogen Population

For most common pathogens, excluding them from planting material is helpful but generally temporary in preventing disease, because they are often introduced into the pathogen-free area at some point. Also, the potential always exists for infection by pathogens that are already present in the growing environment as dormant propagules, as saprophytes, or as pathogens of weeds and nearby crop plants. Thus, the second general approach to management of the pathogen is eradication, which is aimed at reducing the survival of the pathogen.

Eradication is usually not complete, since eliminating the pathogen is economically and biologically impossible, in most cases. Total eradication may be the goal when a quarantine fails, and extensive eradicative measures may delay the spread of a pathogen that has bypassed quarantine restrictions. In fact, history has shown that total eradication is difficult and not likely permanent. As mentioned in earlier chapters, several previous efforts at eradication have failed, including those aimed at citrus canker in Florida, coffee rust in South America, and fire blight in Europe.

The purpose of some seed treatments is to eradicate pathogens before planting. Historically, seeds were soaked in hot water or chemicals and other harsh substances—such as horse urine and salt water—to destroy pathogens. These attempts met with limited success, however. The current use of systemic fungicides, available since the 1960s, has been far more successful. These chemicals can eradicate certain important seedborne pathogens without harming the plants' embryos. Using a small amount of a chemical at the time of planting can eliminate the need later to spray a large quantity of fungicide on growing crops.

Eradication also can include the destruction of infected plants that have a lower economic value than the crop plants to which the pathogen may be transmitted. A good example is the elimination of weed species infected by plant viruses that also infect crop

Science Sidebar

Disinfested or disinfected?

Plant pathologists use the term disinfested to describe the removal of pathogens from the surfaces of objects, such as pruning shears and potting bench tops, and from the surfaces of seeds and other plant parts. Likewise, storage areas should be carefully cleaned and disinfested to eradicate any pathogens left from previously stored plant materials. This eradication can be accomplished by using steam heat or chemicals.

Soil and plant debris may be described as infested with pathogens at the end of a growing season, because neither soil nor dead organic materials can be said to be infected. The term disinfected is used to describe elimination of a pathogen from within infected living plant tissue. The most common example is when infected seeds are disinfected with heat or systemic chemicals.

The terms disinfested and disinfected are used differently by medical personnel and by manufacturers of so-called disinfectants for living areas and medical facilities.

species. Virus transmission from weed hosts to crop hosts can occur when the plants rub together or, more commonly, when sucking insects (such as aphids and leafhoppers) transmit viruses during feeding. This is discussed in more detail in Chapter 13.

Similar attempts at eradication are made with other types of pathogens. For example, hawthorns are eradicated in Europe, because they serve as a source of fire blight bacteria. Hawthorns are closely related to many species of fruit trees and grow wild in hedgerows, which means the bacteria may be carried to nearby apple and pear orchards by insects or windblown rain. Many rust fungi require two unrelated plant species, known as the alternate hosts, to complete their life cycles. This amazing biological phenomenon, explained in Chapter 11, can be exploited for practical disease management. When one host is economically important and the other is not, the less-valuable host species can be removed, thus eradicating the rust fungus by preventing completion of its life cycle.

Eradication of diseased crop plants or their infected parts is commonly practiced. Many growers inspect newly emerged plants and remove diseased individuals. The removal of young, virus-infected plants early in the season, called "roguing," eliminates them as sources of infection for healthy neighboring plants. Also, cankers in woody plant parts serve as important reservoirs of bacteria and fungi that produce inoculum for new infections of healthy plant tissue. Pruning out cankers not only removes sources of inoculum for new infections, but it also eliminates the continued destruction of infected woody tissues by expanding cankers. Pruning in the dormant season is often most practical, for two reasons. The branches of deciduous plants are best exposed during winter and early spring for the inspection of perennial tissues, and most pathogens are in an inactive state and unlikely to spread to new infection sites during the pruning activities.

A multitude of other agricultural practices—the purpose of which is pathogen eradication—can be grouped under the heading "sanitation." Careful growers remove soil from tools and machinery before transporting them between farms (or sometimes even between fields) to reduce the movement of soilborne pathogens. Tools used for pruning, grafting, and other activities that cut plant tissues should be sterilized or disinfested between cuts to prevent pathogen spread.

The careful removal and destruction of plant material at the end of the growing season is often effective for eradicating existing pathogens that might threaten plants in the next planting. This is obviously not a practical means of eradication in a large field, but crop debris and stubble can be burned or plowed under to encourage decay of the remaining plant material. In moist soil, with its teeming microbial population, pathogens find themselves in an intensely competitive soil environment after the plant debris has decayed. Without the safe haven of the host plant, the pathogen population will quickly decline, because most pathogens are far less competitive than the saprophytic microorganisms of the soil. Farmers can exploit this natural competition to help eradicate pathogens before planting the next crop.

Although residue destruction and microbial competition can be exploited for crop protection, two aspects of modern agriculture limit farmers' ability to use them. The first is reduced soil tillage, or plowing. Stubble burning and soil tillage have been greatly reduced in some areas to help prevent soil erosion, to store carbon in the soil, and to preserve soil moisture. Although the effects of reduced tillage vary, depending on the crop and the pathogen, an increase in the diseases caused by pathogens that survive best in crop debris can generally be expected. Increases in fungal foliar diseases in corn, soybeans, and wheat grown under reduced tillage have already been observed. The conservation of valuable topsoil is critical in many areas, however, and many disease problems can be minimized by paying careful attention to plans for crop rotation.

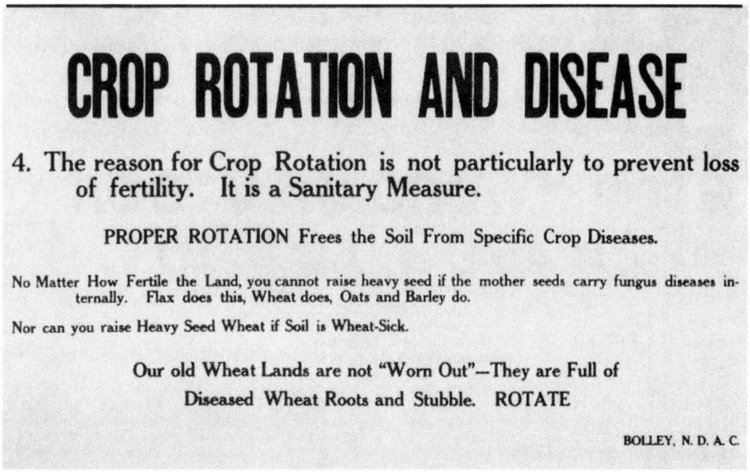

CROP ROTATION AND DISEASE

4. The reason for Crop Rotation is not particularly to prevent loss of fertility. It is a Sanitary Measure.

PROPER ROTATION Frees the Soil From Specific Crop Diseases.

No Matter How Fertile the Land, you cannot raise heavy seed if the mother seeds carry fungus diseases internally. Flax does this, Wheat does, Oats and Barley do.

Nor can you raise Heavy Seed Wheat if Soil is Wheat-Sick.

Our old Wheat Lands are not "Worn Out"—They are Full of Diseased Wheat Roots and Stubble. ROTATE

BOLLEY, N. D. A. C.

Figure 7.4. A 1909 poster by H. L. Bolley, of the North Dakota Agricultural College, explaining that the benefits of crop rotation for wheat result from a reduction of soilborne pathogens.

A second aspect of modern agriculture that limits use of microbial competition to eradicate pathogens is the pressure to limit crop rotation. As mentioned earlier, many soilborne pathogens are host specific to some degree. Thus, in choosing a crop to plant in a particular field, it is important to select the one that is least susceptible to any pathogens that might still be present in the debris from the previous crop (Figure 7.4). Rotation recommendations often involve alternating between crop species from different plant families. For example, a grain crop might be followed by a crop of sunflowers or a forage legume, such as alfalfa. The reason for alternating species is that distantly related plants tend to have different pathogens.

Unfortunately, economic constraints have encouraged many farmers to abandon traditional rotation patterns in favor of more intensive monoculture. Farmers feel pressure to grow high-value crops more often because of the greater earnings these crops bring. Additional incentive comes from the sizable investment many farmers have made in specialized planting and harvesting equipment. As a result of reduced tillage and monoculture, farmers are less able to exploit the ability of competitive soil microbes to eradicate pathogens from the soil.

For some high-value crops and special disease situations, rotation and tillage do not sufficiently eradicate pathogens for economic crop production. In such situations, more rapid and complete pathogen eradication from the soil is necessary.

In greenhouses, relatively small amounts of soil are used in plant production, so chemical or heat pasteurization methods are used to kill pathogens in the soil (Figure 7.5). Soil can be heated to a temperature that kills most pathogens without killing all of the beneficial saprophytic competitors, so some protective competitors will still be present in case a pathogen is accidentally reintroduced. Pasteurization is the term used to describe the heating of soil to kill pathogens without destroying all of the microorganisms. (The same term is used to describe the process of heating milk and other dairy products until most but not

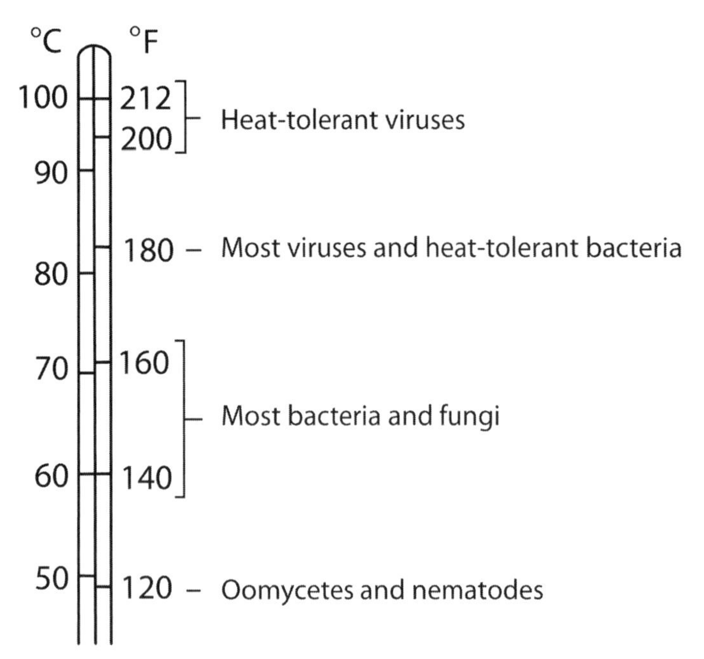

Figure 7.5. Temperatures necessary to kill plant pathogens. Soil is usually heat treated for 30 minutes under moist conditions.

all of the bacteria present have been killed.) When soil is heated to a very high temperature that kills all of the organisms (sterilization), ammonia and salt toxicities arc common problems. In many modern greenhouses, plants also are grown in soilless mixes containing peat moss and perlite and even in liquid nutrient solutions (hydroponics) in rock wool to avoid soil pathogens.

Outdoors, soil pasteurization is expensive and temporary, because only small areas of soil can be treated and pathogens can then recolonize treated soil from the edges of the treatment area. Such soil treatment is done only for high-value crops or in soil where a pathogen is present that can prevent an acceptable harvest if not eradicated. Heat treatment with steam is time consuming and difficult in field situations, although it is sometimes done in nurseries and seedbeds. Outdoor soil is more commonly sterilized with chemical fumigants.

In a climate in which intense sunlight is available, the pathogen population in soil also can be reduced by using clear plastic tarps to trap sunlight and heat the soil in a process called solarization. In a hot and dry climate, repeated tillage may be practiced on fallow land to eradicate certain soilborne pathogens by exposing them to the harsh environmental conditions.

Protection: Helping the Plants

We have seen that management of a pathogen can be approached in two general ways: exclusion, which manages the distribution of the pathogen, and eradication, which manages the survival of the pathogen. Now, we consider what might be done to protect the host plant, assuming the pathogen has not been completely excluded or eradicated.

Certainly, the use of pesticides to kill the pathogen comes quickly to mind. For some diseases, such as late blight of potatoes, frequent protective sprays with fungicides are necessary to produce a crop in a humid climate. However, the use of pesticides for disease management is decidedly ineffective for many diseases, such as those caused by bacteria, viruses, and many soilborne pathogens. Another factor is that economic constraints limit the use of pesticides for plant disease management to relatively few crops. The complex and controversial subject of using pesticides to manage plant diseases is discussed in Chapter 8. This chapter considers other approaches to plant protection.

One approach is to protect plants against infection by pathogens through genetic resistance. Genetic resistance is pathogen specific, so it is unlikely that a particular cultivar will have genetic resistance to all the possible pathogens that can attack it. However, at least 75% of all agricultural crops have been improved by incorporating genetic resistance to at least one pathogen, and many crops are resistant to several pathogens. For example, 98% of all grain and forage crops have genetic resistance to one or more diseases. All modern wheat cultivars have resistance to at least one pathogen, and some have resistance to as many as ten different pathogens. Genetic resistance may be race specific, or it may be generally effective against all races of a single pathogen.

Another approach is biological control, which involves the use of living microorganisms—particularly bacteria and fungi—to help reduce plant diseases. Sometimes, these microorganisms function to eradicate spores or other pathogen survival structures. Most examples involve protecting plant surfaces from infection, as described previously for fire blight of apples and crown gall on certain crops (such as raspberries). These are discussed in more detail in Chapter 9.

A final approach to disease management that protects the host plant from infection includes methods that can be categorized as "cultural practices." These methods consist of agricultural activities that do not directly exclude or eradicate pathogens but instead serve to reduce plant disease by protecting the plant. In a very general sense, vigorous plants are less susceptible to disease than nutrient-deficient plants grown with suboptimal water, poor soil drainage, or intense weed competition. Therefore, agricultural activities that improve plant growth and vigor will, at least indirectly, protect plants from disease.

Many studies have shown the direct and indirect effects of specific soil elements on plant susceptibility. One important example concerns nitrogen, which results in increased infection by many pathogens when applied in high amounts that stimulate the growth of soft, succulent plant tissues. An insufficient amount of nitrogen, on the other hand, may increase symptoms caused by pathogens that reduce nitrogen uptake (such as root pathogens) or make leaves more susceptible to infection.

The application of irrigation water also can affect plant diseases. Having too little water puts plants under stress and makes them more susceptible to some pathogens. Frequent, shallow watering encourages the development of shallow, poorly formed roots that cannot withstand attack by root pathogens as well as a deep, well-developed root system can. Overwatering—the most common cause of houseplant death—fills all of the soil air spaces with water and prevents root growth by depriving these living tissues of oxygen. Stressed roots are more susceptible to attack by root-rot pathogens.

The aboveground effects of watering can be important, as well, because many foliar fungi require water on the leaves for spore germination and infection. Bacteria also require water for multiplication and infection. Irrigation water should be applied early in the day, when the leaves will dry quickly, or during the night hours, when the leaves are already wet from dew. In either case, the duration of leaf wetness should not be extended. Foliar diseases are much more likely to develop when water is applied to leaves in the late afternoon or early evening, causing the leaves to remain wet throughout the night. Despite a common misconception, applying water to plants in the sunlight does not burn the foliage but actually helps cool the plants. Brief applications of water to the putting greens on golf courses in the heat of the day (called "syringing") help keep the closely mown grass alive.

In row crops, rapid drying of foliage can be improved by orienting the row with the predominant wind direction, properly spacing the rows, thinning the plant density, and removing weeds. In modern greenhouse production, irrigation systems apply water only to the soil surface, keeping the leaves completely dry and preventing many foliar diseases. In field situations, drip irrigation, in which water is allowed to trickle slowly onto the soil surface or directly into the root zone, is used for some fruit and vegetable production. When the foliage is wet, pathogens such as bacteria and fungi can be spread during normal activities, such as pruning and cultivation. It is recommended to wait until the leaves dry before handling or working among the plants.

Monocyclic and Polycyclic Epidemics

The preceding discussion described two approaches that can be used to reduce plant diseases: management of the pathogen, through exclusion and eradication, and protection of the plant, through the use of protective chemicals, genetic resistance, plant vigor, and numerous cultural practices. How can we create an appropriate disease management strategy from the many possible options?

Epidemiology has given us some general principles that suggest a rational approach to this problem. J. E. Vanderplank, a South African plant pathologist, introduced the concept of monocyclic epidemics and polycyclic epidemics in 1963 and provided two simple models that illustrate the effects of these two general types of epidemics.

A pathogen that causes a monocyclic epidemic completes one disease cycle during a growing season. An example of such a pathogen is *Fusarium oxysporum,* which infects the xylem and reduces the ability of a plant to absorb water and minerals, causing Fusarium wilt. Spores are present in the soil at a particular density that results in a certain number of plant infections. At the end of the season, when the infected plants decay, new spores are released into the soil and serve as inoculum for the next season.

Vanderplank used a banking analogy to describe monocyclic epidemics. He used the term "simple interest" to describe how new inoculum is added at the end of the season, similar to how a bank pays interest after money has been on deposit for a year. This type of interest is commonly paid on certificates of deposit (CDs). If two people purchase CDs, the person who invests the most money will have earned more money when the CDs mature. Unfortunately, more disease is not as appealing as more money.

A pathogen that causes a polycyclic epidemic produces new generations of inoculum throughout the growing season, so that infections continue to increase in number. A familiar example is *Phytophthora infestans,* the cause of late blight, which produces new sporangia several days after infection. A graph of disease progress of a polycyclic epidemic will show a slow increase in disease at first, because only a few infections have occurred. Acceleration is rapid, however, because each set of infections results in the production of new inoculum. Disease increase levels off later in the season, if no new plant tissue is available for infection. This S-shaped curve is typical for rapidly increasing polycyclic diseases, such as many rusts, downy mildews, and leaf spots caused by fungi and bacteria.

Again using a banking analogy, Vanderplank referred to polycyclic epidemics as "compound interest" diseases. Some bank accounts earn compound interest (often described as "compounded daily"), in which the amount earned each day or month is added to the principal and the entire amount continues to earn interest. This compounding is similar to the increasing amount of inoculum that occurs with each generation of the pathogen. Unfortunately, the rate of increase with plant pathogens is much greater than the interest rate paid by a bank!

How can these two models be used to predict the amount of disease that might occur in a single growing season? For a monocyclic epidemic, we should be able to predict how much disease will occur if we know how much primary, or initial, inoculum is present (that is, inoculum that existed at the beginning of the growing season). The amount of disease at the end of the season should be proportional to the amount of primary inoculum. When the amount of inoculum is high, there will be more disease. Conversely, less inoculum should result in less disease.

Can the same relationship be applied to the primary inoculum in a polycyclic epidemic? In most cases, the relationship is less direct between the primary inoculum and the final amount of disease. The increase of the pathogen during the growing season is governed by factors such as temperature, moisture, and crop plant resistance, which play a much more important role in the final amount of disease than the amount of primary inoculum. The length of time between generations of the pathogen, called the latent period, is governed directly by these factors. Inoculum produced during the growing season, called secondary inoculum, is responsible for the explosive development of some polycyclic epidemics. Pathogens such as *P. infestans* multiply so rapidly that even a low amount of primary inoculum can result in the total destruction of a potato field.

How to Make the Management Choices

How can understanding the differences between monocyclic and polycyclic epidemics be useful in choosing appropriate management measures? Let us first consider the effects of reducing primary inoculum. A reduction of primary inoculum has a direct effect on the amount of disease in a monocyclic epidemic. Management measures that reduce primary inoculum include many of the pathogen eradication methods discussed previously.

How does reducing primary inoculum affect a polycyclic epidemic? The epidemic may be delayed in the beginning, because less inoculum is available, but the exponential increase of the pathogen may still cause severe disease. If some inoculum is present, the rate at which the epidemic develops is governed by factors unrelated to primary inoculum and will be relatively unaffected by its reduction. This means that methods that reduce primary inoculum of the pathogen may not be sufficient to protect the crop. Reducing the rate at which the epidemic develops is usually necessary as well. Thus, for a polycyclic epidemic, reduction of primary inoculum is important and helpful but not usually sufficient.

In the case of potato late blight, primary inoculum is reduced by the destruction of infected tubers and by the planting of pathogen-free tubers. If oospores become widespread in the soil, they, too, will have to be reduced in the future, perhaps by crop rotation. The rate at which the epidemic develops is reduced in several ways: by timing irrigation to prevent extending leaf wetness, by frequently applying fungicides, and by using general resistance that is effective against all races of *P. infestans.*

Coffee rust, downy mildews, and South American leaf blight of rubber are additional examples of diseases that result in polycyclic epidemics, for which rate-reducing methods of disease management are necessary. Of course, if all the primary inoculum can be excluded or destroyed, a polycyclic epidemic cannot occur. South American coffee growers must now approach rust management with rate-reducing methods, because the quarantine failed to totally exclude the coffee rust pathogen, *Hemileia vastatrix.* So far, total exclusion of primary inoculum by quarantine has protected Asian rubber plantations from South American leaf blight.

In conclusion, disease management is most effective when comprised of methods appropriate for the type of epidemic that results from a particular pathogen. Most of the practices that reduce primary inoculum were described previously in the sections on managing pathogens (that is, exclusion and eradication). Obviously, complete pathogen exclusion is desirable for all plant pathogens but not usually possible.

Let us consider these two management strategies using a diagram (Figure 7.6). In most cases, reducing primary inoculum is most effective for managing monocyclic epidemics. Management methods that protect the host plant—such as cultural practices, protective chemicals, and genetic resistance—are primarily effective in reducing the rate of an epidemic, essentially "putting on the brakes" (Figure 7.6, curve C). Thus, these methods are most effective for managing polycyclic epidemics. Reducing primary inoculum may delay a polycyclic epidemic, but it has no effect on the epidemic rate, which is caused primarily by environmental factors that occur during the growing season (Figure 7.6, curve B). If both the primary inoculum and the epidemic rate can be reduced, the epidemic is both delayed and slowed (Figure 7.6, curve D). This is the most desirable situation and explains why a complete disease management strategy for rapidly reproducing pathogens includes methods that reduce both the primary inoculum and the epidemic rate.

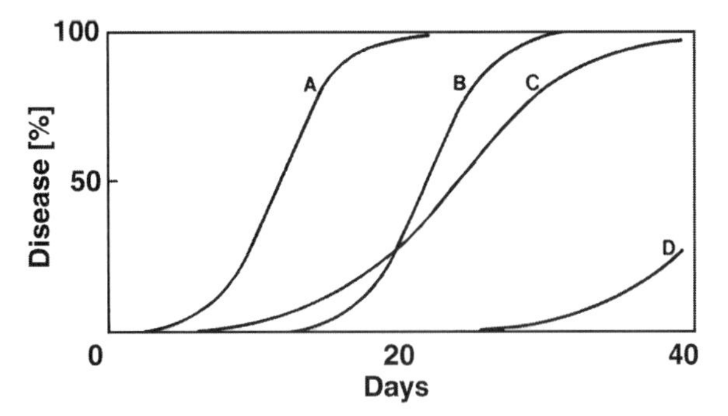

Figure 7.6. Disease progress curves of plant disease epidemics. A polycyclic epidemic under different management strategies: **A,** no management; **B,** reduction of primary inoculum, which delays the epidemic but does not reduce the epidemic rate; **C,** reduction of the epidemic rate; **D,** a combination of B and C (the epidemic is delayed and the rate is reduced).

Although these epidemiological principles are extremely useful, we must not oversimplify the complexities of an agricultural ecosystem. Despite the comfort we get from using categories, few things in nature can be divided so neatly into two types. Many epidemics are not easily labeled as monocyclic or polycyclic. Some pathogens produce only two or three generations of new inoculum during a growing season. Some pathogens that produce more than one generation of inoculum are located in the soil, where inoculum is unlikely to spread to new plants. Sometimes, the categorization of a disease epidemic changes if it is observed over several years, rather than during a single growing season. For example, Dutch elm disease could be called monocyclic in a single year, but its behavior is polycyclic when observed over several years (see Chapter 12). In each case, it is necessary to determine whether reducing primary inoculum is sufficiently effective or possible or using rate-reducing activities is more appropriate.

In addition, we should not conclude that polycyclic epidemics are necessarily more threatening or more destructive than monocyclic epidemics. Some so-called minor diseases result when the host plants have enough general resistance that pathogens cause little economic impact, despite being polycyclic in nature. In contrast, some soilborne monocyclic epidemics cause great losses. *Verticillium* species are fungi that cause monocyclic vascular wilt diseases. If they are present in the soil at a level that causes 5–10% of the plants to become infected, they may reach a concentration capable of causing a 80–90% infection rate in just 3 years.

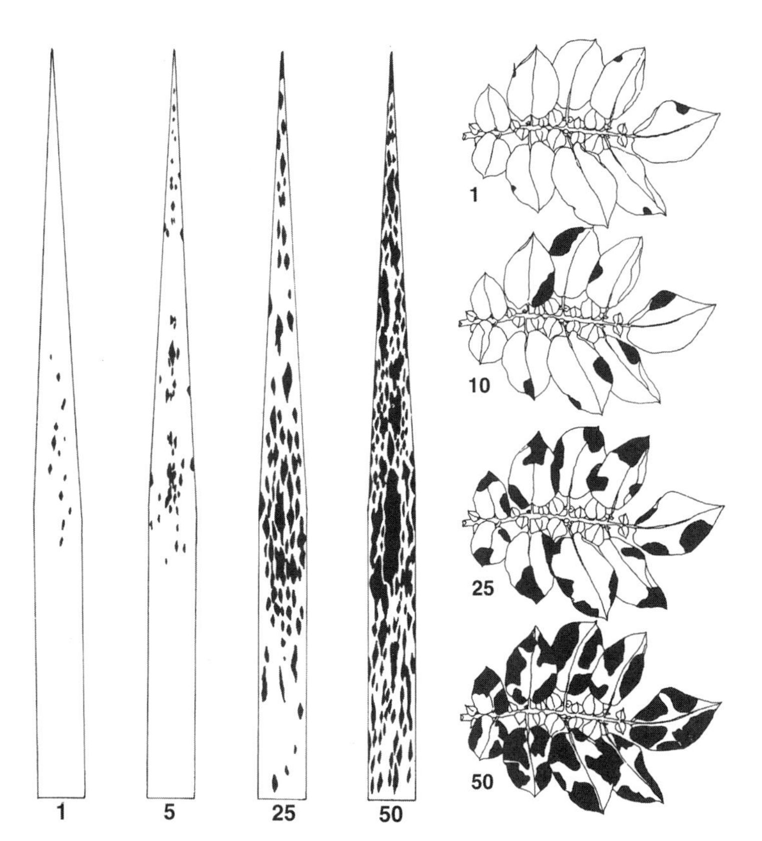

Figure 7.7. Disease assessment keys. These keys are used to make visual comparisons with diseased plants to improve the accuracy of evaluation of resistance, fungicide control, and cultural practices that reduce disease. **Left,** Septoria leaf blotch of cereals. **Right,** late blight of potatoes.

Economic factors, including the cost of the management practice and the value of the crop, always play an important role in making management decisions. Economically sound decisions must be based on accurate disease assessment that can be correlated with predictable losses. Figure 7.7 illustrates how difficult accurate disease assessment can be. Cover up the numbers next to each set of images, which indicate the percentages of diseased tissue, and guess how much tissue is diseased. It takes some practice to accurately assess the percentage of diseased tissue on an infected plant. To improve speed and accuracy, computer programs can be used to assess a single plant or an entire field.

The 1970 Southern Corn Leaf Blight Epidemic: An Important Field Crop Epidemic of the 20th Century

We conclude this chapter with descriptions of two extremely destructive plant disease epidemics. The first is the southern corn leaf blight epidemic that occurred in the United States in 1970, causing more than $1 billion losses in the corn crop.

The pathogen responsible was a leaf spot fungus that causes polycyclic epidemics and is commonly known by the name of its asexual or conidial stage, *Bipolaris maydis* (Figure 7.8). The fungus produces dark, multicellular conidia on the surfaces of infected leaves

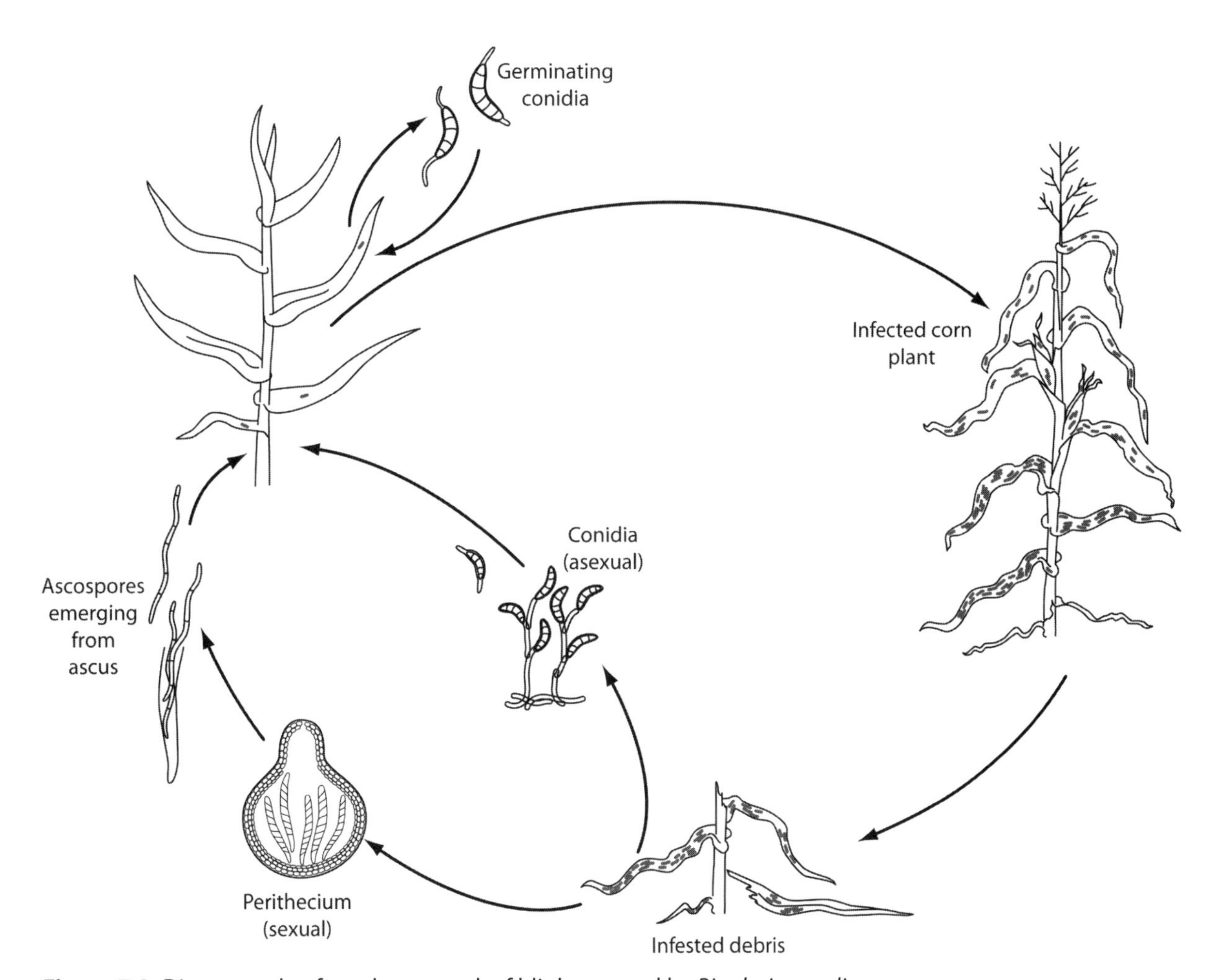

Figure 7.8. Disease cycle of southern corn leaf blight, caused by *Bipolaris maydis.*

several days after initial infection. These conidia are then blown or splashed to new plant tissue. Infection is most common on the leaves and husks of the developing ears of corn. This fungus causes spots on corn leaves in almost any year in all but the most northern states of the United States.

What is the usual management strategy for such a pathogen? Theoretically, this pathogen is best managed by a combination of methods that reduce primary inoculum and the rate of the epidemic. The fungus overwinters primarily in stubble left from the previous crop, so stubble can be plowed under to enhance decay. Corn should not be grown in the same field in consecutive years.

If no-till is in use, this first management option is not feasible. Then what means are available to reduce the rate of the epidemic? Cultural practices that reduce leaf wetness consist of row orientation and spacing. Protection of the foliage with fungicides cannot be justified, because the value of the corn crop is usually insufficient to offset the expense of chemical applications, except possibly in seed production fields. Genetic resistance seems the most economically feasible means of management. The available cultivars of corn appear to have high levels of general resistance to this pathogen, because infections regularly occur but rarely cause significant loss.

In 1969, southern corn leaf blight was one of several corn foliar diseases that together accounted for a loss of approximately 2.3% each year. We must ask: What changes occurred that led to the tremendous loss in 1970?

This corn disease epidemic is an excellent example of the problems that can arise from genetic uniformity and a narrow view of an agricultural ecosystem. To understand what occurred, it is first necessary to review the production of seed corn.

Early in the 20th century, breeders discovered that inbreeding corn did not increase yield but rather reduced it. When two high-yielding parental lines were crossed, the initial progeny produced the highest yields, but the continued propagation of seeds from these progeny produced disappointing harvests. From these studies came the concept of hybrid vigor (also called heterosis), which predicts that the highest yields are produced in progeny from the crossing of two inbred parental lines with many genetic differences. Similar observations have been made for many other types of plants, so seed catalogues offer hybrid seed of many crops. The production of such seed requires that genetic crosses be carefully made each year. If seed from hybrid plants is saved and planted, the resulting plants will not have the vigor of the parent plants.

Figure 7.9. Hybrid corn seed production in which the seed-producing plants have been detasseled to prevent self-pollination.

To produce hybrid corn seed, the tassel that produces pollen at the top of one type of corn plant must be prevented from fertilizing ears of corn on the same type of plant, which serves as the female parent of the hybrid cross. The plants are detasseled by hand to prevent them from being self-pollinated. In a seed corn field, rows of a second type of corn plant with tassels serve as the male parental line and alternate with rows of detasseled corn that serve as the female parental line (Figure 7.9). Seed is harvested only from the detasseled plants, because this seed is certain to be the progeny of the hybrid cross. Obviously, the process of hand detasseling before pollen shed requires a lot of labor, which is expensive, and the work must be completed in a short time. Seed producers constantly worry that weather or other problems might prevent complete detasseling before pollen shed begins.

In the 1960s, a genetic factor was discovered in Texas that caused plants to become male sterile, meaning that the plants produced no pollen. Plants from the female parental line no longer had to be detasseled to prevent self-fertilization of their ears, which greatly reduced labor costs. The genetic factor involved was governed by genes in the cytoplasm, which were later found to be associated with the mitochondria. The so-called Texas male-sterile (TMS) cytoplasm is inherited only through the female parent, because the mitochondria are contributed to the zygote by the egg, as described in Chapter 5. In 1970, nearly 80% of all hybrid field corn produced in the United States contained the breakthrough TMS cytoplasm.

Genetic uniformity was the first step on the road to disaster. It represented an increase in the susceptibility of the host, one of the four important factors necessary for an epidemic to occur.

In the meantime, a genetic change occurred in the pathogen population as well. A new race of the pathogen appeared that was particularly virulent on corn with TMS cytoplasm. The new race was named Race T to differentiate it from the common Race O, which caused only a minor leaf spot disease.

It soon became apparent that the disease caused by Race T was quite aggressive. New generations of inoculum could be produced in as little as 51 hours after infection. In addition, the fungus infected the leaves and husks, and it even destroyed the developing ears (Figure 7.10). In the southern states, development of the epidemic was so rapid that brown paths could sometimes be seen in fields following the prevailing air movement. Fungicides were applied by airplane, despite the high cost, to try

Science Sidebar

Potential dangers in new cultivars

Plant breeders, working with plant pathologists, continually test new cultivars to determine their susceptibility to pathogens. Despite these tests, it is impossible to detect all of the genetic changes in pathogen populations and all of the new genetic interactions between existing pathogens and new cultivars. Situations similar to the southern corn leaf blight epidemic have occurred before.

One of the most famous epidemics caused severe losses in the oat cultivar Victoria (named for England's former queen) and genetically related cultivars in the 1940s. The Victoria oat and its relatives were very popular, because they contained a resistance gene against crown rust, the most important oat disease. Unfortunately, these cultivars turned out to be particularly susceptible to the foliar fungus *Bipolaris victoriae,* named after its best-known victim. This fungus had previously been considered a minor pathogen, but it produces a potent toxin that specifically causes severe leaf blight in Victoria oats and other cultivars with Victoria parentage.

In 1945, when approximately 75% of U.S. oats contained the Victoria crown rust resistance gene, a severe epidemic struck the northern states. A second epidemic struck the southern states in 1947, and after that, growers turned to other oat cultivars.

Figure 7.10. The 1970 southern corn leaf blight epidemic developed very rapidly. The heavily infected corn leaf on the right became infected about 10 days before the corn plant on the left, which has only a few leaf lesions.

Figure 7.11. In the southern United States, the 1970 southern corn leaf blight epidemic completely destroyed the corn crops on many farms.

to save some of the corn crop. In spite of these efforts, in many southern states, entire fields were destroyed and losses of 80–100% were common (Figure 7.11).

The new race caused much less severe disease on corn with normal cytoplasm. Today, researchers know that Race T produces a toxin that specifically binds to the mitochondria of TMS-cytoplasm corn. Even though Race T could cause disease on normal-cytoplasm corn, the T-toxin did not poison the mitochondria. This was the first example of the effects of cytoplasmic genes on disease susceptibility.

Then, the third component of the disease triangle came into play: environment. Race T grows quickly in hot and moist conditions, as commonly found in the southern states. Fortunately for the farmers in the northern states, environmental factors were on their side, so losses from the disease were much less in these cooler states. The year 1970 also was not particularly wet, which helped reduce the incidence of infection. Corn prices doubled late in 1970, so northern farmers—who lost only 10–20% of their crops—came out ahead financially.

In the scramble to take advantage of the male-sterile genetic factor for hybrid seed production, was there no warning of the possible dangers that lay ahead? Reports of increased susceptibility of TMS corn to *B. maydis* were made as early as 1961 in the Philippines, where the climate is hot and humid, but the information was not widely read and its implications were not realized. Corn breeders and seed producers observed the increased susceptibility of the TMS corn in 1969, but that was too late to prevent the use of TMS seed the following growing season. In the meantime, inoculum was increasing, resulting in the explosive epidemic of 1970.

What happened in the hybrid corn industry? By the following year, the connection between TMS cytoplasm and susceptibility to Race T was clear, and seed producers quickly returned to normal cytoplasm for winter seed production in the South and in Hawaii. Farmers planted normal-cytoplasm seed where it was available, blended it with TMS corn, or temporarily switched to other crops. Breeders quickly switched away from the TMS cytoplasm, and many teenagers in the Midwest once again had summer jobs detasseling corn in the seed production fields. Plant pathologists and agronomists again became acutely aware of the dangers of genetic uniformity.

Brown Spot of Rice: Endemic Diseases Sometimes Cause Great Harm

In 1942, the world's attention was focused on World War II. India, then part of the British Empire, was under the threat of attack from Japan.

The staple crop of India then, as now, was rice. Three rice crops—two minor and one major—were sown each year in the province of Bengal in northwestern India. Near the end of June, one of the minor crops, the Aus crop, was sown by broadcasting seeds that grew throughout the monsoon season and were harvested in late September or early October. The major crop, the Aman crop, was transplanted into paddies in July or August and harvested near the end of the year. Another minor crop, the Boro crop, was grown in low-lying areas after the water from the monsoon season had receded.

Most years, the monsoon season in Bengal began in the second half of June, and rainfall continued through July and August and tapered off in September. The total monsoon rainfall ranged from 80–100 inches (30–40 centimeters) across Bengal. After the monsoon season ended, environmental conditions were usually excellent for grain fill and ripening in the Aman crop.

Figure 7.12. Brown spot on rice, caused by *Bipolaris oryzae.*

Brown spot, caused by a fungus commonly known as *Bipolaris oryzae,* had been a common but minor disease in the Bengal rice crop (Figure 7.12). This fungus has the same basic life cycle as *B. maydis,* which causes southern corn leaf blight. Typically, the Aus crop became infected late in its growing season, and conidia from this crop then spread into the Aman crop. Infected seeds and alternate hosts also served as sources of primary inoculum. However, in most years, the level of infection in the Aman crop was not high enough to significantly reduce the yield of this critical crop.

Unfortunately, 1942 was not a typical year. The monsoon season continued through September and October and into November, rather than ending in late August or early September. An additional 15–25 inches (6–10 centimeters) of rain fell, and conidia moved from the heavily infected Aus crop into the Aman crop, causing yield losses that ranged from 70% to more than 90% across Bengal.

With this production shortfall, the price of rice began to rise early in 1943. The Indian government was focused on war, and soon the price of rice was so high that many Indians could not afford it. People were forced to sell their farms and to head to the cities in search of jobs. Sadly, many of them died of starvation and related illnesses. In all, more than 2 million Indians died in the Bengal famine of 1943, making this epidemic as calamitous as the much more famous Irish potato famine that had occurred a century earlier.

This story illustrates that an endemic disease that has been present in a locale for many years can cause a significant epidemic if conditions change. In the case of brown spot of rice, the pathogen and the host did not change. The only aspect of the disease triangle that did change was the environment. The extended monsoon season led to the brown spot epidemic and subsequently to the suffering and death of millions of Indians.

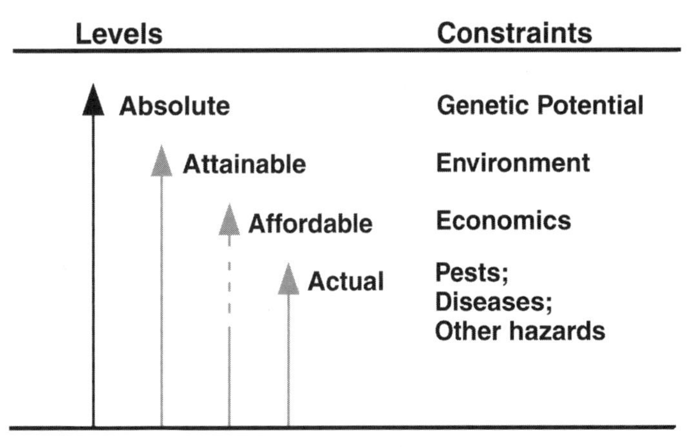

Figure 7.13. Constraints to crop production.

Conclusions

We have now provided a framework of epidemiological principles for disease management. We use these principles in the following chapters to illustrate the means by which humans are learning to coexist with pathogens and still obtain a bountiful food supply (Figure 7.13).

For the past 100 years, and especially since World War II, chemical pesticides—the subject of the next chapter—have become an important means of crop protection. Furthermore, since the 1960s, people have begun to understand some of the ecological effects associated with the use of these chemicals. Ecological principles apply to agroecosystems, just as they govern the natural world.

CHAPTER 8

Chemicals to Protect Plants: Pesticides

Few topics provoke a more emotional response than the use of pesticides in agriculture and landscapes. Some people believe that pesticide use should be banned completely, arguing that the negative environmental and health effects it creates cannot be justified in a time of food surpluses. Others believe that only "synthetic" or "chemical" pesticides should be prohibited, arguing that "natural" pesticides can be used safely. Still others claim that pesticides are a necessary input to modern agriculture, without which our food supply would become expensive and unreliable. Some people who work in agriculture resent government regulations, which they find poorly defined and overly restrictive. Others feel the government has a responsibility to protect the health of its citizens and the environment through pesticide regulation.

An appropriate response to such a complex issue requires becoming informed about its various components. The controversial aspects of the pesticide issue can be divided into two categories: costs and risks.

Before we discuss these topics, it is important to think about how we define the word pesticide. In this book, the word means any chemical used to manage one or more kinds of pests, including plant pathogens. If the word chemical raises alarm, remember that everything on Earth is made of one or more chemicals. Our unfamiliarity with chemistry and its language makes us afraid of an unfamiliar chemical called dihydrogen monoxide, but we feel comfortable about the same chemical when it is called by its common name: water.

According to a broad definition of pesticide, the chlorine used in swimming pools and drinking water—without which many of us would not be alive—is a pesticide. The antibiotics and fungicides used in medicine (for instance, to treat athlete's foot, bronchitis, yeast infections, etc.) are very similar and in some cases identical to those used in agriculture to protect crops. We might willingly use these pesticides to treat infections of our own bodies. Would we feel differently about the fungicides and other pesticides used on plants if they were called "plant medicines"?

Costs and Risks of Pesticide Use

The manufacture, distribution, and application of pesticides cost money. However, costs also are associated with crop losses and plant damage by pests—namely, increased production costs and higher food prices. There also are potential short-term environmental and health costs related to pesticide use, which can be studied and reduced as part of the cost of pesticide regulation. Long-term environmental effects and chronic health problems are more difficult to predict and describe in economic terms.

The second controversial aspect of pesticide use is risk. Evaluating risk has become a dominant concern of modern life. Although living risk free is impossible, some people are very risk averse. This aversion may prevent them from using products that offer great benefits because they also have a slight potential for harm. The prevalence of lawsuits in modern life also affects the availability of certain products.

Certainly, the chance of being injured or killed in an automobile is relatively high, yet people ride in cars on a regular basis because of the transportation benefits they provide. These same people might not be willing to accept exposure to extremely low concentrations of certain chemicals in their food or water. Is this response appropriate? What are the relative risks? How can we evaluate them?

The purpose of this chapter is not to present correct answers to these questions. Rather, the purpose is to give some background information about pesticide use and regulation that can be used to develop an intelligent opinion about this complex issue. We present information about the pesticides used to manage plant diseases, including their historical development and current status, as well as the regulation of pesticide use and applicator training and certification. Finally, we discuss attempts to reduce pesticide use through a better understanding of the agroecosystem, which involves using programs of integrated pest management.

Pesticides for Plant Diseases: What Are You Trying to Kill?

The general term pesticide refers to a chemical that will kill a pest. In agriculture, a number of more specific terms are used. Antibiotics are used primarily against bacteria, and fungicides are used to kill or inhibit fungi and often oomycetes as well. Herbicides kill plants but are most commonly used to manage weeds. Insecticides kill insects, and nematicides kill nematodes (tiny soil roundworms).

Fungicides are the type of pesticide most commonly used to manage plant diseases (Table 8.1). However, pesticides other than fungicides are sometimes used in plant disease management. The use of antibiotics for bacterial plant pathogens has been limited by efficacy problems, expense, and the rapid development of resistant strains of bacteria, as discussed in Chapter 5.

Herbicides are occasionally used to kill parasitic plants or weeds that harbor pathogen-transmitting insects, viruses, or other pathogens. The most common use of herbicides, however, has nothing to do with plant disease management but rather with eliminating weed species that compete with the plants we value.

Insecticides are applied to some high-value crops to kill the insect vectors that transmit plant pathogens. Insecticides may provide effective management of diseases caused by viruses and other pathogens that are transmitted by sucking insects, such as aphids. This subject is discussed in Chapter 13. In many cases, killing vectors is not practical or effective. For example, in the case of fire blight, it is impossible to kill all of the insects that may carry the bacteria to the flowers of the apple and pear trees. Also, insects are involved in pollination, which is essential for fruit production.

Soil fumigants are used in special circumstances for soilborne pathogens, but their use is limited to crops with values that can justify their considerable expense. Nematicides, most of which are highly toxic chemicals originally developed as insecticides, are applied to soil to kill nematodes, but once again, the cost of application must be justified by the crop's value. The use of nematicides and soil fumigants is discussed further in Chapter 9.

Table 8.1. Types of Pesticides Used in Plant Disease Management

• Antibiotics (bacteria)
• Fungicides (fungi)
• Herbicides (plants, weeds)
• Insecticides (insects)
• Nematicides (nematodes)
• Soil fumigants (broadly toxic biocides)

The History of Fungicides: Choose Your Poison

Agriculture was practiced for many centuries without the direct application of pesticides to crops, so the history of pesticide use is relatively brief. Before we knew that microorganisms cause disease and understood something about their life cycles, farmers were helpless to protect crops from pathogens, except by the general means discussed in the previous chapter. As humans discovered toxic compounds, they began to use them to poison the pests that infested their animals, their crops, and themselves. Most early pesticides were inorganic compounds of various toxic heavy metal elements.

The scientific definitions of the words organic and inorganic differentiate between molecules that contain carbon (organic) and those that do not (inorganic). The word organic also is popularly used to describe farming that does not use synthetic chemicals or fertilizers, which has caused confusion.

The elements copper (Cu), arsenic (As), mercury (Hg), and sulfur (S) are toxic to all living organisms. When Bordeaux mixture is applied to leaves, toxic copper ions are absorbed by the vulnerable germinating spores. Only a tiny amount is needed to kill the spores, whereas plant and human tissues are relatively protected from copper's toxic effects by the epidermises that cover them. Arsenic and mercury compounds were commonly applied as pesticides until the 1960s. In fact, until quite recently, arsenic could be purchased at many drugstores for use as an insecticide for household and garden insects.

Because the toxic components of such pesticides are chemical elements, they cannot be degraded after application. The residues that wash off treated plants onto the soil sometimes accumulate in plant products or soils. In some areas, such as the citrus groves in Florida, repeated applications of inorganic copper fungicides actually led to levels of copper in the soil that caused toxicity symptoms in the citrus trees. Mercury fungicides are banned in the United States, because mercury can accumulate in food chains.

As discussed in Chapter 2, the earliest foliar fungicides were sulfur and copper compounds discovered in the 19th century. Elemental sulfur, applied as a dust, was an early insecticide and also was used to manage powdery mildew on plants (Figure 8.1). However, it was not generally applied for plant diseases. Powdery mildews are caused by ascomycetes, which are foliar pathogens. These fungi do not permeate the tissues of the host plant, as most parasitic fungi do. They remain exposed to the environment and to the toxic effects of sulfur if it is dusted on leaves.

Attempts were made to control coffee rust and potato late blight with sulfur, but they were not particularly effective. Sulfur is difficult to apply so that it completely covers the foliage. Like all surface-acting fungicides, it must kill the pathogen before infection, so application to already infected leaves has no curative effect. Sulfur also must be applied with caution, because it can be phytotoxic and burn foliage, especially when applied in hot weather. As described in Chapter 2, Bordeaux mixture (copper sulfate and lime) was discovered in 1885 by Millardet during the grape downy mildew epidemic in France. It was the first foliar fungicide that allowed farm-

Figure 8.1. Dusting potatoes in Waterville, New York (1911).

ers to reliably protect foliage from invasion by fungi, but like other surface-acting fungicides, it could protect plants only before they were infected. The solution also had to be applied repeatedly as new plant tissue was produced and after rain had washed it from the leaves.

Early attempts to kill seedborne pathogens involved soaking seeds in nearly any strong-smelling or repellent substance available, such as horse urine. Germination of seeds was often harmed by exposure to such harsh chemicals, however, and many of the chemicals had no apparent effects on the pathogens, despite their strong odors.

The benefit of copper for seed treatment was discovered accidentally. In 1807, Isaac-Bénédict Prévost, of France, found that grain soaked in water in copper containers produced smut-free plants. Further investigation showed that even the low concentrations of copper ions that dissolved from the container into the water killed the smut fungus spores when they germinated. News of Prévost's discovery spread slowly, but the use of copper for grain treatment had become common in Europe by the mid-1800s. It is interesting to note that it took nearly a century for people to discover that the chemical that killed fungi on plant seeds also could kill them on foliage—and even then, the discovery was accidental.

Application of early fungicides to the foliage of plants was limited by the phytotoxicity that many caused and by the lack of effective spraying and dusting equipment. The inorganic fungicides rely primarily on copper and sulfur to kill fungi. They are broad-spectrum fungicides, effective against all fungi and oomycetes when properly applied. Inorganic fungicides were widely used in the United States until the mid-1940s, when the use of organic compounds started to become more prevalent.

Organic Fungicides: Biodegradable!

Fungicides and other pesticides can be categorized or classified in several different ways (Table 8.2). Beginning in the 1940s, a new classification system came into being with the development of organic fungicides.

In contrast to inorganic fungicides, organic fungicides are toxic because of complex molecules, which include a number of different elements—primarily carbon, hydrogen, oxygen, nitrogen, and sulfur. In some cases, the exact mode of action, or the means by which the compound poisons the fungus, may not be known. Organic fungicides are generally toxic to many fungi and much less toxic to humans and plants. In contrast to inorganic fungicides, these compounds are degraded by sunlight and soil microorganisms when they are washed into the soil, so residues do not accumulate. Despite these advantages, organic fungicides are synthetic molecules and thus usually ineligible for use by organic growers.

The early fungicides, both inorganic and organic, are strictly protective and are sometimes known as contact or protectant fungicides (Table 8.3). A contact fungicide does not enter plant tissue but remains on the surface, which means it can kill the fungus only as it attempts to germinate and penetrate the leaf. If a fungicide is needed to protect a crop, that need must be anticipated, and the spray must be applied before infection occurs. Once infection has occurred, the mycelium of the pathogen continues to colonize the leaf tissue, even if fungicide is applied to the leaf surface. A large volume of fungicide must be applied for full coverage of the

Table 8.2. Categories of Fungicides

• Inorganic	• Organic
• Contact	• Systemic
• Protectant	• Eradicant
• Narrow spectrum	• Broad spectrum

Table 8.3. Contact Fungicides

• Must be on plant surface before infection
• Broad-spectrum activity
• Resistance problems unlikely
• Repeated applications required as plant grows or rain dilutes the chemical

foliage, and repeated applications are necessary as the fungicide is degraded or washes off or as plant growth occurs. Large amounts of applied fungicides are essentially wasted and may create environmental hazards at nontarget sites, such as soil and water.

Systemic Fungicides: Longer Lasting with Less Exposure

Since the late 1960s, the use of systemic fungicides has become more prevalent (Table 8.4). Systemic or eradicant fungicides are absorbed into plant tissue and can have some limited after-infection or curative action. (Of course, no fungicide can bring dead tissue or a dead plant back to life.) Less of this type of chemical needs to be applied compared to a contact fungicide, because it is absorbed into the tissue and does not wash off after application. Systemic fungicides also are less likely to reach the environment outside the plant because they are absorbed into plant tissues (Figure 8.2).

There are several categories of systemic fungicides. Some are locally systemic in plant tissue and move only short distances from where they are applied. Others enter the xylem and move upward in the plant. This type of systemic can be redistributed to new growth, so it does not have to be reapplied frequently. Phloem-transported systemic fungicides can move both up and down in a plant, but very few chemicals can do this.

Systemic fungicides are specific enough to kill or inhibit fungi without causing toxic effects to plant tissues, which often means they are quite specific in their toxicity to fungi and less toxic to other organisms, such as people. For example, some systemic fungicides are effective against basidiomycetes but not ascomycetes. Some kill oomycetes but not true fungi—and vice versa. These systemics are called narrow-spectrum fungicides. In contrast, broad-spectrum fungicides are toxic to many fungi, including important saprophytic soil inhabitants. The narrow-spectrum systemic fungicides that reach the soil may cause less ecological disruption than broad-spectrum fungicides.

Table 8.4. Systemic Fungicides

Table 8.4. Systemic Fungicides
• Do not wash off
• Fewer applications needed
• Some limited curative (after-infection) activity
• Narrower spectrum of disease control
• Resistance may develop with repeated use

Once the mode of action of a fungicide has been discovered, modern organic chemists can then modify the molecules to make them more useful. As a result, many different systemic fungicides have the same mode of action. Fungicide molecules can be modified to inhibit fungi better, to be more biodegradable, or to be less toxic to nontarget organisms.

The use of systemic fungicides has some significant disadvantages. One disadvantage is that many of these fungicides do not move well throughout the plant after application, and those that are transported move primarily upward. This is unfortunate, because it is usually not possible to protect plant roots with foliar applications.

In addition, although specific toxicity may be an advantage, it also may have some practical limitations. For example, some systemic fungicides are effective against oomycetes, such as *Phytophthora infestans* (late blight). However, an oomycete-specific fungicide will not protect potato plants from

Figure 8.2. Modern fungicide application in a potato field.

early blight, a fungal disease caused by *Alternaria solani,* which is an ascomycete. In years past, farmers usually did not deliberately apply fungicides for early blight, because the broad-spectrum contact fungicides were effective against both discases. When an oomycete-specific systemic fungicide was used for late blight, farmers discovered that early blight was serious enough to require an additional fungicide. For ecological reasons, one might wish to specifically attack a primary pathogen, but in terms of crop protection, this same specificity may be a disadvantage if other diseases cause significant losses.

Resistance: Overuse Leads to Ineffectiveness

A second important problem arises from the specificity of systemic fungicides: the development of pathogen populations resistant to the chemicals. The early fungicides were broad spectrum in their ability to kill fungi, because their modes of action usually involved the inhibition of major enzyme systems or a range of toxic effects that interfered with numerous metabolic processes. In these situations, it is highly unlikely that resistant fungal populations might arise. This was true not only for the inorganic compounds but also for most of the early organic ones. Thus, for many years, farmers had the convenience of using these fungicidal chemicals repeatedly without resistant pathogen populations arising.

This was not the case for insect problems, because many organic insecticides had a specific mode of action that involved inhibition of a specific nerve enzyme: cholinesterase. When these insecticides are applied repeatedly, resistant insect populations develop, and the sprays become ineffective. A famous example of this is the repeated use of DDT against malarial mosquitoes and its subsequent loss of efficacy against resistant mosquito populations, first detected in 1959 in India. Since that time, insect populations have arisen with resistance to products with other modes of action as well.

A similar situation has occurred in plant pathology with the advent of the site-specific mode of action (mostly systemic) fungicides—and for the same reason. In 1960, only one fungicide-resistant fungus genus had been discovered. Since then, the number has increased dramatically, and there are now reports of resistant populations of most fungi and oomycetes for which systemic fungicides are used frequently. A number of pathogens have resistance to more than one systemic fungicide group. The modes of action of these compounds are much more specific than those of the older contact fungicides. Thus, even a small genetic change in the fungus population can lead to widespread resistance and sometimes to failure of the chemical to manage the pathogen.

One of the important challenges in pesticide science is to prolong the effectiveness of these chemicals. Chemical companies face the dilemma that increased sales of pesticides lead to profits, but repeated use of some compounds makes them become ineffective because of the development of resistant pathogen populations. In most cases, resistant individuals are already present in a pathogen population at the time a chemical is first applied. With repeated applications, sensitive individuals are killed while the resistant types continue to reproduce. In a rapidly reproducing pathogen population, this can occur quite quickly, and eventually, most of the pathogen population becomes resistant to the pesticide.

Because the discovery and marketing of a new fungicide is expensive and because most major groups of organic compounds have already been screened for their activity as fungicides, it is important that fungicides subject to resistance be used judiciously. People who use fungicides must be aware of the active ingredient and mode of action of each one. Fungicide manufacturers and plant pathologists work together on the Fungicide Resistance Action Committee (FRAC). This group studies the modes of action of the various fungicides and determines which chemicals are cross-resistant. All the chemicals are assigned individual FRAC codes to help applicators use these valuable products effectively while minimizing the potential for the development of resistant pathogen populations.

Fungicide Use: Which Crops Are Sprayed the Most?

Many different fungicides are available for the reduction of plant diseases. Some are contact fungicides (protectants), and some are systemic (eradicants). Each type has relative advantages and disadvantages, and both involve some expense. Systemics tend to be more expensive than broad-spectrum contact chemicals, but their application often requires smaller amounts of the active ingredient at longer time intervals.

Even though the costs of fungicides are relatively low compared to those of most insecticides and herbicides, many crops are not treated regularly with foliar fungicides. This may seem surprising, but the value of many field or agronomic crops does not justify the considerable cost of labor and equipment required to apply these chemicals. In 2002, the National Pesticide Use Database of the Environmental Protection Agency (EPA) tallied how many pounds of active ingredients were applied in the United States in herbicides, insecticides, and fungicides. The highest amount of active ingredient was applied in herbicides (61%), followed by insecticides (22%) and fungicides (17%).

Only a small percentage of the total acreage of crops receives repeated fungicide applications. Fungicides are applied primarily to manage polycyclic epidemics caused by foliar pathogens. Fungicides are not generally used for soilborne pathogens, because it is difficult to effectively deliver fungicides to roots.

Based on discussions earlier in this book, it should be clear that potatoes require fungicide protection against *Phytophthora infestans* until greater genetic resistance can be developed. Apples, citrus, other fruits, and some vegetables are sprayed regularly to meet consumer demands for unblemished produce. In fact, approximately 95% of the fungicides applied in the United States are applied to fruits and vegetables. More fungicides are applied in the eastern half of the United States than in the western half, because the eastern states have greater rainfall and higher humidity—both conditions that support fungal infections. Fungicides are not applied routinely to many field crops, such as barley, corn, oats, and wheat. However, farmers are under increasing economic pressure to produce higher yields, so fungicides are starting to be used more often on some of these crops.

Application of a fungicide is primarily an economic decision. The cost of application (including the chemical, equipment, and labor) must be offset by the increased value of the crop when disease is reduced. For some crops, such as potatoes and peanuts, applying fungicides increases yields by reducing foliar disease, whereas for other crops, such as apples and tomatoes, applying fungicides improves both yield and quality by reducing

Science Sidebar

Avoiding resistance

Some fungicide-resistant fungi will continue to survive in nature, making certain fungicides useless. Others do not compete well with non-resistant fungi. Alternating or mixing fungicides with different modes of action can prolong their usefulness for many years. If fungicides are composed of different chemicals but have the same mode of action, alternating or mixing them will not help, because they kill fungi the same way. This is known as cross-resistance. Alternating or mixing two fungicides does not reduce the development of resistance if the genes that govern resistance to the chemicals are the same.

This situation can be complicated by the fact that the same active ingredient is sometimes sold under different trade names. Applicators may think that they are using different fungicides to avoid resistance, when they are really using the same chemical.

spots and blemishes on fruits as well as foliage. On average, every $1.00 spent on fungicides returns $14.60 in production value. For chemical-intensive crops, the total amount of fungicide used can be significantly reduced by eliminating even a few sprays per season. We return to this idea at the end of the chapter.

Pesticide Use Today: What Do We Want?

Even though Bordeaux mixture was discovered in 1885, the era of agricultural pesticide use really began in the 1940s, when a relatively inexpensive variety of organic pesticides became available. Until the 1940s, the major fungicides used were inorganic compounds based primarily on sulfur and copper. At that time, approximately 300 million pounds of fungicides were used each year, primarily on apples, peaches, peanuts, and potatoes—crops that are still difficult to grow without fungicide protection.

For a short period, pesticides were seen as miraculous tools in the fight against pests, but soon, the environmental side effects and problems with pathogen and pest resistance to the chemicals began to tarnish their image. The change in society's attitude toward pesticides accompanied the change in perspective from controlling diseases to managing them. Contrary to public perception, less fungicide is applied today than in the 1940s, because new fungicides are more effective in lower amounts. In 2002, 115 million pounds of fungicide active ingredients were applied, most of which were organic, biodegradable chemicals.

What do we expect from a modern pesticide? It should be inexpensive and effective. Even though it must be toxic to pests, it should not endanger humans and other nontarget organisms. It should be persistent enough that farmers will not have to spend a lot of time on reapplication or put a lot of it into the environment. However, it should not persist in the environment and accumulate in food chains. Finally, although the pesticide should be specific to the pest, its repeated use should not result in the development of resistant pathogen populations. These conflicting expectations of pesticides are the source of controversies concerning their use.

Pesticides in Organic Agriculture: It Is Not Pesticide Free

Contrary to public perception, organic farming is not necessarily pesticide-free farming. Organic growers face the same pest and disease problems that conventional famers do. Inorganic sulfur- and copper-based fungicides, which are not biodegradable, are allowed in organic production, along with certain other chemicals. (See the Internet Resources for a complete list.)

Copper-based fungicides must be applied repeatedly in wet weather, because they readily wash off leaves. They are strictly protectant fungicides. For example, potatoes may have to be sprayed with copper-based fungicides 9 to 15 times per season to try to prevent late blight. In 2009, many organic growers lost their potato and tomato crops to late blight. On some farms, a large number of heirloom varieties were lost. Also, contrary to the popular image of organic farming, some of the chemicals that are allowed are not particularly environmentally friendly. For example, lime sulfur, which is often used as a fungicide in organic production, can harm apple trees and reduce their productivity. Apple yields doubled in the years following the introduction of synthetic fungicides.

Many people who buy organic produce want it to be chemical free. The word chemical has taken on an inaccurate negative connotation as something harmful. As noted earlier, everything on Earth is made of chemicals, so no food can accurately be described as "chemical free." Some chemicals are harmful, and others are not. The same chemical can be harmful under certain circumstances and harmless under others. Finally, synthetic chemicals are no more likely to be harmful than natural chemicals. The danger posed by any particular chemical is related to exposure. To paraphrase Paracelsus, "The dose makes the poison."

Pesticide Registration: The Legal Use of Pesticides

To balance the conflicting expectations of pesticide use, complex regulatory legislation has been developed. In 1947, the Federal Insecticide, Fungicide, and Rodenticide Act (FIFRA) was passed. This heavily amended act, which is administered by the EPA, and components of the 1938 Federal Food, Drug, and Cosmetic Act (FFDCA), which is administered by the Food and Drug Administration (FDA), govern the use of pesticides in the United States.

All pesticides must be registered with the EPA. For registration to be granted, the manufacturer must provide extensive information about the chemical to the federal government. The amount of information required has increased so much over the years that it now takes 6 to 9 years and more than $50 million from the discovery to the registration of a new pesticide (Table 8.5).

In addition to data about efficacy and chemical formulation, the manufacturer must provide toxicological information about the pesticide's short- and long-term effects on a variety of organisms, including plants, animals, and microorganisms. The manufacturer also must conduct studies to determine the mutagenic (causing genetic damage), carcinogenic (causing cancer), oncogenic (causing tumors), fetotoxic (causing toxicity to a fetus), and teratogenic (causing birth defects) effects of the chemical. A 1972 amendment to the pesticide regulations requires that studies must include the economic, environmental, and social effects of the proposed pesticide as well. This includes studies of how the chemical affects the air, soil, and water environments.

As part of final registration approval, the manufacturer accepts a pesticide label that is required to be displayed on all containers (Figure 8.3). A pesticide label is a legal document that must contain the following items: (1) the product name and ingredients statement, (2) information about toxicity and treatment for poisoning, (3) directions for storage and disposal, (4) precautionary statements about application, (5) directions for use, and (6) crops for which application has been approved. Any use contrary to the information on the label is illegal. Violations include altering the amount or frequency of application of the chemical; changing how and where it is applied, stored, or disposed of; and applying it to crops not specified on the label.

Because registration is so costly, most chemical manufacturers attempt to get the broadest approved applications possible, so that future sales will offset initial costs. Patent protection for the discovery of a new compound lasts only 20

Table 8.5. 2007 Data Requirements for Registration of Conventional Pesticides

• Product performance
• Data from studies that determine hazard to humans and domestic animals: - Acute, chronic, and subchronic toxicity studies - Chronic studies that take place over the lifetime of test animals to determine carcinogenicity - Studies to determine mutagenicity and pesticide metabolism
• Data from studies that determine hazard to nontarget organisms: - Acute and subacute toxicity studies - Chronic and field studies
• Postapplication exposure studies
• Applicator/user exposure studies
• Pesticide spray drift evaluation
• Environmental fate
• Residue studies

years. Since registration approval takes up part of this time, the manufacturer is left with a relatively short period in which to exclusively market enough of the chemical to earn a profit from it.

Registration of a pesticide for food crops is quite specific, because studies must be made on the possible pesticide residue in each harvested product. Registration for use with nonfood crops is often granted with broader approval, because testing on the hundreds of species of plants used as ornamental and landscape plants is prohibitive.

Many specialty crops, which are grown on small acreages, have specific pesticide needs. They are never included on pesticide labels, because the registration costs are too high to be covered by eventual sales of the pesticide for use on these crops. Chemical companies are generally not interested in developing new products for crops grown on less than 500,000 acres (200,000 hectares). (Companies are similarly disinterested in developing so-called orphan drugs for treating rare diseases in humans.) To help remedy this problem for minor crops, test data for pesticide use are provided by scientists from universities and the Cooperative Extension System of the U.S. Department of Agriculture (USDA).

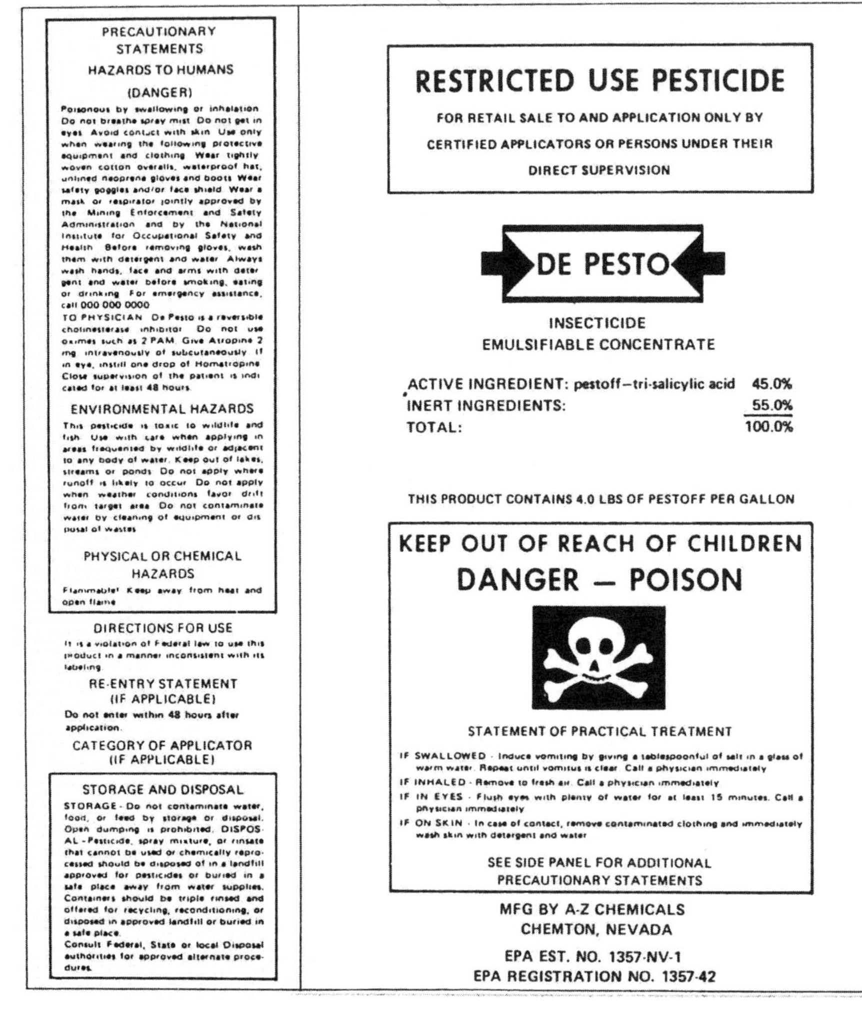

Figure 8.3. Mock pesticide label. A pesticide label is a legal document that must include, among other information, the active ingredient, precautionary statements, directions for storage and disposal, and directions for use.

Relative Toxicity: How Poisonous Are Pesticides?

For most people, an important concern about pesticides is their ability to poison living organisms, which is called toxicity. Since these chemicals are capable of poisoning pathogens and pests, it is reasonable to wonder about their toxic effects on humans.

A pesticide's toxicity is determined by testing its ability to kill organisms at various concentrations. Relative mammalian toxicity is usually determined with rats or mice. When the results of such tests are graphed, the concentration of the pesticide that kills 50% of the test population can be determined. This concentration is called the median lethal dose, or the LD_{50}.

The middle value is chosen to accommodate individuals who are particularly sensitive or resistant to toxic effects. The concentration that causes 50% mortality is considered an accurate representation of the relative toxicity of the chemical, although it is not the exact amount needed to cause death. An LD_{50} is usually expressed as milligrams (mg) of chemical per kilogram (kg, or approximately 2.2 pounds) of body weight of the test organism. Thus, a chemical with an LD_{50} of 10 mg/kg is more toxic than one with an LD_{50} of 1,000 mg/kg. It takes less of a highly toxic compound to cause death.

Insecticides are generally more toxic to mammals than fungicides. In fact, most fungicides have a relatively low mammalian toxicity. For example, 800 to 11,000 mg/kg represents a general LD_{50} range for some common fungicides. One teaspoon is approximately 4,200 milligrams, which means it would take many teaspoons of active ingredient to cause toxicity in a person weighing many kilograms.

This range makes the fungicide with the 800 mg/kg LD_{50} seem a lot more toxic than an 11,000 mg/kg LD_{50} chemical, but an LD_{50} represents relative toxicity for comparison purposes. The LD_{50} of caffeine is 250 mg/kg—about three times more toxic than the most toxic fungicide in this comparison. Yet many people ingest large quantities of coffee every day. This is not to say that coffee is toxic but rather that toxicity depends on the dose, or the amount of a chemical to which an individual is exposed. Many common chemicals, such as caffeine and table salt, are toxic in high doses (Table 8.6). Given the LD_{50} range for fungicides, toxicity is not a major concern for most of them.

An LD_{50} represents toxicity for acute poisoning—that is, from a single oral dose of the chemical. However, studies have shown that some chemicals are actually absorbed more easily through the skin than through oral ingestion. From a practical standpoint, poisoning from dermal exposure during mixing and application of a pesticide is more likely than poisoning from swallowing a dose, under normal circumstances. Inhalation of pesticides is another important means of poisoning, since lung tissue can rapidly absorb many chemicals. The relative toxicity for inhaled chemicals is expressed as the median lethal concentration (LC_{50}) in milligrams per liter (mg/L) of air.

Although acute relative toxicity is an important factor in pesticide risk, chronic exposure to a low level of pesticides is more common and much harder to evaluate. An LD_{50} offers no information about other health risks that may result from exposure to pesticides over many years, such as increased risk of cancer.

Table 8.6. Median Lethal Dose (LD_{50})

Substance	Amount (mg/kg of body weight)
aldicarb (nematicide/insecticide)	1
nicotine (insecticide)	50
caffeine	250
copper sulfate	300
carbaryl (insecticide—e.g., Sevin)	307
aspirin	780
chlorothalonil (fungicide—e.g., Bravo, Daconil)	3,800
fenarimol (fungicide—e.g., Rubigan)	10,000

The protection of pesticide applicators and other workers exposed to industrial chemicals has changed markedly in recent years (Figure 8.4). Today, gloves, boots, hats, goggles, respirators, and moisture-repellent protective suits usually are required (Figure 8.5). For some chemicals, applicators must wear protective clothing as specified on the pesticide label, but prudent individuals prefer to protect themselves from chronic exposure to pesticides whenever possible. Some applicators routinely undergo blood tests for analysis of cholinesterase inhibition. Exposure to carbamate and organophosphate pesticides, most of which are insecticides, causes the blood cholinesterase level to fall. Routine blood testing can detect pesticide exposure before toxicity symptoms are noticed.

Careful applicators and people who do not work with pesticides are only infrequently exposed to pesticides, and then the exposure is at a low level. The human body is capable of detoxifying many poisons, including alcohol and nicotine, and may or may not be harmed by low-level exposure. Negative effects from chronic exposure are determined by age, dosage, frequency, health, and other biological factors. Because of these and other complications, relative toxicity is based on the LD_{50} value derived from an acute oral dose of the chemical, even though chronic toxicity is probably the more common form of exposure.

Pesticide labels use specific signal words to indicate relative toxicity. The word "Danger," along with the image of a skull and crossbones, is used for the most toxic chemicals. "Poison, Warning" is used for moderately toxic chemicals. Most fungicides and herbicides and various pesticides that can be purchased by homeowners have the word "Caution" on their labels to indicate that they are only slightly toxic.

Government regulation of pesticide applications increased in 1973. To better protect

Figure 8.4. Fumigating soil for nematode control. This photograph illustrates an accepted method of pesticide application in 1962. Note that no protective clothing or respirators were worn. The use of this chemical by homeowners is no longer legal.

Figure 8.5. The recommended protective clothing for pesticide applicators includes gloves, water-repellent clothing, boots, goggles, hat, and respirator.

the environment, the general public, and the applicators, certain pesticides were placed on a "restricted use" list. These pesticides can be purchased and applied only by certified applicators who have been trained and passed a written test and who receive continuing education in pesticide application. The federal government sets general standards for certification of pesticide applicators, and individual states administer the licensing, sometimes adding their own stricter standards.

Monitoring Pesticide Residues in Food: Is It Safe to Eat?

Many people are concerned about pesticide residues in food. No one wants to ingest pesticides if doing so can be avoided. EPA registration requires that the acceptable level of pesticide present on a crop at the time of harvest, called the tolerance, be set for each food crop use listed on a pesticide label. The tolerance is the legal residue allowed.

How does the government determine this level? The tolerance is governed by the FFDCA, mentioned earlier. This act was amended in 1958 to include the famous Delaney Clause, which specified that no food additive—including colors, flavorings, and pesticide residues (even though they are not deliberate additives)—may be used if it induces cancer in animals at any concentration. This restriction resulted in the banning of certain food colors and artificial sweeteners. These and other substances had been shown to cause cancer in studies of the maximum tolerated dose in rat and mouse feedings, The scientists who conducted these studies were trying to quickly determine what might cause cancer, even though most people are exposed only to low doses of these chemicals over long periods of time.

Recent studies by toxicologists have shown that it is not easy to determine what materials in our diets may cause cancer and at what levels. It is clear, however, that our diets contain many chemicals, both synthetic and natural, that appear to cause cancer at high doses in animal studies. Does this mean these chemicals are dangerous at any level or only at some threshold level? The answer appears to depend on the specific chemical. Many chemicals that cause cancer at high doses appear to have little to no negative impact at low doses.

Bruce Ames, a biochemist at the University of California, Berkeley, did some of the initial work on this problem. He originally did research that led to the banning of a chemical fire retardant used on children's pajamas after it was learned from high-dose rat studies that the chemical caused cancer. As his research developed, however, he became aware that all kinds of chemicals—including many natural compounds in foods—would be banned based on such tests. Ames has had great influence on the understanding of toxicology and how to evaluate carcinogens in the human diet. He developed the Ames test, described in Chapter 6, to help quickly flag mutagens that require further study.

One of Ames's key contributions was to demonstrate that synthetic chemicals pose no greater risk to our bodies than natural chemicals. The risk is determined by how each chemical affects the human body, which reacts to the chemical, not its source. More detailed information on this subject was compiled in a 1996 report from the National Research Council (NRC) of the National Academy of Sciences, entitled *Carcinogens and Anticarcinogens in the Human Diet.* Ames also pointed out that some mutations—such as those of the p53 gene, which works to suppress tumor growth—have greater health significance than others. Chemicals that cause mutations in such genes should be regulated very strictly.

A final factor that must be considered in modern toxicology is the ability to detect smaller and smaller amounts of various chemicals in the laboratory. When the Delaney

Table 8.7. Modern Laboratories Routinely Detect Chemicals at ppm, ppb, and ppt Levels

Parts per million (ppm)
• 1 drop in the gas tank of a compact car (13 gallons [49 liters])
• 30 seconds in a year
Parts per billion (ppb)
• 3 drops in a railroad tanker car
• 3 seconds in a century
Parts per trillion (ppt)
• 1 drop in 20 Olympic-sized swimming pools
• 3 seconds in 100,000 years

Clause was written, pesticides were usually detected in parts per million. Today, laboratories routinely detect pesticide residues at parts per billion, parts per trillion, and even lower concentrations. As noted in Table 8.7, three drops of a contaminant in a railroad tanker car is equal to one part per billion.

This increased ability to detect chemicals has resulted in many news reports of cancer-causing chemicals being detected in foods. However, using modern laboratory analysis, "detectible" does not mean "cancer causing" for most chemicals. In fact, despite the use of many pesticides in modern agriculture, the rates of cancer deaths have remained fairly steady over the years, with the exception of lung cancer. The rate of deaths from lung cancer has fallen significantly for men, because of decreased smoking, and is still rising for women, because of increased smoking (Figure 8.6).

Because of public concern about carcinogens in foods, the NRC published a report in 1987 in which it tried to establish which pesticides were of greatest concern. The report summarized the results of a conservative study that determined which pesticides increased oncogenic risk, including both benign and cancerous tumors. Another worst-case aspect of the NRC study was that it assumed that the entire crop received all legal pesticide applications and that all residues were at the maximum legal tolerances, which is not likely in most cases.

Surprisingly, the NRC found that 80–90% of the risk was caused by only 10 compounds and that 60% of the risk was caused by certain contact fungicides. Insecticides tend to be more toxic than fungicides, but they are not generally carcinogens. Herbicide residues are rarely detected in foods, because they are used to kill weeds early in the season. In sum, much of the apparent dietary risk was caused by a few compounds on approximately 15 kinds of food. This finding put the focus on fungicides—primarily contact fungicides, which are on the surfaces of foods and thus easily washed off.

Clearly, a lot has been learned about human health and pesticide residues, toxicities, and carcinogens since the Delaney Clause, which had good intentions, was passed. In 1996, the Food Quality Protection Act (FQPA) was passed. The goals of the FQPA are to apply modern science to pesticide regulation with uniform standards. Under this legislation, all previously registered pesticides were required to be reregistered.

Although the new scientific information has reassured many people about the safety of our food supply, the goal of pesticide regulation is to have no residues present at the time of harvest. How does the government plan to reach this optimal result?

The FDA tolerance standards use several important measurements, such as the acceptable daily intake (ADI) of a chemical and the no observable adverse effect level (NOAEL). These figures are determined by physiological studies of the effects of a given chemical on test animals and by studies that determine how much of a particular food is normally eaten. For instance, the tolerance for a pesticide in a commonly eaten food, such as wheat or peanuts, might be lower than the tolerance for a more exotic or unusual food. The tolerance generally reflects a safety factor of 100 times less than the NOAEL. In certain situations, the FQPA allows the tolerance to be 1,000 times less for infants and children.

The FQPA also requires the EPA to examine all uses of groups of chemicals that cause the same health effects, rather than issue a tolerance for each chemical. Another change is that data must be provided on all exposures to a chemical, rather than just the dietary exposure studied previously. Additional types of exposure include the water people drink and the pesticides used on lawns and in homes (for instance, to kill termites). These combined uses of a chemical group are described as a "risk cup" of exposure. When the EPA determines that all of the exposures to a chemical group exceed the acceptable level of risk, some uses must be eliminated. For cancer, the threshold of risk is generally an increase in tumor production of 1 per million persons exposed over a 70-year lifetime.

The manufacturers of a chemical are allowed to make voluntary cancellations of certain uses. For example, many fungicides are no longer available for use on home lawns, because the potential for human exposure is relatively high. When voluntary-use cancellations are not sufficient, the EPA cancels additional uses as necessary to meet the risk-cup limitations. In addition to toxicity and carcinogenicity, the EPA monitors chemicals for additional potential chronic health concerns, such as disruption of the endocrine glands, which secrete hormones.

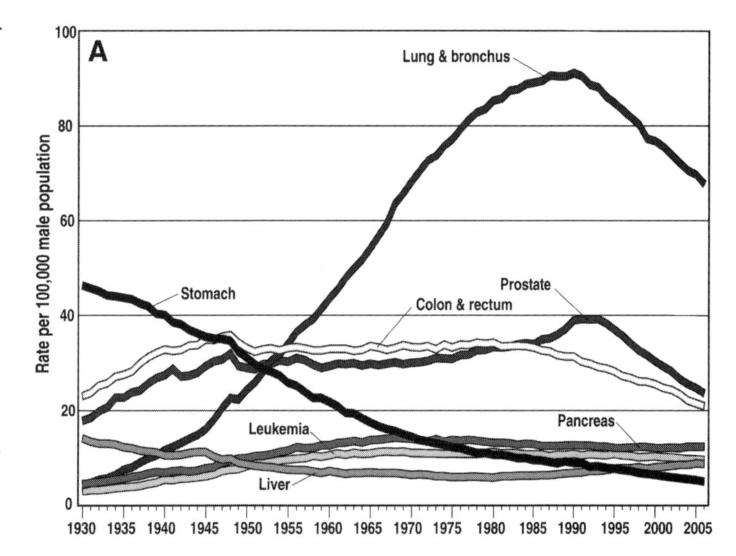

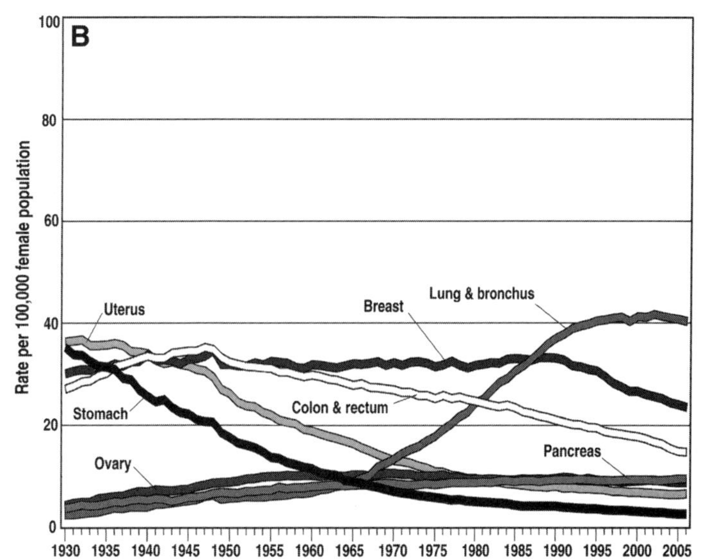

Figure 8.6. Cancer death rates, by type, for (**A**) males and (**B**) females, United States, 1930–2006. With the exception of lung cancer, the rates of cancer deaths have remained fairly constant over the years.

Days to Harvest: What Are You Eating?

After a tolerance has been set, studies are conducted to determine how long it takes for the residue left in a food crop to reach that level after an application. This time reflects normal weathering and degradation of the chemical by wind, rain, sunlight, and microorganisms. An additional safety factor is included so that the time between the final application and harvest should result in no residue or a residue level well below the legal tolerance. The pesticide label indicates this time as the days to harvest after the final application. For pesticides that degrade rapidly, the application can be made within a day or two of harvest. Other chemicals, such as some systemic insecticides, remain in plant tissues for longer periods and cannot be applied within 90 days of harvest.

EPA inspectors may inspect any harvested crop and seize it if the residue level exceeds the tolerance. Pesticide applicators also are required to keep detailed written records of all pesticide use. Many food processors conduct their own pesticide analyses to make sure crops are safe before they are accepted for processing. Of course, as with all laws, enforcement is expensive, so safety really lies with growers who follow regulations. Training meetings and additional education for pesticide applicators emphasize these regulations.

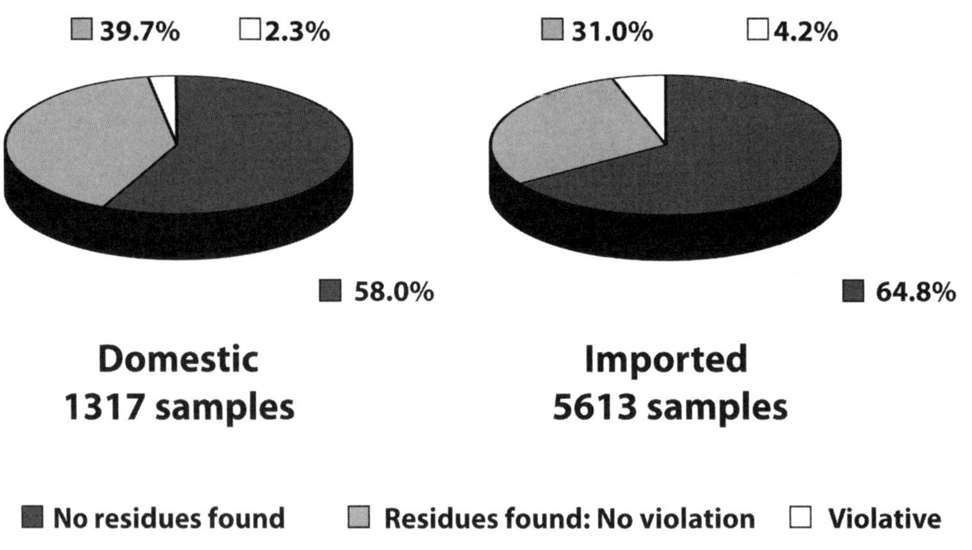

Figure 8.7. Summary of pesticide residues in domestic and imported commodity samples for 2007. Note that violative samples may result from exceeding the tolerance or applying a pesticide to a commodity for which no tolerance has been established.

Each year, market basket surveys are conducted to determine the actual residues present on produce purchased by consumers. These surveys provide data indicating that residues are far below legal tolerances and often undetectable. For example, in 2007, 97.7% of 1,317 domestic food samples and 95.8% of 5,613 imported food samples had no detectable pesticide residues or were below legal tolerances (Figure 8.7). Violative residues may be present because of exceeding tolerances or using pesticides on crops for which they are not labeled. These violations were detected in 0.9% and 2.4% of fruit and vegetable samples, respectively.

Overall, in the real world of foods as we buy them, our exposure to pesticide residues is estimated to be less than 1% of the acceptable daily intake. This means that the safety factors built into the calculations of days to harvest have been very successful. Although the controversy about food safety and pesticide residues will undoubtedly continue for some time, consumers should try to balance these concerns with the overwhelming evidence that fruits and vegetables are important sources of nutrients and anticarcinogens. Healthy produce keeps longer and does not contain harmful chemicals produced by decay bacteria and fungi. Fungicides allow farmers to produce abundant supplies of these healthy foods at reasonable prices.

Reducing Fungicide Use: No One Wants to Spray

What would be the consequences if growers were not able to use fungicides to manage plant diseases? In many cases, no adequate substitutes exist.

Eliminating the need for fungicides is the long-term goal of plant pathologists, but that goal has not yet been reached. In humid environments, crops such as apples, beans, carrots, grapes, onions, oranges, peaches and other stone fruits, potatoes, raspberries, and tomatoes are difficult and often impossible to grow without fungicides.

The application of pesticides costs money and has associated risks. Although food residues in the United States typically pose no significant health hazards, applicators and farm workers are exposed to pesticides at higher concentrations than consumers. Concentrated amounts of pesticides are manufactured, produced, shipped, and stored. Spills sometimes occur, and chemicals have potential nontarget effects on the environment.

In the late 1960s and early 1970s, the research emphasis shifted from developing pesticides to finding ways to reduce their use. This change in emphasis resulted in part from pressure from the environmental movement, which was very concerned about pollution. Another concern was that producing pesticides and applying them both involve fossil fuel products, which began to grow increasingly expensive at this time. A third change came from the continuing study of the ecology of agriculture. Pesticides were not the miracle cure for agricultural pests that many researchers and farmers had hoped for. New information from two main areas—economics and pest life cycles—is critical to the reduction of pesticide application.

Integrated Pest Management: Using Economic and Biological Information

Detailed studies of the relationship between the cost of pesticide application and the resulting increased value of the protected crop demonstrated that application at set intervals might actually cost more than the value gained at harvest time. Thus, the concept of an economic damage threshold became a guiding principle in the new approach to pest management. To justify a pesticide application, the cost of the potential loss had to meet or exceed the cost of management.

The economic threshold concept was dramatically successful for many insect pests. Rather than spray weekly, farmers were advised to wait until the insect pest population reached a predetermined threshold. This meant the pest population was monitored, and insecticide was applied only when necessary. Over the years, insect monitoring has become much more sophisticated. Life cycle studies have made the timing of sprays more precise and even more effective. For many crops, the number of insecticide applications has been greatly reduced while insect damage has been maintained at an acceptable level.

Similar studies have been applied to plant diseases. However, one of the major limitations in the use of economic thresholds for many plant diseases is that plant pathogens damage plants very differently than most insect pests. Although it may be possible to monitor insect populations and apply insecticides when thresholds are reached, similar activity with plant pathogens may lead to massive losses, because significant damage has occurred by the time visible disease symptoms appear. To reduce fungicide use, detailed information is needed about pathogen life cycles, damage, and yield losses resulting from diseases. For example, life cycle information can reveal when the pathogen is likely to cause infection, allowing the timing of fungicide application to be more precise. Economic studies demonstrate how much disease can be tolerated before a fungicide application will offset the loss.

In addition to allowing an accurate economic and biological justification of pesticide application, these new studies emphasize a more complex approach to pest management than was used in the past. Rather than relying on pesticides alone, various cultural practices, genetic resistance, and other management techniques are used. This approach is called integrated pest management (IPM), and programs based on it are common in agriculture and other industries affected by pest problems. Plant pathologists have always promoted the concept of IPM, because so many diseases cannot be managed by pesticides alone. Chemical management must be accompanied by sanitation, crop rotation, and other cultural practices that reduce pathogen populations.

Science Sidebar

Research leads to new management options

Scientists have been conducting research for more than 50 years to reduce the need for pesticides, especially those based on toxic elements such as arsenic, copper, and mercury. The following examples demonstrate that pest management is a much broader enterprise than simply trying to kill pests with chemicals that also may be toxic to people and a threat to the environment.

As scientists have gained better understanding of the physiology of plant pathogens, we have been able to take advantage of some differences found between fungi and other eukaryotic organisms, such as plants and animals. For example, one important group of fungicides inhibits the production of sterols, which are major components of the cell membranes of certain fungi but not of plants or animals. These sterol biosynthesis–inhibiting fungicides can reduce fungal spore germination and hyphal growth and are widely used to kill fungal pathogens of both plants and animals.

Chitin is the major component of the cell walls of all true fungi, so enzymes called chitinases can be used to break down fungal cell walls. Chitinases are produced naturally by bacteria, and chitinase-coding genes have been isolated from plants, including cucumber, rice, and tobacco. Approaches to using chitinases in plant disease management include introducing them into irrigation water, incorporating them as seed coatings to protect germinating seedlings, and genetically engineering chitinase genes into plants or into rhizosphere bacteria.

Chitinases also are useful for insect management, because chitin is a major component of insect exoskeletons. Some commercially available chemicals inhibit the ability of insects to produce chitin, so they are unable to molt and grow into reproductive adults. Juvenile hormones also can be applied to eggs and early immature stages to reduce the ability of insects to mature.

Insects produce pheromones, which attract mates or cause insects to gather in a location. Pheromones can be used to disrupt mating and for trapping the apple codling moth, the cabbage looper, the oriental fruit moth, and the peach tree borer. Pheromone traps for Japanese beetles are available at many garden stores to reduce damage to ornamental plants and to help reduce the larvae ("grubs") that feed on the roots of lawn grasses. The traps are so effective that you best give them to your neighbors, because they will attract too many beetles to your own yard.

Interesting examples of such environmentally friendly pest management tools are described on the websites of many integrated pest management programs at land-grant universities across the United States.

Unfortunately, IPM programs will always be somewhat limited in their ability to reduce fungicide use, because most fungicides function in a protective manner. In most cases, it is not possible to wait until disease reaches a certain threshold before application begins. Treatment with fungicide is still a necessary aspect of IPM for many crops.

The amount of fungicide used for plant protection in modern farming has been greatly reduced because of the use of synthetic fungicides. In the 1930s, 10–60 pounds (4–27 kilograms) of active ingredient was applied per acre (hectare), on average. Today, that range has been reduced to 3–6 pounds (1–3 kilograms).

Management of Apple Scab: How to Grow Healthy Apples

One successful example of the reduction in fungicide use over the decades is the management of scab disease, the most damaging disease of apples and crabapples. Heavy infection by *Venturia inaequalis,* an ascomycete, can cause severe distortions and scabby lesions on fruit, as well as leaf spotting and leaf and fruit drop. Infected apples also are subject to decay by other organisms in storage.

In years past, the intensive application of fungicides resulted from the market demand for unblemished fruit. Small corky scabs on apples cause no harm to the people who eat them but greatly reduce the market quality and therefore the profit margin of apple growers. Could growers reduce their fungicide applications and still produce healthy apples?

V. inaequalis overwinters as mycelium in fallen apple leaves on the orchard floor. During the spring, the primary inoculum—ascospores—mature and are forcibly discharged from the sexual fruiting bodies (pseudothecia) that are produced in the fallen leaves. Some ascospores land on newly expanding apple leaves, germinate, and grow into mycelium between the cuticle and upper epidermal cells (Figure 8.8). Soon, the secondary inoculum—conidia—are dispersed by water to other leaves and developing fruits, causing new infections. Infections on the fruit result in superficial, corky scabs, giving the disease its name (Figure 8.9).

Figure 8.8. Leaf symptoms of apple scab, caused by *Venturia inaequalis,* on crabapple.

Figure 8.9. Symptoms of apple scab disease on apple fruit.

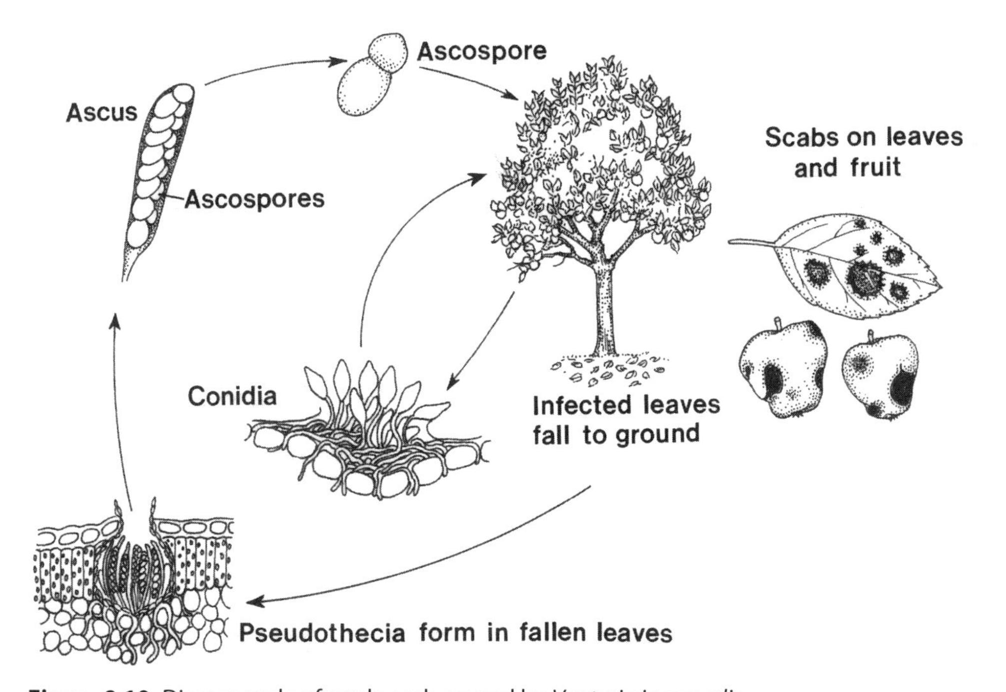

Figure 8.10. Disease cycle of apple scab caused by *Venturia inaequalis.*

In the past, apple orchards received fungicide applications on a regular schedule—from the time of leaf bud expansion in the spring throughout the growing season—to prevent polycyclic epidemics. Leaves must be protected from infection, because early scab lesions provide secondary inoculum for the later infection of developing fruits. Infection by *V. inaequalis* can be divided into primary infections by ascospores coming from the orchard floor and secondary infections from conidia on new leaves (Figure 8.10).

Plant pathologists determined that if fungicide was used to protect plants during the time of primary infections from ascospores, applications later in the season could be reduced because secondary inoculum would be very limited. Cooperative Extension personnel monitor the fruiting bodies in fallen leaves and distribute ascospore reports to warn growers of the onset, peak, and end of the ascospore season in the spring, when primary infections can occur. If an orchard is relatively scab free when ascospores have all been released (usually, about 2 weeks after the fall of flower petals), the apple trees require much less fungicide protection for the rest of the season. The apples are just beginning to develop at this time, so potential fungicide residues on fruit at harvest are greatly reduced.

After determining that fungicide applications for apple scab could be limited to the ascospore season, plant pathologists tried to reduce applications even more. They determined that besides knowing when primary inoculum was present, it also was helpful to know how successful infection would likely be. This could be predicted by determining the average temperature and the number of hours leaves had to remain wet for spores to germinate. The combination of these factors is called an infection period.

Table 8.8. Apple Scab Temperature and Wet-Period Guide, Based on the Mills Table

Average Temperature (°F)	Hours of Moist Foliage Necessary for Leaf Infection
78	13
77	11
76	9
61–75	9
60	9
57–59	10
56	11
55	11
52–54	12
51	13
50	14
49	15
48	15
47	17
46	19
45	20
44	22
43	25
42	30
33–41	More than 2 days

Source: New England Cooperative Extension.

An American plant pathologist, W. D. Mills, in tedious studies, provided this important information, which has been compiled in the famous Mills table (Table 8.8). Commercial apple scab predictors are now available that monitor the environmental conditions and then analyze the data with a computer program to warn growers when fungicide applications are necessary. If the weather is dry, there is no need to apply a fungicide. The computerized predictions take the guesswork out of deciding when to spray and give growers the confidence to skip sprays when they are not needed.

Systemic fungicides, which have a certain amount of curative action, are particularly useful in stopping the superficial apple scab pathogen using such disease predictions. The total number of applications during the ascospore season can be greatly reduced by coordinating sprays with infection periods. When spraying is done only during the ascospore season and then only when necessary, the need for fungicide can be reduced by 80%.

Some attempts have been made to reduce the primary inoculum coming from the fallen leaves. Theoretically, the total removal of fallen leaves eliminates the fungus, and so this practice is recommended for homeowners with a few apple or crabapple trees. Apple trees also are pruned to enhance drying of the leaves on the tree, thus making infection by the fungus more difficult. In commercial orchards, fungicides with after-infection activity are sometimes applied before the leaves fall to reduce overwintering mycelium and prevent formation of ascospores.

In relatively scab-free commercial orchards, applying fungicide during ascospore production has reduced the season-long need for scab fungicides, although summer applications are sometimes necessary to treat other apple diseases. Unfortunately, fungicide resistance has limited the application of disease prediction for apple scab and other crops that require fungicides. An understanding of disease cycles is still critical in determining when fungicides will be needed, but reliance on systemic fungicides to extend spray intervals has resulted in scab fungi that are resistant to some chemicals. To prolong the effectiveness of these valuable chemicals, contact fungicides that are not subject to resistance are used in mixtures or as alternates in spray schedules. This is particularly true in humid regions, where disease pressure can be severe.

Many scab-resistant apple and crabapple cultivars are available. Genetic resistance is the most effective way to reduce fungicide applications, but replanting is expensive and consumers are often slow to change their preferences in fruit. In the long term, the apple scab pathogen is likely to overcome the resistance genes. Plant pathologists are considering using a few fungicide sprays each season, even on genetically resistant trees, to help prevent the establishment of strains of *V. inaequalis* that can overcome the resistance genes.

Pathogens that cause polycyclic epidemics, especially those with a sexual stage, will continue to challenge people who are competing with them for food. Controlled, responsible use of valuable, biodegradable fungicides is important to keep them effective. Just as doctors prescribe the antibiotics that are most effective in treating people, Cooperative Extension advisors work with growers to create optimal fungicide combinations and application schedules. IPM programs integrate cultural, genetic, and chemical tools to create sustainable crop production systems.

Turning on Plant Defenses: Plant-Produced Pesticides

The use of genetic resistance to protect plants from pathogens was described in Chapter 7. This section discusses another kind of plant defense that has been observed since the 1930s.

If a local infection occurs on a leaf, the plant generates a signal that induces resistance throughout the plant and even in new growth. Like animals activating their immune systems, plants can turn on their defenses and protect themselves from further infection. Unlike the specificity of animal immunity, this defense reaction is general and effective against a variety of diseases. This type of resistance has been named systemic acquired resistance (SAR). A cascade of defense compounds is produced in plants with SAR, including phytoalexins and pathogenesis-related proteins, most of which are enzymes. SAR is a biological response that develops best in healthy plants. This explains why stressed plants are often more susceptible to disease.

Until recently, SAR was impractical for applied plant disease management. However, molecular studies of the interactions of plants and their pathogens have identified some specific pathogen chemicals that bind to plant receptors. These chemicals help pathogens initiate disease but also allow plants to realize they are being invaded. Some of these chemicals have been developed into commercial products called "plant activators," because they induce SAR. If applied before disease develops, plant activators may reduce the amount of pesticide necessary to protect plants.

Are plants with SAR safe to eat? In fact, we have been eating them since agriculture began. Will plant activators be eligible for use in organic agriculture as an alternative to so-called chemical pesticides, if these natural chemicals are mass produced in factories? If individual plant defense compounds are tested using the same toxicological methods used to test standard pesticides, which will appear safer?

These are some of the questions that remain to be answered. As with pesticide residues, we are exposed only to very low levels of SAR-induced chemicals when we eat plants. However, to avoid overexposure to any particular potential toxins, we should consume a variety of plant foods.

The Future of Pesticide Use: Responsibilities of Applicators and Consumers

Each time a pesticide application is eliminated, costs are reduced and so are possible risks for applicators, consumers, and the environment. Pesticides will continue to be necessary to protect certain crops, but they must be applied in the safest ways possible. State and federal governments regulate the storage, application, and disposal of these chemicals. They also regulate the applicators and the residues allowed in agricultural products.

Farmers should choose the most effective, least toxic, and most environmentally sound pesticides based on scientific information, rather than on whether a pesticide is natural or synthetic. Farmers can reduce the use of pesticides by practicing IPM and using multiple means to reduce diseases, insect pests, and weeds. Knowledge of pathogen life cycles can help focus these efforts, so fungicides are applied only when infection periods occur. Disease prediction systems using environmental monitoring are available for many crops, including celery, grapes, onions, potatoes, peaches, peanuts, and tomatoes (Figure 8.11). Global positioning systems (GPS) can be used to customize weather predictions for more accurate results.

Figure 8.11. Computerized environmental monitoring equipment. Environmental data such as air and soil temperatures, rainfall, relative humidity, and leaf wetness are determined and stored. Computer models predict disease outbreaks based on environmental data. Predictions and management recommendations can be accessed by crop managers.

Not only can these predictions be used for timing fungicide applications, but they also can be used for predicting when biological control agents will likely be most effective. The application of a large population of living organisms is not sustainable in the field, but it may be effective if applied exactly when needed. The newly available plant activators also provide an additional tool in the battle against pathogens.

Consumers can help in the effort to minimize pesticide use by changing their expectations. Having colorful displays of blemish-free fruits and vegetables often requires excessive pesticide applications. Are you willing to buy an apple with a few harmless scabs if it means fewer fungicide applications?

Consumers also should consider the public health importance of making fruits and vegetables affordable. In 2011, new U.S.

government dietary guidelines significantly increased the recommended daily intake of fruits and vegetables. Buying produce directly from growers at farmers' markets and through consumer-supported agriculture (CSA) farms allows consumers to talk to producers about how they practice IPM and minimize pesticide use. In addition, consumers should consider buying fruits and vegetables in season. Not only do these foods taste better, but they also can be produced with fewer pesticides when they do not have to be shipped thousands of miles before being sold.

Another important area of concern about pesticide use is the environmental hazards they may cause. Two topics of particular interest are the nontarget effects of these chemicals and the eventual damage they may do to the environment. Most pesticides find their way into the soil after running off plants during application, washing off plants during rain and irrigation, or being left as residue on plant debris at the end of the growing season. That means we need to consider what happens to pesticides in the soil.

So far, our discussion of pesticides has focused on aboveground issues. In the next chapter, we "go underground" to consider soilborne pathogens, biological management of pathogens, and groundwater contamination by pesticides.

CHAPTER 9

The World Belowground: Soilborne Pathogens

Healthy plant roots are necessary to support the aboveground parts of plants (see Chapter 4). Roots need oxygen, water, minerals, and suitable environmental factors, such as temperature and pH, for continual growth. The physical environment of the soil directly affects plant roots.

However, plant roots also affect the soil. Roots are a food source for many herbivores and omnivores, including small mammals and insects that chew and burrow into the moist vegetation. Even dead roots are an important food source for many animals, as well as for the microorganisms responsible for decay of these tissues. This chapter explores a world that you cannot see and probably do not often think about: the world belowground.

Root Exudates and the Rhizosphere: Useful Leaking

Many chemical compounds leak from plant roots during their growth. These root exudates include chemicals such as amino acids, sugars, organic acids, and various growth factors. They may serve as nutrients, inhibitors, and stimulants for various soil organisms.

Several important sources of these compounds occur on the growing root. Cells are continuously sloughed off from the root cap, contributing their contents to the soil environment. Elongating cells just behind the meristem leak nutrients and other chemicals. Some cells are torn open when lateral roots exit through the cortex. Compounds also are lost in considerable quantities from root wounds caused by cultivation, insects, nematodes, and rocks. Root exudates create a special environment in the thin "halo" area surrounding the actively absorbing roots and extending several millimeters from the root surface.

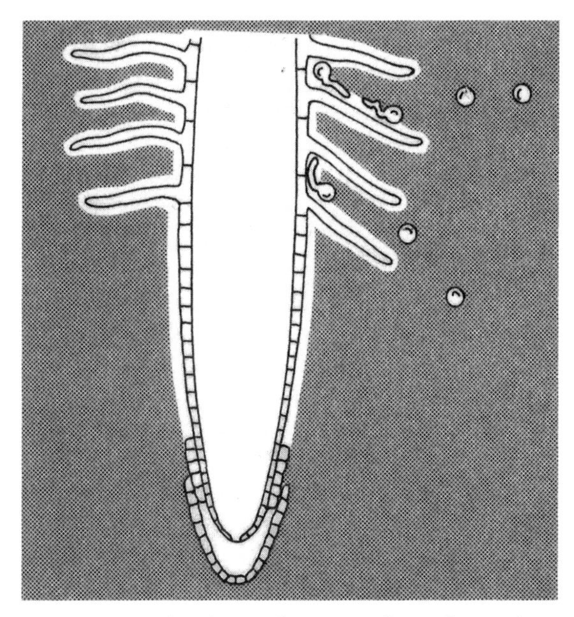

Figure 9.1. The rhizosphere. Exudates from plant roots contribute to a nutrient "halo" near the surfaces of plant roots. Many fungal spores remain dormant until influenced by the nutrient-rich rhizosphere. This phenomenon is called fungistasis.

This halo, influenced strongly by the presence of root exudates, is called the rhizosphere (Figure 9.1). It plays a significant role in the activity of soil microorganisms near root surfaces. Very different numbers and species of microorganisms are found in the rhizosphere than exist in the soil just beyond it. Significant thermal, pH, and gaseous differences also exist. Plants affect the microenvironment of the soil by contributing roots, other underground plant parts, and eventually, dead aboveground plant parts. In addition, plants contribute root exudates to the soil microenvironment.

It may seem wasteful and even dangerous for a plant to leak so much of its photosynthetic products into the soil environment, potentially stimulating the activity of the pathogens lurking there. Studies of the rhizosphere have begun to explain this curious phenomenon. The nutrient-rich environment produced by the plant's roots allows microorganisms to mineralize organic matter so that important elements become more available to the plant. The active rhizosphere microbial community also protects the plant against the activities of root-attacking pathogens. The plant is essentially supporting its own population of biological control organisms by feeding them well.

Some rhizosphere bacteria seem not only to protect the plant but also to increase its yield and growth. These growth-promoting bacteria appear to compete with pathogenic organisms for iron without reducing its availability to the plant. For many years, root pathogens have been reducing crop yields without causing obvious disease symptoms—more than anyone had imagined. Yield increases of 20–30% and more have been obtained in several crops that were "bacterized" with beneficial bacteria before planting.

Soilborne Pathogens: Underground Invaders

Healthy roots are just as important to plants as healthy leaves and stems. The roots of plants are vulnerable to pathogens that live in the soil and attack them belowground.

Soilborne fungi are some of the most frustrating pathogens for plant pathologists, because protecting roots with fungicides is difficult to impossible. Also, disease is not apparent until it is advanced enough to cause symptoms on the aboveground plant parts. Many of the management options available for soilborne pathogens must be applied before or at the time of planting. Some important management methods include improving soil drainage, modifying soil pH, and rotating to a nonsusceptible crop when a known pathogen is present in the soil. It is easier to reduce the movement of soilborne pathogens than airborne ones. Spreading them can be avoided by not moving contaminated machinery and tools. You have already seen some examples of important soilborne pathogens in the oomycetes. Many *Pythium* and *Phytophthora* species attack plant roots and crowns.

Fusarium wilt is a disease caused by the soilborne pathogen *Fusarium oxysporum.* This disease is called a vascular wilt, because the fungus invades the plant through the roots and then colonizes the water-conducting tissue (xylem), disrupting the flow of water and minerals to the rest of the plant. The fungal pathogen is an ascomycete that produces only asexual spores; no sexual stage is known.

The mycelium produces three kinds of asexual spores, each with a different function. The first kind of spore infects the roots of plants and invades the xylem. The second is smaller than the first and aids in spreading the fungus through the xylem. The third kind has thick walls for survival in adverse conditions or when no susceptible plant is available.

Once the xylem of the plant has been invaded, the plant turns yellow and wilts from the bottom up. *F. oxysporum* remains trapped in the xylem until the plant dies. It then produces spores on the surface of the dead plant, which contaminate the surrounding soil. These spores can survive for many years. The nature of this disease cycle explains why Fusarium wilt was mentioned in Chapter 7 as a classic example of a monocyclic pathogen.

Fusarium wilt is a well-known plant disease of some major cultivated crops, such as bananas, cotton, cucurbits (cucumbers, squash, watermelon), and tomatoes. Each host-specific subspecies of *F. oxysporum* is called a special form, or *forma specialis,* abbreviated f. sp. Different *formae speciales* are responsible for different diseases. This means that if a farmer has a problem in a field one year with Fusarium wilt on tomatoes, cucumbers can be planted in that field the next year, because the *F. oxysporum* f. sp. *lycopersici* that infected the tomatoes cannot infect the cucumbers.

Genetic resistance to Fusarium wilt is available for many plants. Many tomato cultivars are labeled with the letters VFN, which indicates the plants are resistant to Verticillium wilt (another vascular wilt disease), Fusarium wilt, and root-knot nematodes. A combination of genetic resistance and crop rotation to nonsusceptible host plants is used to manage Fusarium wilt. Fungicides cannot be used to prevent infection or to eradicate *F. oxysporum* from the xylem.

Panama disease is the name used for Fusarium wilt of bananas caused by *F. oxysporum* f. sp. *cubense.* The wild progenitors of the cultivated bananas are fertile diploids. Bananas originated on the islands around Southeast Asia and western Melanesia. Most cultivated bananas are grown as sterile triploid plants, in which one parent contributed two sets of chromosomes. Because banana fruit contain no seeds, they are propagated vegetatively, like potatoes. New plants grow from the buds of upright, underground stems called corms. Like potatoes, these plants are genetically identical and thus equally susceptible to pathogens.

Banana plants are not trees but rather large herbaceous plants without woody tissue. Within one year, the inflorescence, or flower stem, emerges from the plant and produces a cluster of banana fruit (Figure 9.2). Fruit production occurs only once on a plant, so new plants must be cultivated after each crop. As cultivation of the banana crop continues in the same land year after year, soilborne pathogens build up in the soil. If corms are transplanted to new soil, they will likely carry the pathogen with them in the infected xylem.

Figure 9.2. Banana plant (*Musa* species).

Figure 9.3. Panama disease (*Fusarium oxysporum* f. sp. *cubense*) of banana. Note the numerous wilted lower leaves of the diseased plants.

Despite the massive production of bananas for export, an estimated 90% of the world's bananas are produced by subsistence farmers throughout the tropics for their own use. Many of these bananas are starchy cooking bananas, including plantains. Bananas are the number-one fruit consumed by Americans, and the United States is the world's largest importer of bananas. Costa Rica and Ecuador are the biggest exporters. For many years, the most popular fresh fruit banana export cultivar was Gros Michel. Corms of this popular cultivar were distributed to tropical regions around the world.

As early as 1890, Fusarium wilt was recognized in Panama, which resulted in the nickname Panama disease (Figure 9.3). Since then, the disease has been found in all banana-growing areas of the world. Wherever it is found, banana plants begin to die once the pathogen is established in the soil. Because there is no cure or management for the disease, the only solution is to move banana production to new land. During the first 50 years after recognition of Panama disease, more than 100,000 acres (40,000 hectares) of production land were abandoned in Central and South America.

To plant new crops, land is usually cleared from the rain forest at a cost of $1,000–2,000 per acre ($400–800 per hectare). The fungus can be introduced into these new production areas by infected corms, contaminated soil on feet and equipment, and flowing water. Eventually, when the losses from the disease become too great, the new areas also will have to be abandoned.

As the monocultures of Gros Michel succumbed to Panama disease, they were replaced with monocultures of the Cavendish variety—the sweet export banana familiar to people living in temperate regions. The old song "Yes, We Have No Bananas" was written in 1922, during the transition between the two cultivars, when bananas were difficult to find.

In the 1990s, the Cavendish cultivar was discovered to be susceptible to a new race of *F. oxysporum* in Asia. At present, no commercially acceptable banana cultivar is resistant to all known races of the fungus. This presents a great challenge to banana breeders, who want to select wilt-resistant bananas. Since cultivated bananas are sterile triploids, traditional breeding crosses between them or crosses to wild diploid relatives are impossible. Sources of genetic variation may be derived using tissue culture techniques and genetic engineering, but short-term disease management lies in the techniques appropriate for many soilborne pathogens.

F. oxysporum also plays a role in another interesting story. The destruction of drug crops by herbicides has prompted many legal and political discussions. In recent years, studies have been conducted on the use of *F. oxysporum* as a possible biological control for coca, marijuana, and poppy—all potentially illegal drug crops. There are several advantages of using this fungus, rather than herbicides, to kill the drug crops. For example, the commonly used herbicide glyphosate is usually applied by airplane and is toxic to all plants, including the food crops of nearby farmers. Also, the herbicide must be reapplied periodically, as drug crops are reestablished. Because of the specificity of the special forms of *F. oxysporum,* the fungus can be applied and used to destroy only the drug crop. For example, *F. oxysporum* f. sp. *cannabis* infects only marijuana, and *F. oxysporum* f. sp. *erythroxyli* infects only coca.

In years past, proposals were made to use these naturally occurring fungi as a form of biological control in Florida for marijuana and in Colombia for coca. In 2010, the poppy crop in Afghanistan became diseased with symptoms resembling Fusarium wilt. Some of the farmers accused the North Atlantic Treaty Organization (NATO) of introducing *F. oxysporum* to their fields—accusations that were denied. In fact, it would likely be difficult to establish one of these fungi in a drug crop plantation. It would have to be delivered in a sufficient quantity and in a form that could reach the plants' roots. Once there, however, reapplication of these biological control agents would be unnecessary.

The destructive ability of *F. oxysporum* has made some people uneasy about its use as a possible biological agent against illegal drug crops. Although the products of these plants are abused by some users, the plants have beneficial uses as well. In addition, these species and their close relatives play roles in their native ecosystems. Many people feel that destroying the plants is not the most effective means of controlling illegal drug use. Some have questioned whether application of *F. oxysporum* to drug crops outside the United States violates the international Biological and Toxin Weapons Convention. Clearly, there are advantages and disadvantages to the potential use of either Fusarium wilt or herbicides in the war on drugs.

Science Sidebar

Fungistasis

How does the rhizosphere stimulate growth? Many fungal spores exist in a state of exogenous (externally caused) dormancy. This means the spores are not dormant because of an internal control but because they require an external source of nutrients to become active. This exogenous dormancy is called fungistasis.

Nutrients are present in the soil, but the highly competitive populations of saprophytic bacteria and fungi rapidly metabolize them. Thus, only the nutrient-rich rhizosphere provides sufficient nutrients to overcome fungistasis. Inactive spores usually become active only when a potential host is present.

Fungistasis is an important adaptation that many pathogenic fungi have made to ensure they do not germinate too soon. Their storehouse of nutrients is limited, so successful infection is most likely if they germinate only when a root is nearby. As seeds germinate and roots grow, spores eventually find themselves near a nutrient-rich rhizosphere. In most cases, their response is nonspecific: They become active in the presence of nearly any plant root, but they will die if they germinate in response to the nutrient stimulus from a nonhost plant root that they cannot infect.

Some fungi, nematodes, and parasitic plants produce survival structures that require a specific factor from host roots to stimulate growth, and they can survive in a dormant state for long periods. This is a common feature of obligate parasites, which die if they become active when no host is present. During the wait for a host plant, survival structures may be ingested, parasitized, or otherwise destroyed by other soil organisms. Their populations decrease each year if no host plant roots become available for colonization and reproduction.

Soil Microbes and Plant Pathogens: Microbial Wars

Besides pathogens, what other organisms live with plant roots below the soil surface? Listing them all would be impossible, but some important groups include burrowing mammals (moles, ground squirrels, and mice), numerous insects (some of which feed on plant roots), other arthropods (millipedes, centipedes, spiders, mites, and springtails), mollusks (snails and slugs), and small invertebrates (nematodes). Of course, the most numerous "neighbors" of plant roots are microorganisms, particularly fungi and bacteria.

Soil microorganisms have limited ability to move significant distances in the soil because of their small size. Thus, most soil microbes simply wait, essentially, until some organic matter becomes available. The arrival of a growing root tip and its nutrient-rich rhizosphere causes a burst of biological activity. Intense competition develops between organisms. Those that reproduce quickly and produce inhibitory compounds against other organisms have the advantage. Certain organisms are better adapted for competition and survival in the rhizosphere and will predominate there. Because the nutrient-rich environment typically is available for only a short time, microbes that can exist for long periods in an inactive state are better able to survive until more nutrients become available.

Many soilborne pathogens play this "waiting game," as well, trying to survive in the soil until a plant root arrives. In many cases, the pathogens are in an inactive state—especially fungi, which commonly exist in the soil as thick-walled spores or as dark masses of inactive mycelium called sclerotia. When a plant root grows near these survival structures, they are stimulated to begin active metabolism and growth.

Highly competitive soil organisms that can maintain their populations over long periods of time are called soil inhabitants. Soil inhabitants may successfully exist in soil by one or several means. They are often capable of rapid growth to take advantage of available nutrient sources; they may produce survival structures during adverse conditions; and they may produce various substances antagonistic to other organisms. Humans exploit some of the antagonistic substances for use as antibiotics for medical purposes. Most antibiotics have been isolated from various soil microorganisms, primarily fungi and actinomycetes.

Many soilborne pathogens also can exist in a saprophytic state in the soil. However, pathogens seem to sacrifice some of their competitive ability as they become specialized to invade living plants. As discussed in Chapter 7, pathogens are generally more temperature sensitive than saprophytic organisms. Most pathogen populations decline dramatically in the soil if no host plant is available for several years, because they are far less competitive than other soil inhabitants. Most soilborne pathogens are categorized as soil invaders to indicate their relatively short existence in the soil environment as saprophytes. Unfortunately, a few important soilborne pathogens are very competitive saprophytes and maintain their populations in soil for long periods of time. *Fusarium, Pythium,* and *Rhizoctonia* species are important pathogenic soil inhabitants.

Biological Control: Microbial Protection

An understanding of the rhizosphere and the biology of soilborne pathogens has important practical applications. Soilborne pathogens can survive in soil as saprophytes or as survival structures. These adaptations allow the pathogens to survive despite the absence of a host plant. The presence of the rhizosphere most commonly results in renewed pathogen activity.

In an agricultural situation, it would be most desirable to plant a crop having roots that would stimulate pathogen activity but not serve as a host. The rhizosphere of nonhost roots can "trick" the spores into germinating, but then they starve because they cannot infect the plant. Many soil pathogens parasitize botanically related plants, so common crop rotations recommend a planting cycle that varies the type of crop—for example, potatoes followed by an unrelated crop, such as corn or a legume. When this rotation is followed, most pathogen populations are reduced by both starvation and competition in the absence of a host plant.

A common misperception is that the primary purpose of crop rotation is related to soil fertility. Rotation with crops that can fix nitrogen, such as legumes, may add nitrogen to the soil, but most crops use a similar variety of minerals. Soilborne pathogens cause aboveground symptoms—such as stunting, yellowing, and wilting—that can easily be confused with nutrient deficiencies. These symptoms can cause an observer to conclude that the "tired soils" are depleted of minerals. When a crop is planted that is not susceptible to the predominant pathogens present, however, normal plant growth can be expected.

Crop rotation is our most effective means to prevent soilborne pathogen populations from reaching damaging levels. It is actually a means of biological control, because it depends on rhizosphere stimulation of spore germination and the reduction of pathogen populations by competitive soil inhabitants. In the past, the presence of soilborne pathogens forced farmers to stop growing certain crops and caused good farmland to be abandoned. Crop rotation can make both of these actions unnecessary.

However, in modern times, many forces discourage farmers from using crop rotation as a biological control for soilborne pathogens (see Chapter 7). Besides economic constraints, there are biological complexities. For instance, crops vary in their susceptibility to pathogens, so rotation with one crop may result in a decline in one kind of pathogen but an increase in another. Another factor is weed control. Some weed species are often hosts to important pathogens, so poor weed control can negate an otherwise appropriate rotation plan. Rotation of perennial crops, of course, is impossible. Thus, the levels of soilborne pathogens of asparagus, grapes, tree fruits, turfgrass, and other perennial crops increase each year and can eventually necessitate replanting with other kinds of plants. For high-value crops—such as apples, strawberries, and tobacco—soil fumigation is often necessary to reduce pathogen populations before replanting is possible.

Study of the rhizosphere and its effects on soil organisms has led to a better understanding of how to effectively foster biological control in the rhizosphere to protect roots in the soil. Plant pathologists have been isolating antagonists from the soil for more than 100 years. In the laboratory, these organisms seemed good candidates for biological control, but when they were added to the soil, the results were disappointing.

Recently, however, bacteria and fungi isolated from the rhizosphere, called rhizosphere competent, have seemed to hold more promise as agents for biological control. Many of the successful agents are bacteria that are competitive or antagonistic (Table 9.1). These bacteria have been most effective for short-term protection, such as preventing damping-off of seedlings or infection of plant roots by crown gall bacteria at the time of planting.

Commercial rhizosphere-competent biological controls are now available as seed treatments to protect against soil pathogens. Both rhizosphere-competent bacteria and fungi are commercially available. In many cases, they are not only as effective as chemical seed treatments, but they also offer the possibility of extended protection because they may multiply in the rhizosphere and continue to colonize and protect the growing root surfaces.

Besides competition and antagonism (or antibiosis), another successful method of biological control occurs in certain fungi that parasitize other fungi. Some species of *Trichoderma* wrap their hyphae around the hyphae of plant pathogens and use them for food—a process called hyperparasitism. A hyperparasite is a parasite of a parasite.

The success of these biological controls has confirmed three important principles derived from rhizosphere research. First, it is not necessary to introduce huge quantities of antagonists into the soil when only the rhizosphere requires protection. Second, rhizosphere competence is critical to the success of the biological control agent. Finally, organisms that compete with pathogens should be established on the seed or root surface at the time of planting.

Table 9.1. Mechanisms of Biological Control for Plant Diseases

• Antagonism (antibiosis)
• Competition
• Hyperparasitism

Biological control of aboveground plant parts has not been as successful as biological control of belowground plant parts. The environment on leaf and flower surfaces is much less stable than the root environment. There are extreme variations in temperature, and aboveground surfaces are often dry for many hours. Neither of these conditions is conducive to microbial growth and survival.

One successful example of aboveground biological control is short-term protection of apple flowers against invasion by fire blight bacteria. In addition, commercial preparations of bacteria are available to prevent frost injury. These bacteria displace natural surface bacteria, which normally serve as ice nucleation sites and cause damaging ice crystals to form. In some cases, microbial applications are used to reduce the need for fungicide applications. Several websites list the various products available for management of both soilborne and foliar pathogens. These organisms do not pose health concerns at low concentrations in natural environments. However, commercial biological control products include safety information for applicators to prevent dermal (skin) exposure and inhalation of high concentrations of microbial agents.

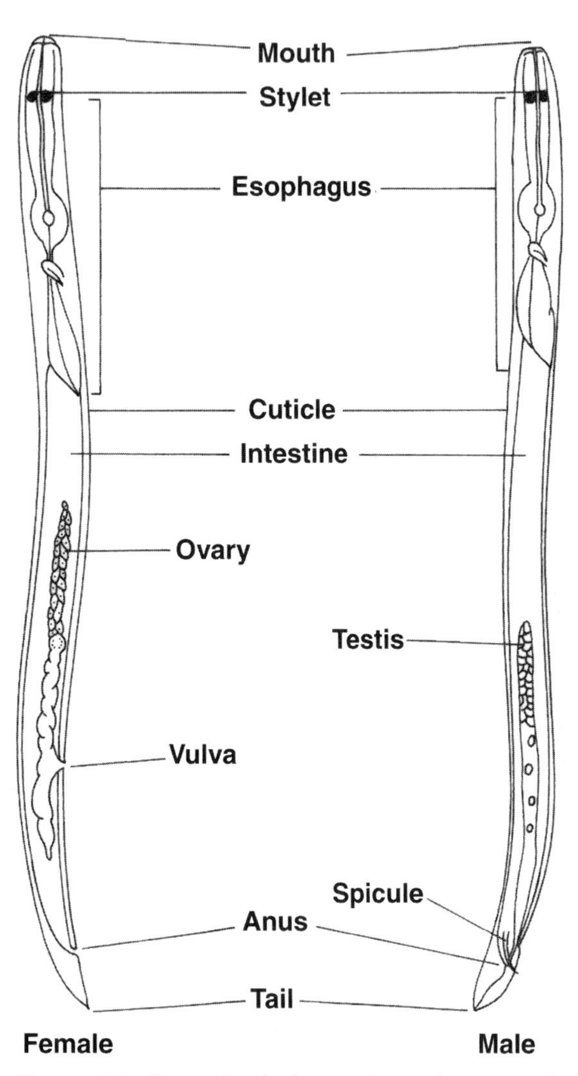

Figure 9.4. Generalized plant-pathogenic nematode anatomy. Note the stylet in the head, with which a nematode feeds.

Nematodes: The World of Worms

Rhizosphere microorganisms can help protect plant roots against infection by bacteria and fungi, but they do little to protect roots from attack by one important pathogen group: the nematodes. Most free-living nematodes are tiny, translucent roundworms that are found in all soils and bodies of water. In fact, they represent the largest number of multicellular organisms found in nature. Most nematodes feed on bacteria or organic matter. Some are specialized pathogens of animals, humans, or plants.

It is difficult to overestimate the significance of nematodes in the world's ecosystems. As Nathan Cobb, the "father" of U.S. nematology, wrote in 1915:

> *In short, if all the matter in the universe except the nematodes were swept away, our world would still be dimly recognizable, and if, as disembodied spirits, we could then investigate it, we should find its mountains, hills, vales, rivers, lakes, and oceans represented by a film of nematodes. The location of towns would be decipherable, since for every massing of human beings there would be a corresponding massing of certain nematodes. Trees would still stand in ghostly rows representing our streets and highways. The location of the various plants and animals would still be decipherable, and, had we sufficient knowledge, in many cases even their species could be determined by an examination of their erstwhile nematode parasites.*

Most nematodes are free-living (nonparasitic) aquatic organisms. In soils, they live in the water in soil pores and in the narrow layer of water on soil particles. Because nematodes must remain in water, they move most easily through sandy soils, where they can go from one large particle to another in the thin layer of water that coats each one.

Nematodes have a simple life cycle. Each begins as an egg and matures into an adult after molting its outer cuticle, just as many insects do. Between these stages, nematodes exist as juveniles of increasing size. In most cases, nematode juveniles look essentially the same as adults but smaller. Only adults are sexually mature. Although sexual reproduction between males and females is common in nematodes, some of them possess both male and female sex organs (hermaphroditic), and others are capable of reproducing without males (parthenogenetic) (Figure 9.4).

Pathogenic Nematodes: Problems for Animals (Including People)

Pathogenic nematodes have plagued human beings throughout history. Well-known human pathogens include pinworms and the nematodes responsible for diseases such as elephantiasis, Guinea worm, hookworm, and trichinosis. Trichinosis is caused by *Trichinella* nematodes, which are pathogens of pigs, bears, and several other carnivores. Humans usually become infected by eating meat that has not been thoroughly cooked. The existence of trichinosis likely explains why at least two major religions forbid the eating of pork.

Many readers of this book live in temperate climates and have access to modern sanitation and clean water. People who live in tropical areas—especially those where sanitation is lacking—suffer from some terrible diseases caused by nematodes.

Pets and farm animals are frequently infected by intestinal nematodes and can be "de-wormed" by medication. Heart worm of dogs, for example, is vectored by mosquitoes and can be prevented by medication. The largest nematode ever found is a parasite of the placenta of a sperm whale. It can be 26 feet (8 meters) in length and has a very descriptive scientific name: *Placentonema gigantissima.*

Some insects have nematode pathogens that have been exploited for biological control. The nematodes swim into openings on insect larvae in the soil, carrying with them bacteria that cause a massive infection and kill the larvae. A commercial example is the application of nematodes to lawns to kill Japanese beetle larvae, which feed on the roots of the grass.

Pathogenic Nematodes: Problems for Plants

The nematodes that are plant pathogens are all obligate parasites and thus cannot live without their host plants. Estimates of nematode damage to crops worldwide approach $100 billion, and this figure does not include the cost of

Science Sidebar

C. elegans *in biological research*

One of the most famous free-living nematodes is *Caenorhabditis elegans* (commonly known as *C. elegans*). Scientists have observed and counted every cell division of its 959 cells, from the egg to the adult. *C. elegans* was the first multicellular animal to have its entire genome sequenced. Scientists discovered that almost half the genes identified in the human genome match those identified in *C. elegans,* so much can be learned from studying this simple nematode.

C. elegans has a short life cycle, going from egg to egg in only 3 days. It can be grown easily in a petri plate containing agar and *Escherichia coli* bacteria as a food source. All of its organs can be observed through the cuticle, and many mutant versions have been produced in laboratories.

Today, *C. elegans* is a real workhorse for scientists studying development and genetics. Several Nobel Prizes have already been awarded for research using *C. elegans.* In 2002, a prize was awarded for advances in the understanding of genes that regulate organ development and programmed cell death. In 2006, a prize was awarded for work that demonstrated RNA interference (RNAi) and gene silencing (described in Chapter 6). In 2008, a prize was awarded for research on the use of a gene that codes for green fluorescent protein (GFP) discovered in jellyfish. When this gene is inserted into a cell, all of the cells and tissues that it gives rise to can be tracked, because they will glow under ultraviolet light. Although first accomplished with *C. elegans,* this technique is now used to study the development of other kinds of animals, including mammals. Without a doubt, *C. elegans* will continue to play a role in groundbreaking biological discoveries.

Science Sidebar

Tropical scourges caused by nematodes

Elephantiasis is caused by nematodes that infect the lymphatic system and produce grotesque distortions of parts of the body. The nematodes are spread by mosquitoes in tropical areas, such as some of the Caribbean islands.

A second tropical disease caused by nematodes is Guinea worm disease. Guinea worms were first described in ancient Egyptian scrolls in 1550 B.C. They were described later in the Bible as "fiery serpents." The genus name *Dracunculus* means "little dragon."

Guinea worms can be up to 3 feet (1 meter) long. They develop inside the human body after an individual drinks water contaminated by copepods (small crustaceans) carrying the immature nematodes. Eventually, a worm breaks through the skin, usually in the leg, causing burning pain. Sufferers seek relief from their pain in cool water, but ironically, exposure to water completes the nematode's life cycle with the copepods. Guinea worms must be drawn out of the body slowly over several days to prevent them from breaking and causing infection.

For more than 20 years, former President Jimmy Carter has led a program to rid the world of this terrible pathogen. That program finds infected people and cares for them in medical clinics. When Carter started this program, 3.5 million cases of Guinea worm disease were reported annually in 20 countries. In 2009, fewer than 3,200 cases were reported in four countries: Ethiopia, Ghana, Mali, and Sudan. Carter hopes to eliminate the problem in his lifetime.

A third disease is caused by hookworms, which parasitize approximately 800 million people in tropical countries. Hookworms enter the body through the feet, which makes many people in infested areas vulnerable because they have no shoes. The hooklike structures in the nematode's mouth are used to feed on blood in the victim's intestine, causing anemia. Because of poor sanitation, nematodes in feces contaminate areas where people walk.

President Barack Obama started a global health initiative to manage such "neglected tropical diseases" as elephantiasis and hookworm.

managing the nematodes. Like human pathogens, plant-pathogenic nematodes generally cause the most damage in warm, humid regions, where their populations are not reduced by dry, frozen winter soils. These nematodes cause the most damage in sandy soils.

Second-stage juveniles hatch from eggs and move to host plants to feed. Some feed from the surfaces of plants (ectoparasites), while others penetrate the tissues of plants (endoparasites). Plant-pathogenic nematodes have a sharp, hollow mouthpart called a stylet, which allows them to pierce plant cells and ingest their contents. These nematodes cannot infect people or other animals. In fact, we probably eat a few nematodes each time we have a salad or fresh fruits or vegetables (Figure 9.5).

Historically, the first nematodes discovered to damage plants were the seed gall nematodes, which parasitize many grains and grasses. These nematodes swim from the soil to the developing grain head in a layer of water on the plant surface and burrow into the developing flowers. Then at harvest time, seed galls are harvested along with healthy grains. Until the life cycle of this nematode was understood, farmers planted the shriveled, blackened, nematode-filled galls along with the healthy seeds. This returned the nematodes to the soil, and new plants then became infected. Proponents of spontaneous generation used nematode-infected grains as evidence of their idea before the nematode's true life cycle was understood. Foliar nematodes of the genus *Aphelenchoides* also swim in water on plant surfaces, feeding and damaging leaves and other aboveground parts of many plants, including ornamentals such as begonias and chrysanthemums.

Two important kinds of plant nematodes never live in the soil. They are tree pathogens and are carried to new host plants by insect vectors. One of these nematodes, which is vectored by the palm weevil, causes a tropical disease called red ring of coconut. The disease is named for the ring of reddened tissue observed when a dead palm tree is cut down after nematodes have killed it.

A second tree pathogen, the pinewood nematode, infects and kills pine trees. Recently, it was reported that bacteria carried by the nematode play a role in the disease. Further research is needed on the relationship between the nematode and the bacteria and on the etiology of the disease. The pinewood nematode is carried from tree to tree by sawyer beetles. Pines native to the United States are not usually killed by these nematodes. However, Scotch (or Scots) pine—widely planted in the United States in landscapes and on Christmas tree farms—is highly susceptible to pine wilt disease, and infected trees often die rapidly. Dead trees can be replaced with Norway or blue spruce, Douglas fir, cedar, or hemlock, all of which are virtually immune to the disease.

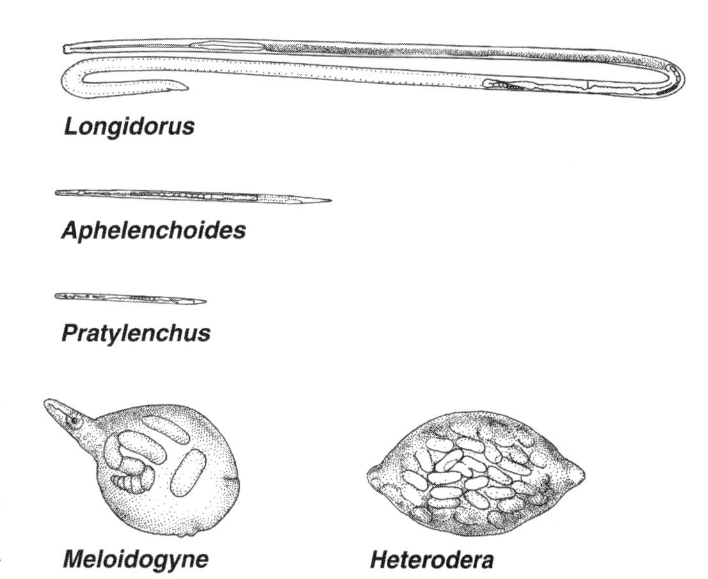

Figure 9.5. Appearances and relative sizes of some important plant-pathogenic nematodes.

Japanese red and black pines are very susceptible to the disease, and huge areas of pine forests in Japan have been killed since the pinewood nematode was accidentally introduced from the United States. It has since spread to China and other areas of pine forests in Asia, causing extensive mortality. Recently, the nematode was introduced to Portugal, where strict quarantines have been established to prevent its spread into the pine forests in the warm, dry areas of Europe. Restrictions on the export of North American lumber to Europe and Asia because of the pinewood nematode have caused increased expenses for the lumber industry. Most lumber must be treated with chemicals or heat before it can be exported to areas that do not yet have this nematode.

In contrast to these aerial nematodes, most plant-pathogenic nematodes remain in the soil and attack roots and other belowground plant parts. As previously mentioned, the nematodes may be ectoparasites or endoparasites. Three kinds of endoparasites cause the most economically important problems: lesion, root-knot, and cyst nematodes (Table 9.2). Lesion nematodes, in the genus *Pratylenchus,* penetrate and feed on root cells, leaving behind streaks of dead and dying cells that become brown. These lesions

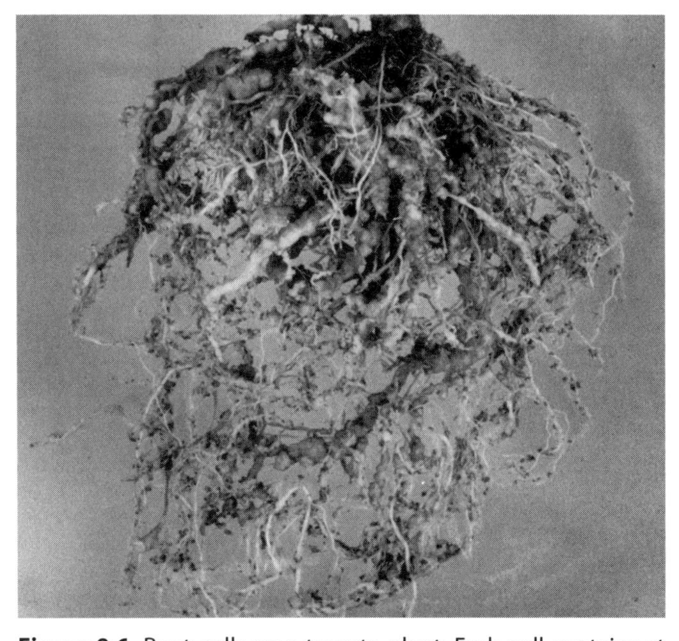

Figure 9.6. Root galls on a tomato plant. Each gall contains at least one female root-knot nematode (*Meloidogyne* species).

Table 9.2. Some Examples of Plant-Pathogenic Nematodes

• Ectoparasites
• Endoparasites
– Cyst
– Lesion
– Root-knot

are often invaded by other microorganisms. Lesion nematodes are pathogens of hundreds of host plant species.

Root-knot and cyst nematodes are the other two most economically important plant pathogens. Unlike lesion nematodes, they induce plants to create permanent feeding sites in roots through plant growth regulators produced in their saliva. Specialized multinucleate cells form near the vascular system. Root-knot nematodes cause roots to develop small, visible knots, or galls, each of which contains a female nematode and a permanent feeding site. Root-knot nematodes are in the genus *Meloidogyne* (Figure 9.6). A mature female root-knot nematode loses its wormlike shape, and its swollen body can lay a mass of several hundred eggs (Figure 9.7). Similarly, female cyst nematodes (in the genera *Heterodera* or *Globodera*) lose their wormlike shape, but they retain many of their eggs and actually become a protective sac (or cyst) for the living eggs when the female dies. Although cyst nematodes induce multinucleate feeding sites in plants, they do not cause galls on roots.

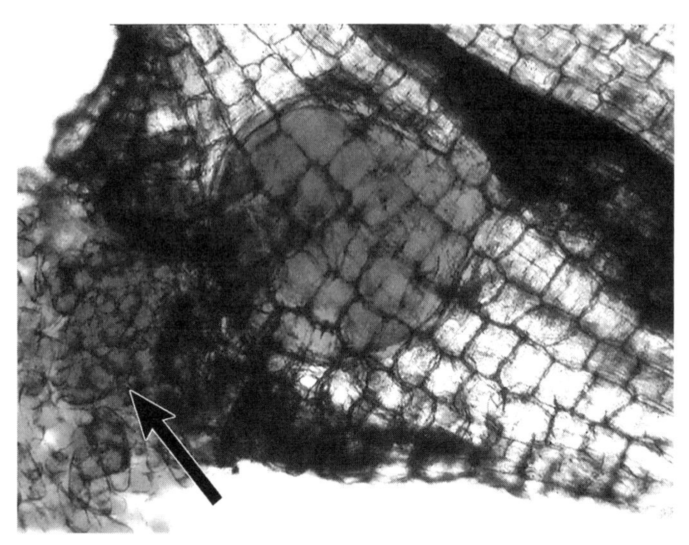

Figure 9.7. Swollen female root-knot nematode in root gall. Note that the head is placed near the vascular tissue of the root for feeding. A mass of eggs (arrow) in a gelatinous matrix has emerged from the gall (light micrograph).

One other important biological difference between root-knot and cyst nematodes is their host range. Root-knot nematodes are pathogens of many species of plants, whereas cyst nematodes tend to be quite host specific. Genetic resistance can interfere with development of the required feeding sites for both root-knot and cyst nematodes. This has been an effective management strategy for many economically important crops, such as VFN tomatoes (resistant to Verticillium wilt, Fusarium wilt, and root-knot nematodes) and soybeans (resistant to cyst nematodes). The wide host range of root-knot nematodes makes it difficult to find a nonsusceptible crop for use in crop rotation. It is easier to develop crop rotation plans for host-specific cyst nematodes.

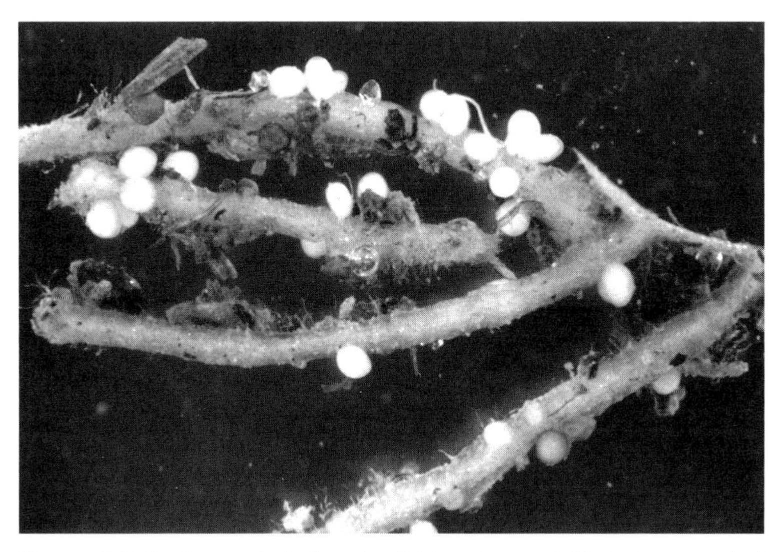

Figure 9.8. Cyst nematode females (*Heterodera* and *Globodera* species) become swollen and eventually become protective containers, or cysts, for eggs.

Soybean cyst nematode (*Heterodera glycines*) is the most economically important pathogen of soybeans, a major world crop. Most of the world's soybeans are grown in North and South America. Soybean crops are used to produce 57% of the world's vegetable oil and 68% of its vegetable protein. Most of the soybeans grown are used for animal feed (80%); only 6% is used for direct human consumption.

Soybean cyst nematodes originated in Asia, along with the soybean plant, and were introduced accidentally to the United States when soil was imported from Japan to distribute nitrogen-fixing bacteria to farmlands. The nematodes' eggs can survive for years in the protective cysts (Figure 9.8), even though a large percentage of them hatch in the first year. These pathogens are now managed with a combination of crop rotation, resistant cultivars, and monitoring of egg populations in fields. Integrated pest management (IPM) prevents soybean cyst nematodes from reaching damaging levels.

The potato cyst (golden) nematodes made their way from their center of origin in South America to Europe, probably in soil carried on the surfaces of tubers. The nematodes were then carried to Long Island, New York, on machinery used in Europe during World War I and transported back to the United States in airplanes. Even today, Long Island potatoes are under quarantine for this pathogen, and farmers must grow resistant cultivars and comply with crop rotation requirements. The quarantine was successful for more than 60 years, but now, potato cyst nematodes are found in a few additional places in New York State and Canada. Like most soilborne pathogens, golden nematodes are spreading slowly, primarily via soil on machinery.

Many genera of nematodes are represented in the ectoparasitic group, which feed on roots but do not actually enter them (Figure 9.9). When relatively inexpensive soil fumigants became available after World War II and their use significantly increased crop yields, plant pathologists became aware of how nematodes impact plant health. Ectoparasites often do not cause obvious damage to roots, but their feeding can result in yield losses, reduced plant growth, and increased infection by other soilborne pathogens in wounds made by nematode feeding. Some ectoparasites may serve as virus vectors, as they move from plant to plant. Genetic resistance to ectoparasitic nema-

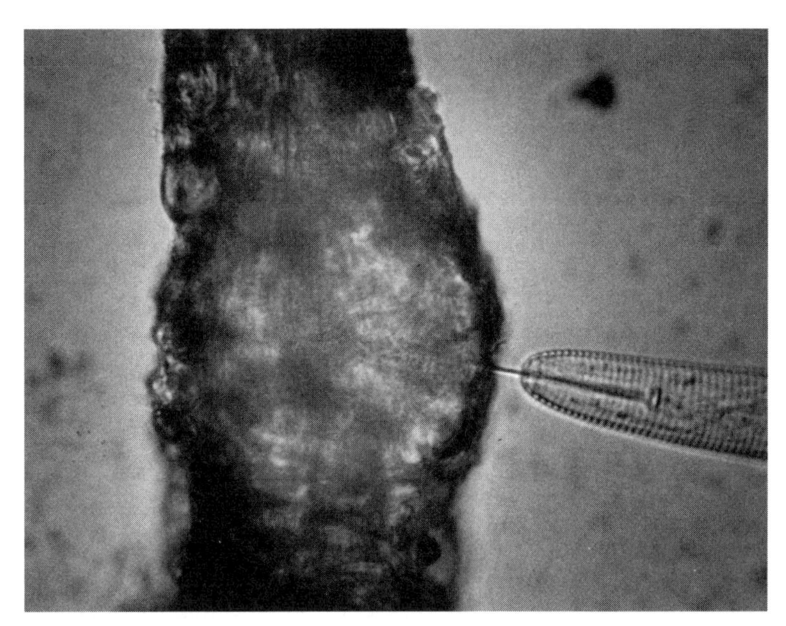

Figure 9.9. Ectoparasitic nematode feeding on a plant root with its stylet (light micrograph).

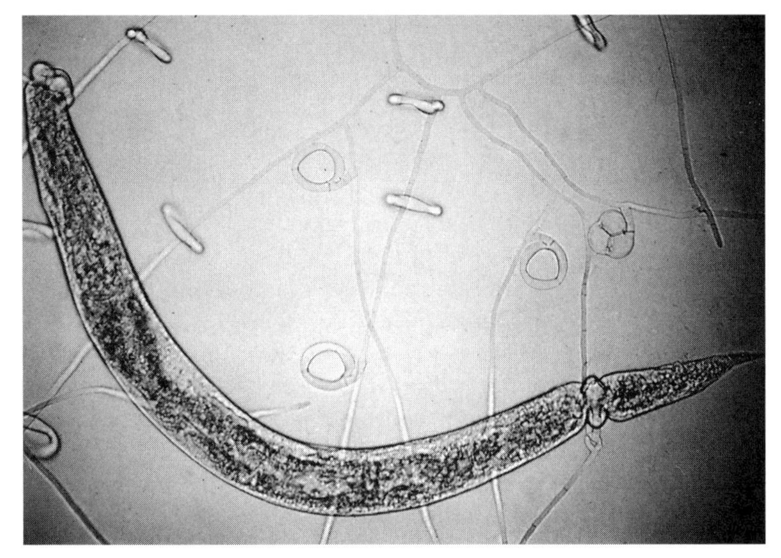

Figure 9.10. Potential biological control for nematodes. This fungus, *Arthrobotrys brochopaga,* actually traps nematodes by inflating a hyphal noose. The nematode is then parasitized and killed by the fungus, which uses it for food (light micrograph).

todes has not been particularly successful, and crop rotations are difficult to recommend because of the wide host ranges of many species.

Numerous potential biological control agents have been discovered for plant-pathogenic nematodes. These agents include predators and pathogens of amazing variety. Some fascinating fungi parasitize nematodes. Some produce a hyphal noose, which quickly inflates to trap a nematode that swims through it (Figure 9.10), while others parasitize eggs and cysts.

Unfortunately, the commercial application of biological controls has not yet been successful. Use of biological controls is limited by their inability to compete with the tremendous killing potential of chemical controls. Some recent research has demonstrated the potential for using ribonucleic acid interference (RNAi) to help control root-knot nematodes. Plants have been genetically engineered to produce double-stranded RNA that silences a nematode parasitism gene and thus reduces the amount of disease (see Chapter 6). By combining a number of alternatives to the toxic nematicides, management of plant-pathogenic nematodes could be more environmentally sound and sustainable in the future.

The Fate of Pesticides in Soils: Where Do They End Up?

The most common pesticides used for soilborne pathogens are nematicides and soil fumigants. Because many plants are vulnerable to a variety of nematode pathogens, plant pathologists searched for a broad-spectrum means of nematode control. Nematicides and soil fumigants are two types of chemicals that rapidly reduce the populations of all nematodes.

Nematicides, mostly carbamates and organophosphates, are nerve poisons, so they are toxic to organisms with nervous systems. Most also are insecticides and were first developed for that purpose. They were chosen for soil application because of their high toxicity and water solubility. Aldicarb (trade name: Temik) is a highly toxic nematicide/insecticide. It was found in groundwater in Florida after application to citrus crops, on Long Island after application to potatoes, and in other places as well. Aldicarb is now a restricted-use pesticide, and its registration is very limited and may be withdrawn.

Soil fumigants are volatile chemicals that are applied to soil to kill pathogens and pests. Many soil fumigants are biocides, which means they are toxic to all living things, including crops, fungi, insects, nematodes, and weeds. Because of their broad-spectrum activity, these chemicals must be applied before a crop is planted and allowed to clear from the soil to avoid crop damage. Soil fumigation is expensive and time consuming, so it is usually done to soils in which high-value crops will be planted. Soil fumigation is especially popular for perennial crops, such as grapes, raspberries, and strawberries. However, over time, pathogens and pests will recolonize fumigated soils.

Methyl bromide is a commonly used fumigant that has been threatened with registration withdrawal for many years because of its potential to harm the ozone layer. Withdrawal has been delayed because there are no effective alternatives for the chemical for certain crops in states such as California and Florida, which have high-value crops and significant soilborne pathogens. Many nematicides and soil fumigants have been banned from use, and those that remain registered are restricted in use and will likely lose registration in the near future.

Pesticides that are deliberately applied to soil and those that reach soil in runoff from foliar applications may pose a threat to groundwater. Because most pesticides are not applied directly into the soil in high concentrations, it was long assumed that they would be degraded by microorganisms (mostly soil bacteria), which use the chemicals as a food source. Groundwater is found in soil pores, sediments, and rock fractures beneath the ground surface. An area from which groundwater can be extracted for human use is called an aquifer (Figure 9.11). Aquifers serve as water sources for some municipalities and many individual homes in suburban and rural areas, where there is also a significant amount of food production. Aquifer contamination by pesticides became an important concern in 1979, after the first discovery of a nematicide in groundwater. Since then, other pesticides—primarily herbicides, soil fumigants, and other nematicides—also have been discovered in groundwater, generally at low levels measured in parts per billion and less.

Since groundwater is the source of water for about half of all U.S. citizens, contamination of this water presents several problems. If pesticides and other impurities that threaten human health are present in groundwater, multimillion-dollar water treatment plants must be built to remediate the groundwater or the site must be abandoned. In addition, piping systems must connect rural homes to a water supply, or individuals will have to resort to using bottled or filtered water. However, if the contamination is extensive, then bottled "spring" water, which is popular with people seeking an alternative to tap water, may not be safe either.

Testing water for numerous organic compounds and their breakdown products is prohibitively expensive in many cases. Also, scientific data are lacking about the health effects of minute amounts of contaminants in water. Recent advances in detection technology allow scientists to measure chemicals at parts per billion, trillion, and even less. Many scientists believe that for the majority of compounds, such low concentrations are not biologically significant. Water also may contain very low levels of naturally occurring elements, such as arsenic, lead, mercury, and other heavy metals. At this time, health standards for pesticide contamination are based more on the technology of detection than on proven health risks.

Still, none of us want pesticides in our water. After the initial panic about detecting pesticides in groundwater, additional studies demonstrated that the problem is less widespread than first thought. The pesticides that have been detected so far have mostly been chemicals applied to bare soil or injected into soil. One of the important benefits of

Figure 9.11. Groundwater sources in aquifers.

Table 9.3. Factors That Increase the Chance of Groundwater Contamination by Organic Contaminants

• Applications to bare soil (less microbial activity than areas with growing plants)
• Chemicals that degrade slowly
• Chemicals with high water solubility
• High water table (the aquifer is close to the ground surface)
• Rainy climate
• Sandy soil (low organic matter)

Roundup Ready crops, which have glyphosate resistance, is that the herbicide is applied to growing plants with an established rhizosphere, eliminating high-risk bare soil applications.

Some soil environments are at greater risk for groundwater contamination than others. High-risk environments include those where the soil is primarily sandy and allows water to penetrate quickly, thus increasing the chance that chemicals will be carried below the biologically active soil layer. Organic matter in the soil has tremendous binding power and holds pesticides tightly until they are degraded. Sandy soils have lower amounts of organic matter than other soils.

Another factor that increases the risk of water contamination is a high water table (Table 9.3). When the water table is high, the aquifer is close to the soil surface. This means microorganisms have little time to degrade contaminants before they reach the water. Areas with shallow water tables—such as Florida, Long Island, and river valleys—have had the greatest contamination problems. The climate affects the likelihood of water contamination too. An abnormal amount of rainfall increases the chance of chemicals washing through the soil and into the groundwater before being degraded.

What happens to pesticides in the soil? Organic pesticides are no different than other organic compounds in the soil. Like organic matter from dead plants, animals, and microbes, pesticides are degraded by microorganisms. The presence of plant roots and their associated microbial populations is one of the most important factors that affects whether a pesticide will be degraded before it reaches the groundwater. How long it takes a pesticide to be degraded depends primarily on its chemical structure. Some compounds are rapidly biodegraded, but others take longer. Most pesticides in use today degrade quite rapidly and have a half-life of only 30 to 60 days in the soil. To protect groundwater, registrations have been withdrawn for many pesticides that do not degrade rapidly.

How can groundwater contamination by pesticides be prevented? Studies of the movement of pesticides in soils have led to the recommendation of planting grass, which has a dense root system, around the edges of row crops and between trees in orchards. This practice helps reduce run-off into surface waters and leaching of pesticides into groundwater. In addition, considerable effort is being made to train pesticide applicators in techniques that will reduce the chance of contamination. For example, pesticide containers must be triple rinsed and, when not recyclable, disposed of properly. Also, waste and spill water must be contained and treated as hazardous waste if it cannot be reused. Wells are provided with special protection from back-siphoning and spills.

Science Sidebar

Other groundwater contaminants

Pesticide professionals have become well educated in approaches to groundwater protection, but pesticides are not the only potential contaminants. Other agricultural chemicals, such as fertilizers, have been found in groundwater at even higher concentrations than pesticides. Some people dump numerous toxic substances (cleaning compounds, motor oil, gasoline, paints, varnishes, thinners, insecticides, and herbicides) into drains that lead to septic tanks and sewage lines. Some people even pour wastes directly onto soils and into streams and ponds. Gas stations, farms, and many homes have underground fuel storage tanks, which are subject to leaking. Industries have deep-well injected wastes, and nuclear wastes are stored deep in the ground. The waste of human beings from sewage sludge, septic tanks, leaking sewage pipes, and landfills seeps into the earth and contaminates water that has been clean for millions of years—until now.

Many of the contaminants found in groundwater reflect ecological abuse by human beings. The cost of protecting groundwater is high, but the cost of cleaning it is even higher. We all participate in this pollution, so we all must consider what we can do to reduce this ecological overload. Without careful stewardship of underground water resources, the time may come when no water can be pumped safely into homes untreated.

CHAPTER 10

Natural Poisons and Gourmet Delicacies: Fungi in Food

Countryside scenes of people eating foods untouched by the contaminating influences of modern life are popular images in advertisements. The public perceives these foods as safe, because they have been eaten over millennia by foraging humans seeking nourishment. Despite this, everyone knows that some plants are poisonous and should be avoided, although the specific folklore about the edibility of natural plants is familiar to few people in industrial societies. Most of us simply choose from among the foods displayed in supermarkets and at farm stands.

Even plants that are commonly eaten are not necessarily nontoxic. Plants, which cannot run away from their enemies, have developed toxins to protect themselves from pathogens and pests. For instance, the leaves and stems of potato plants contain high concentrations of toxic alkaloids. Tubers left in sunlight turn green and accumulate these same alkaloids; therefore, potatoes should be stored in the dark, and green tubers should not be eaten. Cassava, or manioc (*Manihot esculenta*), a staple food for more than 500 million people in tropical regions, contains high levels of cyanogenic glucosides. The starchy root of this plant must be ground and washed thoroughly before being eaten to avoid cyanide poisoning. Many natural components of commonly eaten plants are toxic, and many are carcinogenic. Research by Bruce Ames, at the University of California, Berkeley, has identified mutagenic compounds in familiar foods, including raw mushrooms and many vegetables.

Over the years, people have learned which plants and parts of plants are safe to eat. Our bodies also have developed many detoxification mechanisms that allow us to eat some toxic compounds with little or no harm. Plant breeders, who select plants with resistance to insects or pathogens, test the resistant cultivars for increased levels of toxic compounds before releasing them for human consumption. In fact, many of the compounds used by plants to resist these pests can cause toxic reactions in us as well. Such compounds can be present at high levels throughout the plant tissues, resulting in health risks far greater than the presence of low-level pesticide residues on plant surfaces, where they can be washed away. Despite the potential dangers of these natural foods, many consumers consider the label "all natural" on food products to mean "risk free."

Ergotism and the Holy Fire

After the edibility of various plants had been established, humans began to select plants that were high yielding and suitable for agriculture. As a result, most people today eat a limited number of plants. Only a few species provide the major calorie sources. As this specialization developed, the potential health threat from any particular plant increased greatly because of the disproportionate role these few species play in our diet. One particular risk that is receiving renewed attention in the interpretation of historical records is associated with the cultivation of the cereal rye (*Secale cereale*).

In the Middle Ages, a frightening disease known as "holy fire" or "St. Anthony's fire" was common but unpredictable. Unlike most diseases, which spread quickly among the residents of overcrowded cities, holy fire was most common among the rural poor. The disease also was unusual in that it did not seem to be contagious and could strike one family or even individual members of a family without necessarily affecting their neighbors. Children and people who were weak or frail were most susceptible to the disease. Nursing mothers might see the effects in their babies. A folk cure for these victims was to have the nursing mother eat white bread rather than rye bread, which was coarse and inexpensive.

This scourge of the Middle Ages (500–1400 or 1500) has an unclear history. Medical diagnostics were primitive at the time, and some of symptoms commonly described could be attributed to several different medical problems, making historical records difficult to interpret. Common symptoms included mental aberrations and hallucinations. People also described the feeling of burning skin or insects crawling under their skin. During outbreaks of the disease, women frequently miscarried and fertility was generally reduced. Severe cases resulted in gangrene because of the constriction of blood vessels in the extremities. Many victims lost hands and feet and became permanently crippled.

Figure 10.1. St. Anthony. Redrawn from a woodcut made in Germany by Matthias Grünwald (1475–1528).

St. Anthony, considered the "father" of monasticism, was said to have had his faith tested by demons with flames, so he became a model for disease victims to emulate. Hospitals dedicated to St. Anthony took in these victims, caring for them during their painful and prolonged suffering. The amputated limbs of holy fire victims were hung over the entrances of monasteries devoted to this saint, indicating they would care for patients as well. The limbs were kept in anticipation of the resurrection of bodies expected to precede the Last Judgment. Mandrake roots (*Mandragora officinarum* in the family Solanaceae) were sometimes used as a local anesthetic or twisted into talismans against this dreadful disease. Pigs belonging to the monasteries had bells hung around their necks and were allowed to wander the villages and be fed by the local residents.

Artwork from the period shows symbols of St. Anthony on the clothing worn by people hoping to ward off the effects of the disease. The paintings of Dutch artist Hieronymus Bosch (1450–1516) contain many strange creatures, which may be related to the hallucinations of the diseased people. The German woodcut that has been re-created in Figure 10.1 contains several symbols associated with St. Anthony, including amputated hands and feet, a tiny doll made from a mandrake root, and a pig with a bell around its neck.

Ancient European records of the disease before the Middle Ages cannot be found, and the

Greeks and Romans did not seem to describe the malady. The disease became prominent after rye was introduced as an agricultural grain following the invasion of eastern Europe by nomadic groups such as the Vandals. Outbreaks of the disease were sporadic in the Middle Ages but occurred particularly in years in which a cold winter put the rye under stress and a cool, wet spring prolonged the flowering period.

France was the center of many severe epidemics for two primary reasons. Rye was grown as the staple crop of the nation's poor people, and the cool, wet climate was conducive to development of the fungus that caused the mysterious malady. Today, we know that toxic alkaloids produced by a fungal pathogen on rye caused the various symptoms observed in human beings. Hard, purplish-black, grainlike structures (called ergots or sclerotia) produced by the fungus *Claviceps purpurea* were harvested along with the regular rye grains. *Ergot* is the French word for the cockspurs found on the legs of roosters, which these sclerotia resembled. When the grain and ergots were ground together during milling, the flour became contaminated by the potent alkaloids of the fungus, which persisted even when baked into bread.

The Cause of Ergotism: A Fungus in the Rye

The disease known today as ergotism occurs in places where contaminated rye bread is a staple food. Hundreds of thousands of humans, as well as cattle and horses, died or became seriously ill in the years before the cause of the disease was understood (Figure 10.2). The role of ergots in human and animal disease was not established until a French physician, Dr. Thuillier, finally made the connection in 1670—after thousands more deaths had occurred following the end of the Middle Ages. Ergots were so commonly observed on rye plants that early botanical drawings of the plant included ergots, because they were not known to be produced by an entirely separate organism. The actual role of the ergots and the life cycle of the fungus were not discovered until 1853.

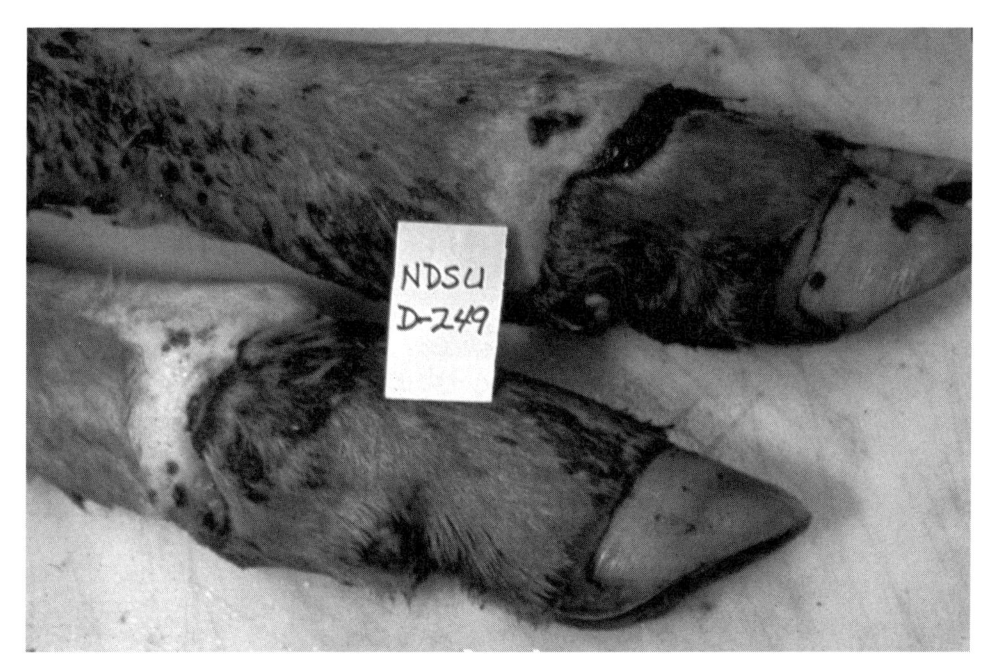

Figure 10.2. Leg lesions of a calf affected by ergotism. Note the sharp demarcation between living (top) and dead (bottom) tissue.

As the rye grain matures, some ergots fall to the soil. These masses of mycelium, which have a dark rind, are survival structures for the fungus, similar to those produced by many other fungi. These structures help ensure the survival of the fungus through the winter. In the spring, small mushroomlike structures are produced from ergots on the soil surface. They forcibly eject the microscopic ascospores (sexual spores) of the fungus upward, to be picked up by wind currents in the field of rye. Rye, like many grasses, is wind pollinated. The stigma of the rye flower is large and featherlike to help trap the windborne pollen. This structure also helps trap the airborne ascospores expelled from the reproductive structures of the fungus on the soil surface (Figure 10.3).

If moisture and temperature conditions are favorable, the ascospores germinate and infect the ovaries of the flowers. Like many fungi, *C. purpurea* then produces asexual spores (conidia) to increase the number of infections in the rye field. This must happen quickly, because the flowers remain susceptible to infection for only a short time. To help spread the conidia, the fungus also secretes sweet, sticky honeydew over the infected flower, which attracts flies and other insects. As they feed on the honeydew, their feet and bodies pick up the conidia. As the insects move among the rye plants searching for more honeydew, they effectively disseminate the conidia. Depending on the weather conditions, this honeydew stage may last for a few or many days.

Rather than produce a rye grain, the infected flowers may be sterile, which directly reduces yield. Other infected grains are of reduced quality, while still others are completely

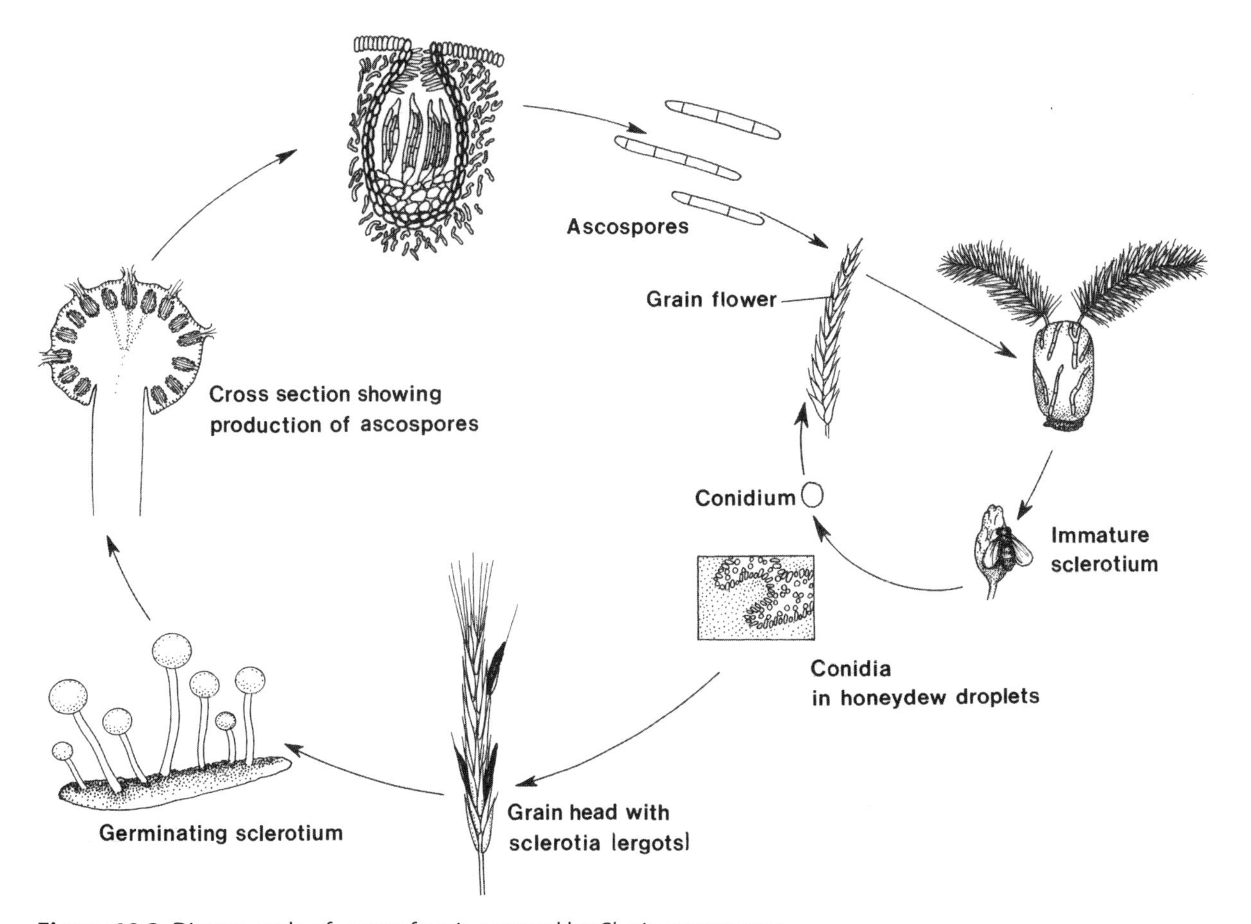

Figure 10.3. Disease cycle of ergot of grains caused by *Claviceps purpurea.*

replaced by the sclerotium, or ergot, of the fungus. Management of the disease is very successful if the stubble is plowed under after harvest to bury the remaining ergots and the rye crop is rotated to different fields. It also is important to mow nearby wild grasses, many of which are susceptible to ergot infection and thus can serve as a source of inoculum in rye crops. Ergots on wild grasses can be found in nature late in the summer, especially in years with wet spring weather.

More Roles for Ergot: In History and Today

Ergotism may have played a role in the establishment of the modern border between France and Germany. Beginning in the 700s, the Vikings settled on the northwest coast of France and conducted many raids on the Holy Roman Empire. Ergotism may have weakened the ability of the Franks to defend their territory. Eventually, in 887, King Charles III was forced to abdicate his throne, leading to the split of the Holy Roman Empire into the two kingdoms that later became France and Germany.

In *Poisons of the Past: Molds, Epidemics, and History,* historian M. K. Matossian builds quite a strong case for the theory that ergot toxicity played a multifaceted role in European history. She challenges a common theory of the role of the potato in the European population explosion between 1750 and 1850 and presents convincing evidence that this growth was directly related to the removal of toxic compounds from the human diet (including ergot alkaloids and other fungal toxins discussed later in this chapter). This occurred when the potato replaced rye and other susceptible grains as the preferred food of poor people. Matossian's population studies of individual countries also provide evidence for the role of ergotism in demographic depression, particularly in France, where rye remained the predominant food of the poor long after the potato had become more popular in other countries.

Matossian also presents evidence that ergot poisoning was a factor in the accusations of witchcraft made in Europe from the 1300s to the 1700s and in the United States in the late 1600s. The author suggests that ergot poisoning in the spring of 1789 affected the mental state of the French peasants and thus contributed to the French Revolution. The hallucinogen lysergic acid diethylamide (LSD) is derived from an alkaloid of ergot.

Accusations of witchcraft have been shown to correlate with historical weather data, including cold, wet weather in winter and spring. These accusations were concentrated in areas of France, central Europe, and the Rhine Valley, where rye was a staple cereal. Robin Cook, writer of medical mysteries, concocted an exciting story in modern times involving a neurobiologist whose girlfriend's ancestor was accused of witchcraft in Salem, Massachusetts, in 1692. He isolates a hallucinogenic fungus from the dirt floor of the old family house—an action that is inaccurate in terms of what we know about the biology of the ergot fungus. Cook's book *Acceptable Risk* is probably the only *New York Times* best-seller with the word sclerotia in it.

G. L. Carefoot and E. R. Sprott, authors of *Famine on the Wind,* also describe the role of ergotism in history. For instance, in 1722, Peter the Great was halted in his attempt to capture Constantinople (Istanbul) and gain access to warm-water ports when his troops and their horses were poisoned by ergoty grain at the mouth of the Volga River. Many serious outbreaks of ergotism also were recorded in Russia. The outbreaks probably would have continued if the potato from the New World had not replaced rye as the major food crop of the poor. Potatoes apparently became plentiful enough that many were fermented to produce vodka, for which Russia is famous.

Midwives have long used ergot to induce abortions and as an aid in childbirth, which explains the ergot's common name of "mothercorn." The fungus produces a range of pow-

Science Sidebar

Endophytes

Fungi related to the ergot fungus have been discovered in pasture grasses, where they have caused serious diseases in grazing animals. The symptoms are very similar to those of ergotism and have been named "ryegrass staggers" and "fescue foot." Investigations have demonstrated that the hyphae of these fungi are found within the aboveground parts of the infected plants. The fungi are endophytes; that is, they exist entirely within the host plant. The mycelium infects developing seeds, so new seeds contain the fungus in the next generation.

Although this is a disadvantage if the grasses are used in pastures, because of their toxic effects on animals, these same fungi have been found to deter insects that feed on aboveground parts of some turfgrasses used for lawns and golf courses. Endophyte-containing seed is now sold as a biological control against chinch bugs, sod webworms, and some other turfgrass insect pests in perennial ryegrasses and fescues (Figure 10.4).

Figure 10.4. Endophyte-infected turfgrasses are used as a biological control against feeding by certain turfgrass insect pests. The perennial ryegrass cultivar Pennant (right) resists billbug damage better than ryegrass cultivars without endophytes.

erful alkaloids, which can be purified and, with the dose carefully measured, used in modern medicine to reduce postpartum bleeding. Other modern pharmaceutical uses of ergot alkaloids include treatment of migraines and mental and neurological disorders, such as Parkinson's disease and restless leg syndrome. Rye crops are deliberately inoculated with *C. purpurea,* and the ergots are harvested for pharmaceutical processing.

Once ergots were known to be the source of disease, they were easily separated from the grain before milling. Careful harvesting and milling practices have made ergot poisoning unlikely in modern times. Animal poisonings are occasionally reported, however, because animal feed is not as carefully monitored as human food and because *C. purpurea* can infect many wild grasses. Ergot poisoning is most common when animals such as horses and cows feed on wild grasses in pastures or in field borders following a prolonged wet spring, when ergot infection is most severe.

Ergot also can infect wheat but is rarely found in this crop. Wheat is self-pollinated, and the flowers are open only for a short period of time, which reduces the infection window for ergot ascospores. Since the 1970s, when male-sterile breeding lines of wheat and barley were developed, there has been concern about increased susceptibility of these lines to ergot infection, because they have an extended flowering period, leaving the plants open to infection longer. Breeding lines are being carefully evaluated to avoid selection of new grain cultivars in which ergot might be an important problem. Other *Claviceps* species infect additional plants, as well, causing problems in corn in Mexico and in sorghum and pearl millet in tropical regions of Africa, in particular.

Today, U.S. government standards prevent the sale of grain containing more than 0.3% ergot by weight for human consumption. Because of the long history of ergotism, a cultural preference for wheat bread over rye bread remains among people in France, in other countries in Europe, and in the United States.

Decay Fungi in Foods: The Danger of Mycotoxins

Fungi have been competing for food with insects and other animals since long before humans existed. We may find the rotten and moldy food produced by fungal digestion objectionable, because the compounds these microbes produce make the food look, smell, or taste bad to us. Or perhaps we react with repugnance to moldy food because of the experiences of our ancestors, who learned that it could be toxic.

Table 10.1. Major Mycotoxins and Toxin-Producing Fungi from Cereals, Corn, Peanuts, Soybeans, and Other Products and Some of Their Effects on Animals

Toxin or Syndrome	Fungal Source	Feed or Foods Affected	Possible Effects on Animals
***Aspergillus* toxins**			
Aflatoxins (primarily B_1, B_2, G_1, and G_2). B_2, G_2, M_1, and M_2 are metabolites; M_1 and M_2 are important regional contaminants in milk	*Aspergillus flavus* and *A. parasiticus*	Cereal grains, corn, cottonseed, peanuts, and other foods	Liver injury; carcinogenic; reduced growth rate; hemorrhagic enteritis; suppression of natural immunity to infection; decreased production of meat, milk, and eggs
Ochratoxins (nephrotoxins)	*A. ochraceus* and *Penicillium viridicatum*	Cereal grains, coffee, grapes	Toxic to kidneys and liver; abortion; poor feed conversion, reduced growth rate, general unthriftiness; reduced immunity to infection
Sterigmatocystin	*A. nidulans* and *A. versicolor*	Cereal grains, coffee	Toxemia; carcinogenic
Tremorgenic toxin	*A. flavus, P. cyclopium,* and *P. palitans*	Cereal grains, peanuts, soybeans, and other foods	Tremors and convulsions
Ergot alkaloids	*Claviceps* species and *Neotyphodium* species	Cereal grains, forage grasses, and sorghum	Tremors and convulsions; gangrene in extremities; abortion; hallucinations
***Fusarium* toxins**			
Zearalenone (estrogenic syndrome)	*Fusarium graminearum* (sexual state *Gibberella zeae*), *F. tricinctum,* and, in a minor way, *F. moniliforme*	Cereal grains, corn	Hyperestrogenism, infertility, stunting, and even death
Emetic or feed refusal factor (deoxynivalenol [DON] or vomitoxin)	*F. graminearum* (sexual state *G. zeae*)	Cereal grains, corn	Food refusal by swine, cats, dogs; reduction in weight gain
Other trichothecenes (T-2, monoacetoxyscirpenol [MAS], diacetoxyscirpenol [DAS])	*F. tricinctum,* some strains of *F. graminearum, F. equiseti, F. lateritium, F. poae,* and *F. sporotrichioides*	Cereal grains, corn	Severe inflammation of gastrointestinal tract and possible hemorrhage; edema; vomiting and diarrhea; infertility; degeneration of bone marrow; death; reduced weight gain; slow growth; sterility
Fumonisins	*F. verticillioides* and other *Fusarium* species, including *F. proliferatum*	Corn	Fumonisin B_1 causes leucoencephalomalacia in horses ("blind staggers"); pulmonary edema; carcinogenic
***Penicillium* toxins**			
Patulin	*P. griseofulvum, P. expansum, P. patulum, P. claviforme,* and *A. clavatus*	Apples, cereal grains	Hemorrhage; edema; paralysis; convulsions; capillary damage
Rubratoxin	*P. rubrum*	—	Liver damage and hemorrhage
Citrinin	*P. citrinum*	Grains	Kidney damage

Source: Shurtleff, M. C., Kirby, H. W., and Eastburn, D. M. 1990. Mycotoxins and Mycotoxicoses. Report on Plant Diseases 1105. University of Illinois, Urbana. Used with permission. Revised by G. Munkvold, 2011.

Many different fungi invade moist food products and contribute to their decay but do not produce toxic compounds. Even though the fungi make the food distasteful, they do not necessarily make it inedible. An example of an unappealing but edible food is a potato tuber rotted by *Phytophthora infestans.* Some of the rot fungi, such as the common secondary invaders of the genus *Alternaria,* do produce toxic compounds. Unless starving, most of us dispose of moldy food, rather than eat it. We may try to salvage some items with a little mold, such as hard cheese. The green, powdery spores on the surface are produced by hyphae that penetrate deep into the cheese, so consider this when you make your cut. Moldy soft cheeses and breads should not be eaten.

Sometimes, we develop a taste for toxic compounds. An example is ethyl alcohol, a product of anaerobic fermentation produced by yeasts, which are ascomycetes, and bacteria. Most human cultures have developed a taste for alcoholic beverages, despite their negative health effects, and people deliberately create conditions conducive to enhanced production of alcohol in various products, such as barley, corn, grapes, potatoes, rice, rye, and wheat. In fact, we are not the only species to find the alcoholic content of foods attractive. Some birds and small mammals ingest fermenting berries and other fruits to the point of intoxication. Even the word intoxication reflects the knowledge that the product is not healthful, despite its deliberate ingestion.

The most important mycotoxins, or poisons produced by fungi, are those that invade grains and seeds. Infections generally occur while these food products are in the field before harvest. Certain conditions increase the chance of invasion, including wounds in the seed coat and a high seed moisture level, which allow more extensive colonization by the fungi. These factors explain why crops such as corn are allowed to stand for weeks in the field until the seeds have dried and hardened. An extended period of rainy weather in the late fall can require expensive artificial drying or result in losses when moldy grain has to be thrown away. Seeds containing high amounts of oil are particularly vulnerable to infection and mycotoxin contamination. Examples include coconut, cottonseed (which is commonly used as a cattle feed), peanuts, pecans, sunflower seed, and walnuts. Barley, corn, grain sorghum, rye, and wheat also are very susceptible to mycotoxin contamination (see Table 10.1).

Grain Mycotoxins: The Invisible Poisons

Fungi of the genera *Aspergillus, Fusarium,* and *Penicillium* produce toxic and carcinogenic compounds that threaten the health of animals and humans who ingest them. Hogs, cows, and other animals often reject moldy grain and feed, which suggests they have developed instincts from unfortunate past experiences. Although animals are fed unprocessed grain or feed and refuse badly contaminated products, humans usually mill or otherwise process grain products until harmful compounds can no longer be detected in the foods. Therefore, we must rely on the careful inspection of grain and the rejection of contaminated grain by federal or state inspectors and food processors.

Of the 300 listed mycotoxins, the government most carefully regulates five major kinds: aflatoxins, fumonisins, ochratoxins, vomitoxin (deoxynivalenol or DON), and zearalenone. Aflatoxins and ochratoxins are produced by *Aspergillus* species, and the others are produced by *Fusarium* species.

Aspergillus flavus produces several of the most potent mycotoxins: the aflatoxins. The term aflatoxin comes from combining parts of the scientific name of the fungus. The word *Aspergillus* comes from the resemblance of the fungus and its spores to an aspergillum: the perforated liturgical device used to sprinkle holy water (Figure 10.5). A closely related fungus, *A. parasiticus,* also produces aflatoxins. These toxins are tasteless, odorless, and colorless. Environmental conditions affect the amounts of aflatoxins produced, but even grain that shows no obvious infection by the fungi in the field or in storage may be contaminated by the poisons.

Aflatoxins are some of the most powerful toxins and carcinogens known. Rats fed a diet containing 1 part per billion (ppb) of aflatoxin B_1 develop liver cancer. Aflatoxins are thought to be 100 times more likely to cause cancer than polychlorinated biphenyls (PCBs). These toxins also are mutagenic, toxic to the liver, and immunosuppressive. Aflatoxins in the diet not only affect the body's immune response to pathogens, but they also affect the success of vaccinations against other important pathogens. Ochratoxins, produced by *A. ochraceus* (and *Penicillium viridicatum*), are primarily toxic to the kidneys, but they also can cause abortions and affect the immune system and general health.

Figure 10.5. *Aspergillus flavus* (electron micrograph).

Historian M. K. Matossian, whose study of the role of ergot and mycotoxins was mentioned earlier, also has studied the "Black Death," or bubonic plague, which occurred in 1348–1350. She suggests that the tremendously high mortality rate from the plague may have been caused by the immunosuppressant effects of mycotoxins in moldy grains. These same mycotoxins also may have killed rats, which fed on the grain. With fewer rats available, fleas looked for alternate sources of blood. As the fleas fed on humans and domestic animals, they transmitted the bubonic plague bacteria to these populations. Many people and animals became sick and died as a result.

Aflatoxins proliferate in peanuts and corn, particularly when the seeds are not well dried before storage, when water leaks into storage areas, and when storage insects feed on the seeds, making wounds that allow easy invasion by fungi. For instance, aflatoxin levels can increase from 200 to 2,000 ppb in just 3 days in high-moisture corn. Aflatoxins may remain even after the processing and baking of contaminated products.

When dairy cows eat contaminated grain, they metabolize aflatoxins into another highly carcinogenic compound that appears in their milk. Because cottonseed is a significant feed source for dairy cows, the occurrence of aflatoxins in cottonseed is of concern. Treatment of cottonseed with ammonia greatly reduces the level of aflatoxins, allowing it to be fed to animals.

Peanut growers in the southeastern United States have strict aflatoxin testing standards because of the continual risk of contamination of peanuts in their humid climate. *Aspergillus* species also can cause a serious lung infection, known as aspergillosis, with symptoms similar to those of tuberculosis.

Knowledge of the health effects of aflatoxins and other grain mycotoxins is relatively recent. The health effects of aflatoxins were first dramatically demonstrated in Great Britain in 1960, when 100,000 turkeys and many other birds died of so-called turkey X disease after eating a shipment of peanut meal from Brazil. The primary organ damaged by aflatoxins is the liver, and ingestion of a high level of aflatoxins has been demonstrated to cause liver cancer. Further evidence comes from epidemiological studies that have found high rates of liver cancer in Africa, where large amounts of peanuts are eaten.

The Food and Drug Administration (FDA) set the first action level (that is, permissible level) for aflatoxins in food products in 1965. Federal regulations on interstate shipments of grain in foods such as cornmeal and peanut butter set the legal limit at 20 ppb, since only part of our diet is potentially contaminated by aflatoxins. The same threshold is used in feed for dairy cows, but higher levels of aflatoxins (100–300 ppb) are allowed in feed for animals raised for meat.

The recent development of immunoassays for rapid detection of aflatoxins in food products has improved inspectors' ability to screen food products for safe aflatoxin levels. However, there is some concern about the 20 ppb threshold. Specifically, some scientists believe the limit should be more restrictive. Japan and some countries in Europe have rejected U.S. grain that does not meet their stricter aflatoxin standards. Controversy also exists about the accuracy of statistical methods used for sampling large quantities of grain, since pockets containing high levels of aflatoxins are common in large storage containers.

It is interesting to note that the Delaney Clause—which was added in 1958 to the Federal Food, Drug, and Cosmetic Act (FFDCA) to protect food supplies from carcinogenic food additives—did not apply to aflatoxins because they are "natural." Today, many scientists believe that the public health risk from aflatoxins is much greater than the risk from the low levels of pesticide residues found in or on food and thus a more pressing food safety concern. This is something to consider when purchasing "all natural" peanut butter.

High aflatoxin levels in corn (maize) crops occur periodically and are mostly associated with hot, dry conditions, which increase the production of *Aspergillus* conidia. Poor husk coverage of the tips of the ears exposes the corn to the conidia, and cracks in the corn kernels increase infection. Ear feeding by insect pests also increases infection by introducing conidia to the corn kernels.

In 1977, aflatoxin contamination of more than 60% of the corn grown in the southeastern United States resulted in losses totaling $200 million. In 1988, the severe drought in the Midwest resulted in aflatoxin contamination of 5–25% of the corn crop. One-third of the 1988 corn crop tested in Iowa and Illinois contained dangerous levels of aflatoxins. Milk had to be dumped in these and other midwestern states, after it was discovered that cows had eaten contaminated grain. A year later, increased rates of abortions in hogs were linked to aflatoxins in feed. In 2006–2007, high levels of aflatoxin contamination of this crop were found again in the hot and humid southeastern states. Since the average American consumes approximately 160 pounds of corn and corn products each year, aflatoxin contamination represents a significant public health concern.

In 1988, speculators bought aflatoxin-contaminated corn and held it until later in the year, when desperate cattle farmers were willing to pay higher prices for the contaminated corn because good feed was no longer available. The U.S. government allows grain that is excessively high in aflatoxins to be diluted with other grain to reach the 20 ppb threshold only under extraordinary situations, such as a summer with an extreme drought. In most cases, contaminated crops must be destroyed.

Aspergillus species are not the only fungi that produce mycotoxins in grains and other foods. *Penicillium* species are familiar causes of mold and decay of leather, books, and other materials stored in basements and damp closets. *P. italicum* and *P. digitatum* cause the common blue and green molds on citrus fruits. A number of *Penicillium* species produce mycotoxins that damage the liver, lungs, brain, and kidneys. Some are carcinogens as well. Several species produce the mycotoxin patulin, which is often found in cider squeezed from damaged apples already invaded by these fungi. *Penicillium* species also invade grains and contaminate them with mycotoxins.

Fusarium species are the third group of fungi that commonly invade grains and may spread under conditions of high moisture in storage. A "rule of thumb" has generally been that aflatoxins are a southern problem and *Fusarium* toxins are a northern problem, but both can be found in both areas. *Fusarium* toxins—including fumonisins, T-2 toxin, vomitoxin (DON), and zearalenone—are common in corn and other grains in late harvests after wet summers. Fumonisins in corn cause a disease of horses called "blind staggers." T-2 toxin was first identified as the cause of alimentary toxic aleukia in Russia in the early 1900s, but it and other toxins have probably been a health threat in moldy, poorly stored grains for centuries.

Another important toxin produced by *Fusarium* is commonly known as vomitoxin. This mycotoxin came to prominence during the rainy 1990s, with the terrible Fusarium head blight (scab) epidemics in barley and wheat fields that occurred in the Red River Valley and the Ohio River Valley, as well as other parts of the United States and Canada. The head blight epidemics were exacerbated by the use of no-till agriculture, which left more *Fusarium*-infested stubble in the fields. Since 1990, head blight has cost farmers more than $3 billion, of which $83 million has been caused by reduced prices due to vomitoxin contamination. In 1995, 16 million tons of dog food was recalled by a single producer, because it contained wheat contaminated with vomitoxin. Bird seed can be contaminated too. People who like to feed birds should take care that the seed does not become moldy in rainy weather and end up poisoning the birds.

Zearalenone, another common *Fusarium* mycotoxin, has estrogenic effects in animals, which reduce fertility. The presence of mycotoxins with estrogenic effects, as well as phytoestrogens found in many plants (including soy products), should be considered when determining the significance of synthetic chemical contaminants with estrogenic effects. The synthetic chemical contaminants get far more media attention.

By the time we purchase processed grain products that may be contaminated by mycotoxins, we can no longer detect mold, an off taste, or a change in odor. We must trust farmers, food manufacturers and processors, and government inspectors to appropriately harvest, store, and test the foods before they are made available for purchase.

Since many of the infections take place in the field before harvest, management strategies revolve around protecting the ripening grains and seeds. These strategies include good growing conditions, sufficient field drying, and harvesting techniques that reduce injuries to the seed coat. Once harvested, grain must be kept cool, dry, and free of rodents and insects, which cause wounds. Treatment with organic acids, such as proprionic acid, reduces the growth of fungi that produce mycotoxins, but treated grain can be used only as animal feed.

Besides careful harvesting and storage, some specific methods have been developed to reduce aflatoxin contamination. Some isolates of *A. flavus* do not produce aflatoxin and have been used in biological control in peanuts to competitively displace the more dangerous isolates. Recent studies of genetically engineered corn with BT insect-resistance genes have demonstrated that mycotoxin levels were significantly lower, because the corn borer larvae were prevented from making feeding holes, which would have allowed entry by *Aspergillus* and *Fusarium* species.

Most scientists agree that improved accuracy and speed in mycotoxin detection will result in improved healthfulness of the foods we eat. Mycotoxins continue to be a concern, however, in countries without funds for the technology needed to test and treat contaminated stored foods. Many developing countries have warm, humid climates, which favor the growth of mycotoxin-producing fungi, and lack the cold air of winter found in temperate climates for storing food.

The Food and Agriculture Organization (FAO) of the United Nation estimates that 25% of the world's food crops are significantly contaminated by mycotoxins. The media frequently report on environmental and food contamination by pesticides and other synthetic chemicals, but they rarely investigate many of these "natural" poisons, which are widespread and clearly threaten our health.

Edible Fungi: Mushrooms and Corn Smut

Humans all around the world eat fungi. It is well known that some fungi, such as the "death cap" mushroom (*Amanita phalloides*), are deadly, whereas others, such as the common meadow mushroom (*Agaricus bisporus, A. campestris*), are cultivated for supermarkets by the ton. A number of fungi are deliberately added to food and food products to enhance their flavor, apparently with no toxic effects.

Figure 10.6. Corn smut.

Knowledge about the edibility, toxicity, and medicinal value of plants and other foods has slowly accumulated over centuries of trial and error by people throughout the world. Mistakes can be deadly, as proven by a group of recently immigrated Southeast Asians in California, who died painful deaths after eating mushrooms that appeared the same as edible types found in their homeland. Other fungi are toxic only to certain individuals. Mushroom hunters who try a new species after its careful identification by an expert are cautioned to eat only a small amount at first. Some mushrooms, like the inky caps of the genus *Coprinus,* can cause gastric upset when consumed with alcohol, such as wine. It's not worth eating a mushroom unless you are certain of its identity. The "cure" for some mushroom poisonings is a liver transplant!

As humans foraged for foods, they sometimes found fungi growing on plants and ate the plants anyway. Just as it is impossible to judge the danger of a compound based solely on its origin in nature or the laboratory, it is impossible to determine the toxicity of fungi and fungal by-products based on their appearances.

Smut fungi are basidiomycetes, in which the entire mycelium turns into a mass of black spores, thus giving the fungi their common name. Perhaps surprisingly, one smut disease is actually prized by gourmets. Corn smut is a disease familiar to many home gardeners, because it affects sweet corn more than field corn. The disease is recognized by the silvery white, translucent galls that can be found nearly anywhere on the corn plant, including the tassels, ears, and leaf whorls (Figure 10.6). Corn smut usually damages the corn yield and quality only when it invades the ear. The galls produce huge quantities of black smut spores, which overwinter in debris and infect corn produced the next year. Yet despite the unappealing appearance of this fungus, it is edible and highly prized in its early stages, when the developing mycelium has colonized the gall tissue produced by the corn plant.

For example, corn smut has been cultivated in Mexico for centuries. The corn smut fungus is called by the scientific name *Ustilago maydis* and was named *huitlacoche* or *cuitlacoche* by the indigenous

Nahua people in Mexico (Figure 10.7). These Nahuatl names may translate to mean "excrement" of various forms, which corn smut resembles. Today, U.S. markets sell the fungus using the more appealing common name smoky maize mushroom. Farmers in Georgia, Pennsylvania, and some midwestern states cultivate corn smut for gourmet food markets, where the fungus sells on the corn ear for a retail price of approximately $50 per pound. Both traditional and upscale Mexican restaurants often feature corn smut in special dishes, and adventurous home cooks can order *cuitlacoche* online.

Figure 10.7. Portion of a mural painted by Guillermo Lourdes in the city of Durango, Mexico, showing the importance of corn in agriculture and in Mesoamerican culture. Young smut galls, which are used as food in parts of Mexico, can be seen in the background. (From a painted copy by Rudy Cruz.)

Truffles are another curious edible fungus fancied by gourmets. Truffles are fruiting bodies of parasitic ascomycetes produced on the roots of host trees. They produce a strong aroma, which attracts certain insects and animals. Truffle fungi rely on animals to detect their aroma, dig them up, and break them open, dispersing the spores. Somehow, people discovered these strongly scented fruiting bodies and began to eat them, usually slicing them thinly on various foods or flavoring sauces with them.

Some people train dogs and pigs to find and dig up the highly prized truffles (Figure 10.8). Pigs love to eat truffles and may not give up their prizes so easily, however. Many truffle hunters now use dogs, which can be trained to find the truffles but willingly settle for another kind of reward for their work. The truffle markets in Italy and France tend to be secretive and highly profitable. Individual truffles can sell for thousands of dollars. Black and white truffles, as well as truffle-flavored oil for seasoning, can be purchased online.

Farmers have tried to inoculate trees to produce reliable crops of these valuable fungi, but apparently the conditions must be precise for success. In the United States, pecan trees produce a species of truffle that sells for about $100 per pound.

Figure 10.8. Woman hunting truffles with a pig trained to sniff out and dig up the underground fungi.

Science Sidebar

Distasteful but safe

Many smut fungi that infect the seeds of cereal grains are not as obvious in appearance as the corn smut fungus. The common names for these diseases include common, dwarf, and karnal bunts. Common bunt also is called stinking smut.

These smut fungi are in the genus *Tilletia,* named for M. M. Tillet, a French biologist who in 1755 demonstrated the contagious nature of stinking smut. He dusted the black spores onto healthy grain because he believed a poisonous substance was associated with the dust. His work was a key step in the early study of fungi as plant pathogens.

Stinking smut spores survive in the soil and infect germinating grain seeds. The spores are released when infected grains break open during harvest, contaminating any newly harvested healthy grain. In years of heavy stinking smut infection, dusty clouds of smut spores are released (Figure 10.9). In some cases, the air has been be so filled with spores that farmers have been killed when sparks from harvesting machinery or at grain elevators have touched off explosions.

T. tritici and *T. laevis*—the fungi that cause stinking smut—do not produce mycotoxins, but smutty grain has an unappealing "rotting fish" taste and smell. Flour made from smutty grain is brownish gray because of the presence of the smut spores. The story is told that gingersnaps were invented to make use of flour produced from smutty grain. The bad taste was covered up with spicy ginger flavoring, and the brownish color was masked by adding molasses.

Figure 10.9. Release of stinking smut spores (*Tilletia* species) at harvest time. The spores contaminate the harvested grain and soil where they land.

Moldy Food: On Purpose!

Most of the fungi just described are large and relatively visible potential foods. It is not difficult to imagine a brave individual trying to eat them for the first time. More surprising is that some people have tried to eat foods contaminated by fungi, which most people would probably describe as "moldy." However, people have always been hungry, and hungry people will even eat moldy food when little else is available. In some cases, eating moldy food has resulted in delicious discoveries.

One very familiar fungus, *Penicillium roqueforti,* produces the blue-green color and distinctive flavor of blue cheese. It apparently originated as an accidental contaminant of cheese ripening in caves in France. The name Roquefort cheese is reserved for cheeses from that area of France, just as the name Champagne can only be used in France for sparkling wines from the province of Champagne. The same fungus is used to produce all of the green and blue cheeses, such as Gorgonzola, Stilton, and plain blue (or *bleu*) cheese.

To produce Roquefort cheese, spores of *P. roqueforti* are mixed with the cheese curds. Then, metal rods are used to make air holes in the mixture, in which the fungus develops. The mycelium and conidia of the fungus colonize the cheese, until the blue-green veins are visible throughout. Because the fungus makes so many spores and they are easily airborne, factories that produce blue cheese usually produce only blue cheese, rather than risk contamination of other kinds. Another species, *P. camemberti,* is used to produce the famous surface-ripened cheese known as Brie or Camembert.

The genus *Penicillium* also contains a number of species that produce powerful antibiotics, known as penicillin, for medicinal purposes. *P. notatum* and *P. chrysogenum* are the most prominent species in commercial antibiotic production. Neither of the *Penicillium* species used in cheese produces penicillin, so people who are allergic to penicillin can safely eat these cheeses.

Aspergillus is another common fungal genus that is responsible for mold and decay of foods and stored materials but also includes species used in food production. One species of the genus *A. oryzae* is added to rice and soybeans and allowed to ferment for the production of sake and soy sauce, respectively. Citric acid, produced by *A. niger* in large industrial vats, is used to flavor cola drinks.

Gray mold is a common disease of small fruits, flowers, and the older leaves of many kinds of plants. The distinctive brownish-gray mold is visible on infected plant materials in humid weather. The fungal pathogen is *Botrytis cinerea,* which produces prolific conidia for wind dissemination and sclerotia for survival in plant debris. This mold is frequently seen on raspberries and strawberries, as well as the aging leaves and flowers of chrysanthemums, geraniums, marigolds, and petunias. The moldy strawberry commonly found in the bottom corner of the box from the grocery store almost certainly is infected by *B. cinerea* (Figure 10.10).

Figure 10.10. Strawberries with postharvest development of gray mold fruit rot caused by *Botrytis cinerea.*

This pathogen also is a common parasite of grapes, and integrated pest management is practiced in many vineyards to minimize Botrytis blight. However, at times, winemakers use

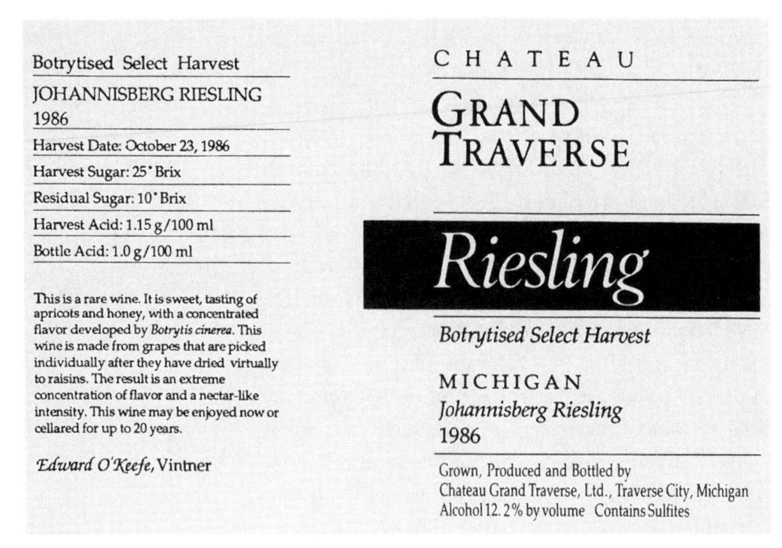

Figure 10.11. Label for a wine produced from grapes infected by *Botrytis cinerea,* the "noble rot."

B. cinerea for a special purpose. Under the ideal conditions of cool night air, humid soils, and sunny, warm afternoons, infected grapes become sweeter, as the natural sugars and tannins become concentrated while the water content of the grapes decreases. Once infected by this "noble rot," the raisinlike grapes are hand picked and used to make special types of wines—particularly dessert wines (golden Sauternes) and Riesling, Sauvignon Blanc, and Semillon.

Only certain areas have the correct environmental conditions for producing infected grapes without developing the destructive gray mold disease. The fungus is endemic in the vineyards and therefore develops naturally in areas conducive to its growth and spread. Today, the fungus is often deliberately inoculated on grapes. The infected grapes must be harvested over a period of time, as different grapes reach the proper stage for wine making. *Botrytis* gives a distinctive taste to the wines that bear the name of the fungus or have the words "botrytised" and "late harvest" on their labels (Figure 10.11).

A Healthy Diet: Relative Risk

In the past few chapters, we have examined a number of aspects of modern food production that must be evaluated for their relative risk, including the use of genetic engineering and pesticides in food production and the presence of mycotoxins in our food supply. As we consider how to address these issues and how much of our limited resources to spend on them, we must be certain to keep them in perspective.

As noted in Chapter 8, the risks posed by modern agricultural production are relatively minor compared to other risks we face in our daily lives. In the United States, a young adult is most likely to die in an accident (especially a motor vehicle accident) or by a homicide or suicide. Among all ages, the most common causes of death are heart disease and cancer. These causes of death are linked to medical and social problems to which time and money should be devoted for studying causes and identifying solutions. For example, certain life-style choices, such as smoking and the use of tanning beds, have been clearly linked to increased cancer risks. An advisory committee for the 2010 Dietary Guidelines for Americans called obesity "the single greatest threat to public health in this century." The committee's report calls for people to cut back on their consumption of added sugars and solid fats and to follow a more nutrient-rich, plant-based diet.

We do want an abundant, varied, and safe food supply. Most media reports about food safety focus on the risks related to what are considered "unnatural" changes in crops. Genetic engineering and pesticides receive much of the attention. Modern crop plants are already very different from their wild relatives, and they change continuously, as plant breeders try to find plants that give us resistance to pathogens and pests and variety in our diet. These natural gene recombinations may potentially affect food safety, just as small changes caused by genetic engineering may. The chemicals that make plants genetically resistant or that result from systemic acquired resistance (SAR) are potentially as toxic or carcinogenic as any detectible pesticide residues.

Media reports rarely give consumers the information they need about which crops have pesticide residues, at what levels, and of what chemical types. Somehow, the word chemical has come to represent the dangerous, synthetic molecule, compared to the safe, natural one. In fact, everything in the world is composed of chemicals, some of which are dangerous and others without which we could not survive. Some natural microbial contaminants of food sicken and even kill people. Both human pathogen contaminants, such as some strains of *Escherichia coli* and *Salmonella* species, and fungal toxins, such as aflatoxins, are important natural threats to our food supply.

Food safety regulation by the federal government is popular with most Americans. Regulations are in place both for compounds in foods that occur naturally in the original plants, animals, fish, and fungi and for potential contaminants, such as mycotoxins and pathogens. Food additives also are regulated, including preservatives, colors, flavors, and pesticide residues. The regulation of these potential health hazards is based on their biological activities as carcinogens, mutagens, teratogens, and toxins. These effects are the results of their molecular structures, not their origins. Some "natural" compounds pose distinctly greater health risks than many "synthetic" compounds. Risk assessment must address the dose, frequency, and biological activity of a given compound.

Many scientists believe that considerably more study and screening should be done of grains and seeds with the potential to be infected by mycotoxin-producing fungi. The concept of relative risk requires evalauting all potential chemicals in our food supply, both natural and synthetic, using the same criteria.

It is interesting that no herbal medicines and few dietary supplements have been regulated by the FDA since passage of the Dietary Supplement Health and Education Act in 1994. The manufacturers of these products do not have to prove they are safe or effective before marketing them. Herbal medicines, specifically, are considered foods. The FDA investigates and sometimes recalls or otherwise regulates a product only after a problem occurs with it.

Certain active ingredients in plants have physiological effects that can improve our health. In fact, many modern pharmaceuticals were discovered in plants. However, ingesting an herbal medicine is very different from taking a pill that contains a measured dose of a particular compound. Plants vary considerably in their medicinal content, which may be more or less than expected. Many herbal medicines consist of dried leaves, in which the medicinal compounds, as well as a wide range of other plant chemicals, have been concentrated.

Herbal medicines and dietary supplements are allowed to make only very broad claims about improving health and also must include this caveat: "This statement has not been evaluated by the U.S. Food and Drug Administration (FDA). This product is not intended to diagnose, treat, cure, or prevent any disease." The lack of regulation of these products is probably related to the fact that so many people believe that "natural" means "safe." Many companies are ready to exploit this fallacy for profit.

Finally, it is possible that reducing the use of some pesticides may actually increase health risks. For instance, foods that are not protected by fungicides are more likely to have spots and blemishes, which are vulnerable to infection by mycotoxin-producing storage fungi. Thus, produce labeled "organic" may not necessarily be safer than commercially grown produce.

Optimal storage conditions and protection against infection by storage fungi are requisites of a healthy food supply. In temperate climates, the cool storage of food supplies is possible during the winter for free. In tropical climates, which have warm, moist weather all year, losses of food to storage fungi are much higher, as is the danger posed by mycotoxin production. Creating dry and well-ventilated food storage structures is a high priority in the development of a healthy reserve food supply in some of the world's most populated areas, including countries in Africa, Asia, and Central and South America.

CHAPTER 11

Ancient and Present-Day Foes: The Rusts

"Nobody is qualified to become a statesman who is entirely ignorant of the problems of wheat." —Socrates

Farmers have battled rust fungi since agriculture began. Long before humans understood the role of pathogenic fungi in plant diseases, and certainly before anyone considered the activities of soilborne pathogens, people had studied and recorded the strange rust-colored eruptions on the leaves of diseased crops. The rust fungi were named for the color of their powdery spores. Like the fuzzy, white growth on potato leaves with late blight, the rust-colored spores on infected plants were believed to be the result, rather than the cause, of the shriveled grains and reduced yields.

Robert Hooke, the 17th-century physicist famous for his theory that all living organisms are composed of cells, spent considerable time studying minute things with the newly invented microscope. He described and drew molds on leather and also published the first illustrations of rust on a rose leaf in 1665. To Hooke, the rust appeared to be tiny plants similar to mosses. Over the next 200 years, many scientists studied the amazing structures produced by rust fungi and began to recognize the clues to their biology and their role as plant pathogens.

Historical Records: From the Beginning

Although nearly every kind of plant important to humans has a rust fungus pathogen, the grain rusts are the ones described most often in history. Archeological digs in Israel discovered wheat stem rust spores on grain dating from 1300 B.C. Wheat (*Triticum aestivum*), an important crop in ancient times, continues to be a major crop in the modern world. At least $5 billion is lost to cereal rusts each year, and even more is lost in years with severe epidemics.

Wheat, barley (*Hordeum vulgare*), and oats (*Avena sativum*) have their centers of origin in Asia Minor, and their associated rust fungal pathogens originated there as well. As the early farmers began to select wheat grains and plant them in plots, the rust fungi took advantage of the feast. Unlike the pathogen that causes late blight of potato, which can leave a potato crop a rotted mass of vegetation, rust fungi are more delicate pathogens. They carefully obtain their nutrients without destroying the entire plant. Yet their presence takes its toll, reducing yield and weakening stems just before harvest. Fortunately, the grain from rust-infected plants is not toxic, so farmers throughout history have accepted their losses and harvested what was left.

Rusts are described in many ancient writings that mention crops and agriculture. Despite some confusion in translation, the Bible seems to mention rusts, as well as blasts, blights, mildews, and smuts. In some cases, the problem may have originated from hail or a storm, but in other cases, it is very clear that a fungus was observed on the plants. The fungus was still considered the result, rather than the cause, of the plague. Fungi that cause powdery mildew, rust, and smut are probably the pathogens that would have been most obvious to early farmers.

The story of the escape of the Jewish people from Egypt and its celebration in the Passover holiday is familiar to many people. In *Famine on the Wind,* Carefoot and Sprott suggest that wheat stem rust may have played a role in how the Jewish people came to be in Egypt. In the Bible, Joseph—of the "coat of many colors"—is sold into slavery by his brothers. He eventually interprets the pharaoh's dreams as a prediction of a coming famine, for which the pharaoh should prepare by storing food. The famine is described as a blast coming from an east wind, perhaps suggesting rainy weather, which would favor rust disease. Because of the famine, Joseph is reunited with his father, Jacob, and his brothers. They are given the land of Goshen and eventually become enslaved by the pharaoh.

The ancient Greek writings of Aristotle and his student Theophrastus also describe rust diseases of crops. In particular, Theophrastus, who is often called the "father" of botany, noted that plant species vary in their susceptibility to rust diseases. He also recorded the role of environmental effects, such as moisture and air movement, in the development of rust epidemics.

In Roman times, rust diseases continued to plague farmers. The rust-colored spores seen in the early stages and the black spores that appeared later in diseased wheat fields were all too familiar to Roman farmers. The god Robigo/Robigus was honored in the Robigalia, a religious ceremony practiced for more than 1,700 years that involved sacrifices of reddish-colored animals, such as dogs and cows. The origin of these rituals may lie in the belief that humans were being punished with rust-infected crops because a live fox had cruelly been set on fire. A traditional fox hunt also was held during the Robigalia, in which torches were tied to the tails of foxes. The red and black colors of wheat stem rust are repeated in many of the symbols associated with these ceremonies: red dogs and cows, foxes, bloody sacrifices, and fire (Figure 11.1).

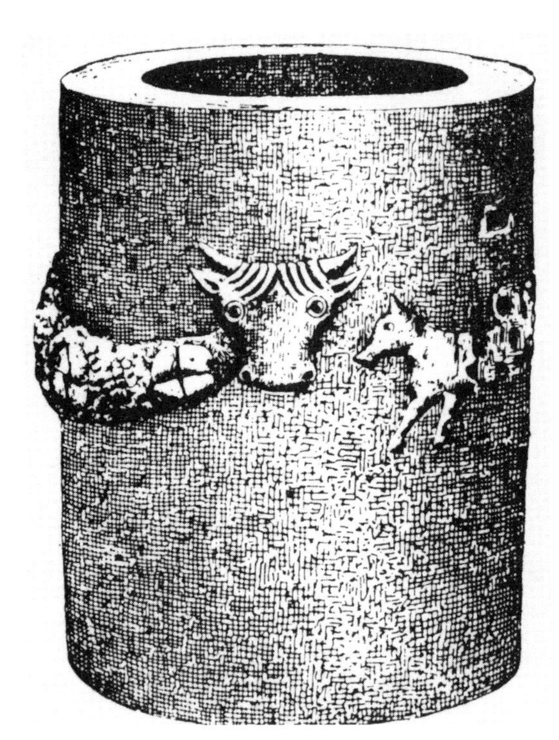

Figure 11.1. Drawing of the Robigo altar found at Castiglioncello, Italy.

These ceremonies probably arose from folk beliefs dating back to the beginning of recorded history, more than 3,000 years ago, as farmers looked for a reprieve from the rust epidemics. The Roman calendar included three agricultural holidays in the spring: the Cerealia (April 12–19), the Robigalia (April 25, about the time wheat was heading and needed protection from the rust god), and the Floralia (April 28). The early Christian calendar, which absorbed the traditions of many ancient holidays, included Rogation Days around April 25 as a time for blessing the crops. Since 1978, Rogation Days have no longer been officially required, but days of prayer for crops are still held in several European countries.

Grain crops have been the focus of agriculture in temperate climates for many centuries. Wheat and rice are the two important grains used primarily for human consumption, rather than as animal feed. Harvested grain is easily stored and can be ground and baked in an enormous variety of forms. The gluten in wheat allows dough to rise when yeast is added. Bread has provided a social and religious symbol for centuries. The word companion, which means "a person with whom one eats bread," is used to describe a friend. The motto of the United Nations' Food and Agriculture Organization (FAO) is *Fiat panis,* which translates to "Let there be bread." Grains also are an important trade item that can be shipped around the world.

Modern-day wheat, the traditional "staff of life," is grown throughout temperate regions in small plots as well as huge fields measured in square miles (kilometers), rather than acres (or hectares). Today, many classes of wheat are grown. They are often designated by the time of planting (winter, which is planted in the fall, versus spring wheats), kernel color (red versus white wheats), and kernel hardness (hard versus soft wheats). The specific uses vary according to the type of wheat. For example, bread in the United States is usually made with hard, red, spring or winter wheats.

Cultivation of the ancient relatives of modern wheat—grasses called emmer and einkorn—began in approximately 8000 B.C. in an area of the Near East called the Fertile Crescent. Cultivation of modern wheat spread slowly throughout southwest Asia and Europe, coming to the Americas with the European settlers. The cultivation of wheat was an important factor in the change from hunter-gatherer societies to farming communities.

Rust epidemics have been part of wheat production since the beginning. As agriculture has become more modern and fields have become larger and more genetically uniform, the risk of loss to rusts has increased (Figure 11.2).

In ancient Italy, which had no natural barriers separating it from Asia Minor, the Romans suffered from rust epidemics whenever weather conditions were conducive. Carefoot and Sprott, in their book *Famine on the Wind,* suggest that unusually wet weather in the early Christian era may have contributed to the decline of the Roman Empire. They

Figure 11.2. Modern-day wheat harvest in the United States.

also note that rust was probably not a major problem in western Europe in the early days of wheat production, because grains were carried over the Alps mountain range and grown in isolation. As travel between the Middle East and Europe became more common, however, a shrub—the common barberry (*Berberis vulgaris*)—was established in western Europe, and wheat stem rust epidemics occurred. The barberry has played an important role as an alternate host in wheat stem rust epidemics, but it took more than 2,000 years for scientists to discover the relationship between barberry and rust.

Barberry and Wheat Stem Rust: A Mystery Solved

Farmers who observe crops firsthand, year after year, are often able to draw empirical conclusions about agricultural problems, even without a scientific education. In the case of wheat stem rust, they looked beyond prayer for a solution to the rust epidemics. Farmers noticed that rust was often most severe near common barberry bushes.

In 1660, in Rouen, France, the first legislative act concerning a plant disease was enacted to help reduce wheat stem rust by eradication of barberry plants. This law was enacted nearly 200 years before the Irish potato famine and more than 200 years before Anton deBary, the famous mycologist, discovered the biological connection between barberries and wheat stem rust.

In the meantime, European colonists had settled in the New World, bringing with them both barberry and wheat plants. Common barberry was grown as a medicinal plant and served as a source of tool handles and yellow dye for the colonists. It also provided a fast-growing hedge to contain animals, and its berries were used in sauces and jellies. The Japanese barberry (*B. thunbergii*), so common in today's landscape plantings, is in the same genus but is not susceptible to the wheat stem rust fungus.

Following enactment of the legislation in France, laws were passed in the American colonies of England in the 1700s forbidding the planting of barberry near wheat fields. Barberry laws were recorded for Connecticut in 1726, Massachusetts in 1754, and Rhode Island in 1766. Attention to these laws faded, however, as grain production moved to the western territories.

Despite discovery of the microscope and passage of the barberry laws in the 1600s, an understanding of rust epidemics was slow to develop. One complicating factor was that a single rust species can produce up to five different spore stages on two unrelated host plants. Scientists began to notice similarities of spore types among the rust fungi found on various host plants, even though all the fungi had been given individual Latin binomials. At this point, scientists did not yet understand how the various spore types were connected in the complex disease cycle.

Around 1860, deBary had been studying a rust fungus that infected garden beans. He discovered four different types of rust spores and, through careful study, determined the order in which they were produced. deBary found that, as with many rust fungi, a dark, thick-walled spore (teliospore) is the survival spore for bean rust. In the spring, four thin-walled basidiospores, the products of meiosis, are produced by each teliospore as primary inoculum. Because rust fungi produce basidiospores, they are members of the basidiomycetes, just as mushrooms and smut fungi are. Shortly after infection of the plant by basidiospores, small, vase-shaped structures (pycnia) are produced. They protrude through the upper epidermis of the leaf. Some weeks later, cuplike structures erupt through the lower epidermis, producing chains of aeciospores, which are dispersed by wind.

During the growing season, one additional type of spore is produced: the urediniospore. It is produced in pustules that break through the leaf epidermis. Each tiny pustule can pro-

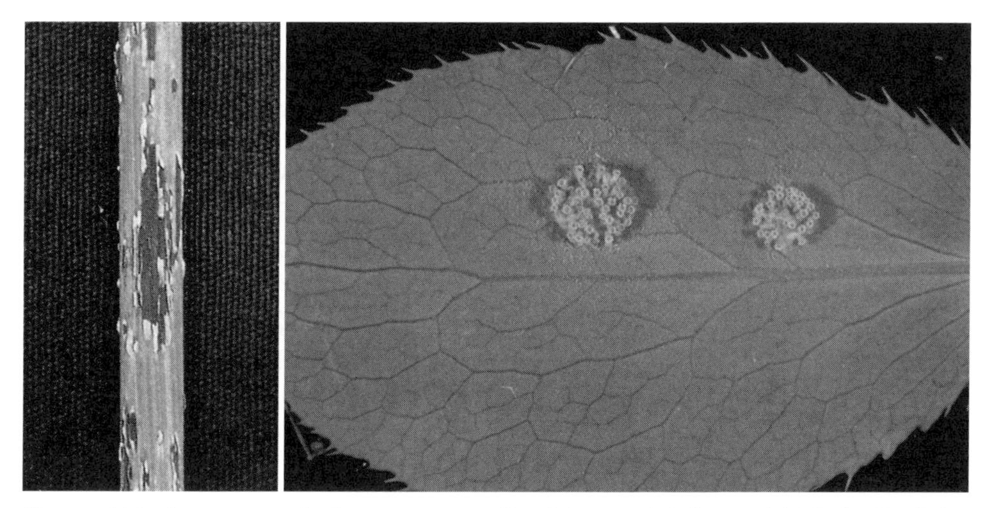

Figure 11.3. Symptoms of wheat stem rust, *Puccinia graminis* f. sp. *tritici*. **Left,** Urediniospores on wheat plants are the repeating stage, which causes polycyclic epidemics in wheat fields. **Right,** Aeciospores appear on the lower surfaces of leaves of common barberry. Aeciospores infect wheat.

duce 2,000 urediniospores per day for two to three weeks (Figures 11.3 and 11.4). The common name for the rust diseases comes from the large number of orange- to red-colored urediniospores that are produced.

Urediniospores are responsible for the terrible rust epidemics, so we will pay special attention to them. The inoculum potential of these fungi is enormous, which makes it easy to imagine how quickly an epidemic could develop. This is one reason that wheat stem rust was investigated for use in biological warfare, as discussed in Chapter 3. Toward the end of the season, the mycelium of the rust fungus stops producing urediniospores and begins to produce dark teliospores for overwintering, completing the life cycle.

After studying the bean rust fungus, deBary turned his attention to the more confusing and economically important disease found on wheat. He harvested the thick-walled black teliospores, allowed them to germinate to produce basidiospores, and inoculated healthy wheat plants, as he had done with his garden beans. He was puzzled by the lack of infection. He then considered the long-time belief held by farmers that common barberry somehow played a role in rust epidemics. In 1865, deBary inoculated barberry leaves with basidiospores that had been produced from the teliospores found on wheat plants. The basidiospores infected the barberry, and the missing spore stages were produced on that host, completing the disease cycle deBary had observed in his bean rust studies. Pycnia were produced on the upper leaf surfaces of the barberry, followed by the production of aeciospores on the lower leaf surfaces.

Figure 11.4. Urediniospore pustules of *Puccinia graminis* (electron micrographs).

The aeciospores, though produced on barberry, infect only wheat, not barberry. For the first time, the amazing phenomenon known as a heteroecious disease cycle had been identified. The parts of the word heteroecious indicate that the fungus has "different houses." Two unrelated host plants, wheat and barberry, are necessary for the production of all spore stages of the wheat stem rust fungus. This discovery led rapidly to the matching of various host pairs, and the alternation between different hosts was found to be quite common for rust fungi.

Biology of Rust Fungi: A Function for Each Kind of Spore

A complete understanding of the wheat stem rust disease cycle was not accomplished until J. H. Craigie's work was published in 1927. Although deBary had observed fungal structures on barberry, he had not understood their function. Craigie, a Canadian scientist, finally discovered their role in introducing variability into the rust fungus population.

Studies of chromosome numbers have demonstrated the following life cycle (Figure 11.5): The basidiospores of all basidiomycetes, including rust fungi, are haploid. After germination, they produce haploid structures called pycnia, which consist of tiny spores (pycniospores) and special receptive hyphae of a single mating type. Cross-fertilization between two mating types on different pycnia is necessary for the life cycle to continue beyond this point. As with the ergot fungus, honeydew is produced by the pycnia to attract insects. Insects or water transfer pycniospores of one mating type to the receptive hyphae in pycnia of the other mating type. Successful fertilization brings two genetically different nuclei into the same hypha.

As we have seen in previous examples from the fungal world, nuclei may exist in the same hypha without fusing. Fertilization in the wheat stem rust fungi results in a dikaryotic mycelium, which grows through the barberry leaf tissue, producing dikaryotic aeciospores on the lower leaf surface. In heteroecious rust fungi, the aeciospores infect the other host plant species, producing more dikaryotic mycelium. Aeciospores are always the spores that move from one host plant to the other species—in this case, wheat. Later, urediniospores, still dikaryotic, are produced on the wheat, but they infect only the plant species on which they are produced. Because urediniospores can cause repeated infection of the same plant or infection of neighboring plants of the same species, they are referred to as the repeating stage. This is the spore stage that causes the damaging, polycyclic epidemics so well known to ancient and modern people alike.

The fungus remains dikaryotic until the black teliospores are produced later in the season. Within this spore, karyogamy (fusion of nuclei) finally occurs, and a diploid nucleus is produced. Meiosis follows karyogamy, and haploid basidiospores are produced in the spring, completing the life cycle. Basidiospores of the wheat rust fungus can infect only barberry.

Heteroecious rust fungi always require two host species. The term alternate host is usually reserved for the plant with the lower economic value. In the case of wheat stem rust, barberry is considered the alternate host, because wheat is the plant that is valued more.

Biology of Rust Fungi: Unique Pathogens

It is easy to become so absorbed in the complex life cycles of rust fungi that we fail to appreciate other aspects of these interesting pathogens, which produce so many kinds of spores. Rust fungi have been the subject of many studies, but this work is challenging. Rust

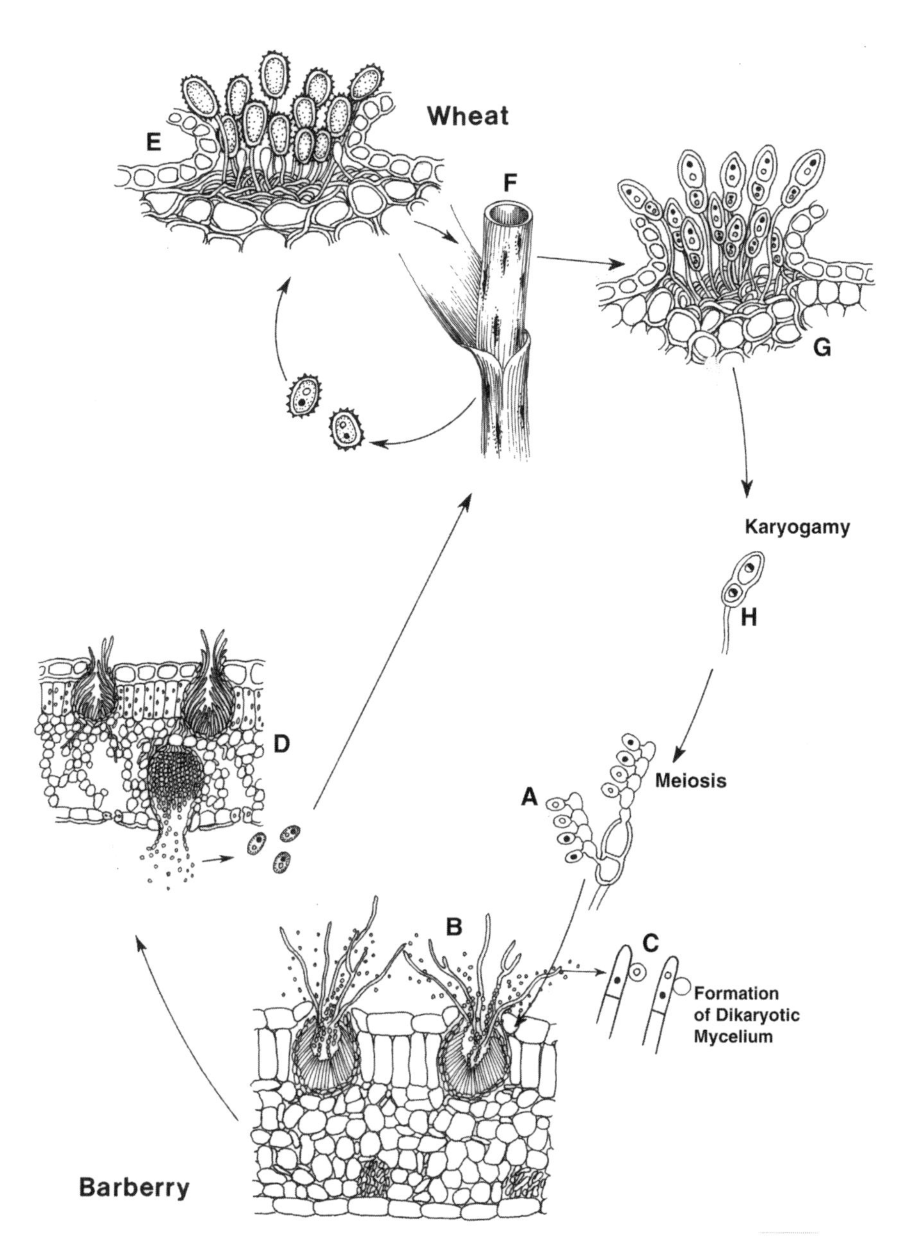

Figure 11.5. Disease cycle of stem rust of wheat (*Puccinia graminis* f. sp. *tritici*). The rust fungus life cycle begins with haploid basidiospores that infect common barberry (**A**). Barberry infections result in pycnia composed of pycniospores and receptive hyphae (**B**). Pycniospores contribute a nucleus to a receptive hypha from a different mating type, which initiates the formation of a dikaryotic mycelium (**C**). Dikaryotic aeciospores are produced on the lower leaf surface. Aeciospores infect wheat (**D**). Dikaryotic urediniospores develop on wheat stems. They infect wheat and are known as the repeating stage, which causes polycyclic epidemics (**E**). Pustules of rust spores break through the epidermis of the wheat stem, weakening it (**F**). Later in the season, the same dikaryotic mycelium forms thick-walled teliospores that are capable of surviving harsh winter conditions (**G**). Karyogamy and meiosis occur in teliospores, which result in the formation of haploid basidiospores in the spring (**H**).

fungi are obligate parasites and quite host specific. Only a few have been grown on complex artificial nutrient media, so most have been studied in the tissue of their host plants.

Rust fungi invade plant tissue intercellularly (between the cells), causing little visible damage in the early stages. They obtain nutrients by absorption via special feeding organs called haustoria, and little necrosis occurs. Because the host tissue is not macerated or degraded to any great extent, histological studies can clearly observe hyphal and haustorial development and subsequent spore production. Few other fungi, except perhaps the powdery mildews, have been studied genetically in such detail and had the corresponding genetics of their hosts determined. H. H. Flor, whose work in the early 1940s led to the gene-for-gene hypothesis, conducted his studies with flax and its rust pathogen, *Melampsora lini*.

Many fascinating studies have been conducted to try to understand how rust spores identify and penetrate their hosts. Urediniospores infect plants through their stomata. The germ tubes appear to locate stomata using both chemical and physical signals. They can locate stomata even on an artificial plastic cast of the host plant's epidermis. Studies also have shown that if the epidermis is stripped away, the lack of physical clues causes germ tubes to wander over the surface of the host plant without penetrating it because they cannot identify their host. Studies using artificial surfaces have suggested that the germ tubes can discern both the distance between epidermal cells and the height of the lip of the stomatal guard cells as signals to guide them to an appropriate penetration site.

It is amazing to consider the genetic information necessary for such complex interactions between host and pathogen. Not only are these studies interesting, but they also may have very practical purposes. If scientists can understand how a fungus finds a penetration site, they may be able to prevent infection by confusing a germ tube, either chemically or through crop-breeding procedures that alter epidermal patterns.

Rust fungi that produce all five spore stages are called macrocyclic. The heteroecious, macrocyclic life cycle is considered the original rust fungus life cycle that evolved early. Without going into great detail, some evidence for this theory is that the rust fungi found on primitive plants tend to be heteroecious and macrocyclic. Examples include those that alternate between conifers, such as firs and larches, and other hosts, such as birches and ferns. Other evidence comes from fossils of rust fungi parasitizing plant tissues. Fossils of soft fungal structures are generally rare. Study of rust fungi in leaf fossils is possible because the fungi are obligate parasites, which leave host plant tissue fairly intact, so that structures of both the host and the pathogen can sometimes be seen.

In the last ice age, which occurred 20,000–60,000 years ago, many kinds of plants were eliminated from the British Isles. A number of birches and ferns eventually returned to Great Britain approximately 12,000–14,000 years ago. Because these plants hosted the repeating stage, their rust fungi continued to maintain themselves over the years. When firs and larches were later reintroduced near sites with birches and ferns, the fungi were able to reestablish a complete life cycle, in spite of the long separation from their alternate hosts. Thus, the host range of rust fungi seems able to remain quite stable.

Managing Cereal Rust Epidemics: Protecting the Staffs of Life

As wheat production in North America moved into the western prairies of both the United States and Canada, the common barberry followed. The edible berries of this shrub were carried by birds to new locales, and people who were probably not wheat farmers planted the shrub, contributing to its spread. Wheat stem rust epidemics occurred every year and were particularly severe when the weather was moist and warm. New generations of urediniospores are produced approximately 7 to 14 days after infection, so these polycyclic pathogens could multiply quickly under favorable weather conditions.

In 1916, 200 million bushels of wheat were lost to wheat stem rust in the United States, and an additional 100 million bushels were lost in Canada. The U.S. government decided that wheat production was important not only to feed the nation but also to provide national security. Wheat was needed to supply the Allies fighting in World War I. In 1918, an intense federal program of barberry eradication was initiated at the urging of E. C. Stakman, a Minnesota plant pathologist. The program continued until 1975–1980, when it was gradually turned over to individual states (Figure 11.6). During the program, more than 3 million barberry bushes were eradicated in Illinois alone. The current cost of the program is approximately one-hundredth of a cent per acre of small grains. The goal of the eradication program is to remove the alternate host of the wheat stem rust fungus, *Puccinia graminis* f. sp. *tritici.*

There are actually three common rusts of wheat. *P. graminis* f. sp. *tritici* causes the most important disease: stem rust or black rust of wheat. Grain yield is reduced in both quantity and quality in infected plants. Worldwide, 1.1 million tons (1 million metric tons) of wheat are lost to stem rust annually. In addition, the numerous pustules that break through the surface of the stem weaken the plant and cause it to lodge, or fall over, which can make harvesting difficult or impossible.

P. graminis exists in more than 300 forms that are identical in morphology (appearance) but differ in host range. The *forma specialis* that infects wheat, barley, and some wild grasses is called *tritici.* There are six other *formae speciales* of this important species, all of which share the same woody alternate host: common barberry. For example, *P. graminis* f. sp. *poae* causes rust on Kentucky bluegrass lawns, and *P. graminis* f. sp. *secale* infects rye.

A second rust disease of wheat and other grasses, called stripe rust or yellow rust, is caused by *P. striiformis.* The name stripe rust comes from the lines of spore pustules that form stripes on infected leaves. The alternate host of *P. striiformis* f. sp. *poae,* which infects Kentucky bluegrass, was discovered in 2009, when it was found that aeciospores of this pathogen were produced on several species of barberry. *Berberis* species also serve as alternate hosts for the wheat stripe rust fungus and may explain the production of new races of the fungus during the sexual cycle. Of further concern is the fact that *P. striiformis* may be able to infect barberry species commonly used in ornamental plantings.

Figure 11.6. Spraying common barberry (*Berberis vulgaris*) with herbicide to eliminate the alternate host of *Puccinia graminis* in Erie County, Pennsylvania (1952).

Stripe rust is the most important wheat rust in Europe, where the climate is cooler and more humid than in the United States. In 1956, stripe rust accounted for a loss of 70% of the winter wheat crop in the Netherlands, and in 2010, stripe rust destroyed nearly one-third of the wheat crop in Syria. New stripe rust strains with increased virulence have been reported in recent years in Australia, Europe, India, and North America. In the United States, stripe rust is a problem mostly in the Pacific Northwest, although serious epidemics have recently occurred in the Great Plains.

A third wheat rust, caused by *P. recondita,* has meadow rue as its alternate host. The fungus causes brown leaf rust, which can be severe in certain years. It is usually the most important wheat rust in the Great Plains, because most wheat cultivars have genetic resistance to stem rust.

Effects of Barberry Eradication: Success or Failure?

What biological and epidemiological effects can be expected from the eradication of barberry? If the spore stages found on barberry are removed, the only spores left to consider are those produced on wheat: urediniospores and teliospores. Teliospores may still be produced, but the resulting basidiospores will have no host to infect after barberry has been removed. Thus, the only functional spores left in the disease cycle are the urediniospores, which are destroyed by cold winter weather in the northern United States and Canada. Does this mean we no longer have wheat stem rust?

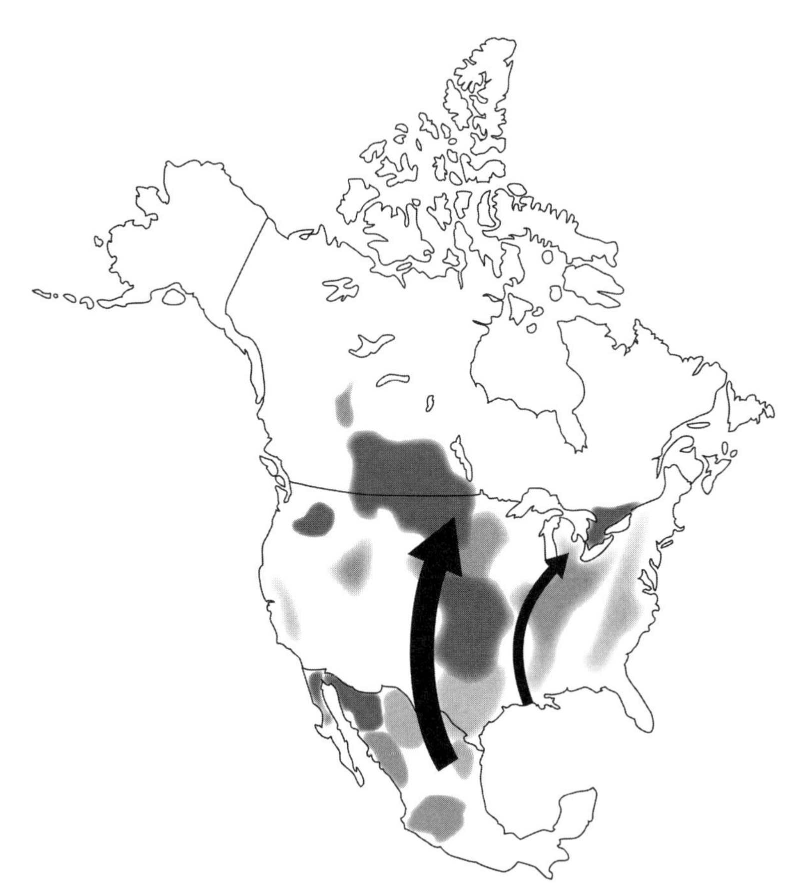

Figure 11.7. The Puccinia pathway. Each year, urediniospores are blown from southern states and Mexico to Canada. Shading indicates amounts of wheat production in various areas, with darker areas having higher production.

Unfortunately, wheat stem rust still exists. Barberry eradication failed to eliminate this rust from the North American continent, because the wheat-growing areas in the southern United States and Mexico—where winters are not as harsh as on the northern plains—serve as an annual source of urediniospores. Each year, the spores are carried in winds from field to field in the long journey north. Thus, rust infections are found throughout the Puccinia pathway, which stretches across grain-growing areas from Texas and Mexico north to the wheat lands of Canada (Figure 11.7).

Was the eradication program useless? It was not a total success, but it greatly improved farmers' ability to manage wheat stem rust. Removal of the barberry delays epidemics for several crucial weeks, since the inoculum must make its way north over many hundreds of miles each year. A single barberry bush can be the source of 64 million aeciospores, so a significant source of inoculum was removed when the barberries were eradicated. Most importantly, the eradication program broke the sexual cycle of the fungus. In the absence of barberry, asexual urediniospores are the only functional spores in the disease cycle.

Based on the number of known resistance genes in wheat plants, it has been estimated that the wheat stem rust fungus has the potential to produce 56,000 races. This potential genetic variation would be overwhelming to any breeding program. Without the normal sexual cycle, however, the major source of genetic change is mutation. Mutation can still be a significant source of variation for a fungus that produces thousands of spores in a single rust pustule. However, in 1918, before the eradication program, 28 different races were detected and seven races predominated, whereas today, approximately five races are commonly found and only one or two predominate. Even though the barberry eradication program did not eliminate wheat stem rust, it produced other less obvious benefits in managing the disease.

Genetic Resistance to Rusts: The Never-Ending Battle

Single-gene resistance remains our main line of defense against the cereal rusts. The long-term goal in breeding programs is to produce a durable, general resistance effective against all races of the rust fungus. This is sometimes referred to as "slow rusting," because it reduces the rate of an epidemic. In the meantime, how can single-gene resistance be effective against a polycyclic pathogen that invades such large expanses of cropland? The general use of single-gene resistance has been a matter of much theoretical speculation and practical experimentation.

Some scientists recommended creating a single cultivar with a number of single resistance genes, since it would be unlikely that the fungus could make all of the necessary genetic changes to cause infection. The accumulation of resistance genes in a single cultivar through multiple crosses is called pyramiding. Although most scientists agreed this would likely be successful, they also recognized that the use of such cultivars would put selection pressure on the fungus, perhaps resulting in the development of a so-called superrace. All of the resistance genes used would then be useless.

Other scientists proposed the geographical deployment of different resistance genes in bands south to north, so that arriving rust spores would be met by varying resistance genes throughout the Puccinia pathway, thus preventing major epidemics. To be successful, this approach would require the complete cooperation of growers, seed producers, and state seed associations and coordination by an efficient and ongoing international bureaucracy. Such gene deployment has been successfully used for oats against the crown rust fungus (*P. coronata*), which has buckthorns (*Rhamnus* species) as its alternate hosts.

In some areas, oats are planted with physical mixtures of seeds of various cultivars that have similar characteristics and vary primarily in a single resistance gene. These mixtures are called multilines. One idea behind the use of such mixtures is that the pathogen will meet a variety of resistance genes, somewhat like what might occur in a natural ecosystem. Another hypothesis behind using such mixtures is that no superrace will arise because of something called stabilizing selection. Evidence suggests that "extra" genes make races less fit and less competitive than the common wild types of races. Thus, the use of multilines keeps disease at a low level and does not select for superraces. Multilines have the added advantage of allowing different resistance genes to be included in the mixtures from season to season.

Although these ideas remain speculative and controversial, each year, wheat plant pathologists maintain plots called "rust gardens," consisting of a number of wheat cultivars with known resistance genes, to determine which rust races predominate. Resistance genes

necessary to complete a life cycle: teliospores and basidiospores. Because of its minimal disease cycle, the hollyhock rust fungus is called microcyclic. Carnation, chrysanthemum, and geranium rusts can be important diseases in greenhouses, where splashing of overhead irrigation water can cause rapid spread of the urediniospores, even though their alternate hosts are thus far unknown.

In 2000, a new rust disease on daylilies, caused by the Asian rust fungus *P. hemerocallidis,* was discovered in the United States. It is a heteroecious rust fungus whose alternate hosts are plants in the genus *Patrinia,* some of which are grown as ornamentals in the United States. In only one year, this rust spread to 30 states, primarily because daylilies are commonly shipped to retail stores. Unfortunately, the urediniospores are found on daylilies, but the spores die in winter conditions. Gardeners should check new plants carefully and remove any diseased leaves, and they also should watch established plants. Daylilies vary in their susceptibility to rust, so it is possible to choose cultivars that are more resistant. Watering of only the soil surface helps prevent the spread of rust and other fungal pathogens.

Tree Rusts: A Perennial Threat

Some important rust fungi infect trees. One commonly seen on apple and crabapple trees is *Gymnosporangium juniperi-virginianae* (Figures 11.8 and 11.9). The disease is known as cedar-apple rust, in which the word cedar refers to red cedar (juniper), rather than white cedar (arborvitae). Aeciospores are produced on apple trees, and they infect junipers, on which a small brown gall develops over a 2-year period. In the second spring after infection, starting just as the apple blossoms are ready to open and lasting for about a month, orange, jellylike protrusions appear on the junipers. These are called telial horns and are comprised of masses of teliospores. They form when moisture is available to produce basidiospores and disperse them on their journey to newly expanded apple leaves and fruit.

This heteroecious rust fungus is different from the macrocyclic rusts already mentioned, because it lacks the repeating urediniospore stage—a disease cycle called demicyclic (Figure 11.10). Thus, there is no secondary spread from junipers to other junipers. Also, rust infections of apple trees do not serve as sources of spores that can cause an epidemic within an orchard. The aeciospores produced on apple trees can infect only

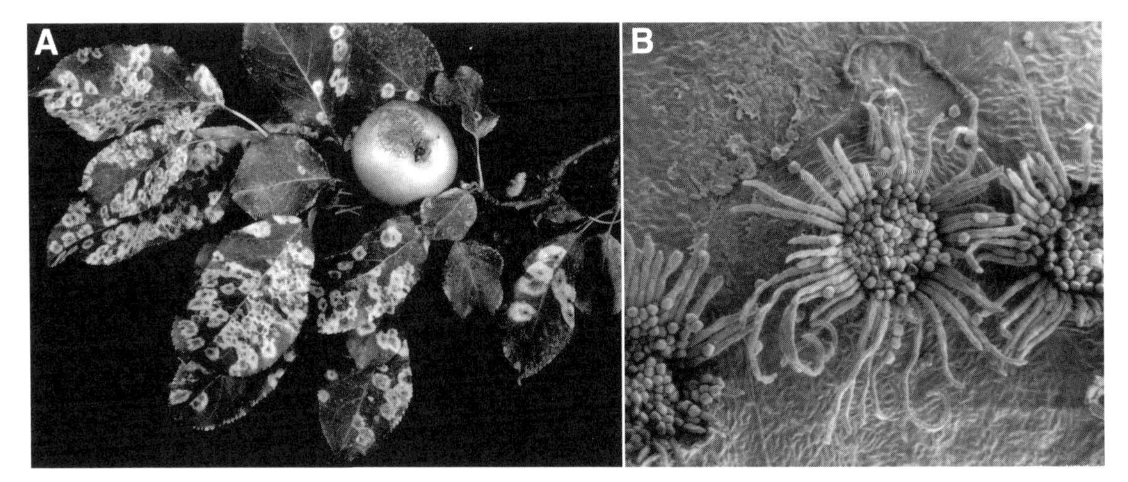

Figure 11.8. A, cedar-apple rust symptoms on apple. **B,** aecium with aeciospores produced on the lower leaf surface of an apple leaf infected with *Gymnosporangium juniperi-virginianae.* The aeciospores infect junipers (electron micrograph).

junipers, and the basidiospores from the junipers can infect only apple trees. Rusts of this type can be completely controlled by removal of one host. Because junipers are less economically important than apple trees, they are removed from areas near commercial apple orchards.

Fungicides can effectively protect apple trees for the short period during which basidiospores may arrive from nearby junipers, but that is usually not necessary. Fungicides applied for the apple scab disease may be sufficient to protect trees from rusts as well. Furthermore, many commercial apple and crabapple cultivars have good genetic resistance to rust diseases.

Figure 11.9. Gall stage of cedar-apple rust on juniper.

A single juniper gall can produce more than 100 billion spores, so severe infections may occur when junipers and crabapples are grown together. In landscaping, either junipers or crabapples should be selected, unless rust-resistant cultivars are planted. As in apple orchards, applying fungicides will protect crabapples in landscapes, but this is not as reliable as genetic resistance. Spores from wild junipers in nearby fields also may be an important source of infection, although the spores are only viable approximately 1 mile (1.6 kilometers) from their source.

In the early 1900s, another rust disease destroyed the white pine forests of the northeastern United States. The white pine blister rust fungus, *Cronartium ribicola,* is macrocyclic and heteroecious, with plants of the genus *Ribes* serving as alternate hosts, as suggested in the specific epithet. This genus includes currants and gooseberries. The rust fungus apparently arrived on white pine seedlings imported from Germany to replant the lands stripped by timber companies.

The widespread devastation caused by the fungus made white pine blister rust one of the diseases (along with citrus canker and chestnut blight) that finally convinced U.S. legislators that the nation's borders needed to be protected. This led to passage of the Plant Quarantine Act in 1912. Unfortunately, the new law could not stop the pathogens that had already arrived and become widely dispersed.

Knowledge of the white pine blister rust disease cycle suggests a useful management strategy. The aeciospores are found on white pine, whereas the urediniospores and teliospores are on the *Ribes.* Thus, the repeating stage that causes epidemics among plants of the same species was, as it turned out, on the host of less economic value.

Once the destructive nature of this rust was realized, many people were hired to scour the woods, cutting and killing wild *Ribes* plants within 1,000 feet (305 meters) of white pines—the distance basidiospores can travel and still be viable. During the Great Depression of the 1930s, as many as 11,000 men in the Civilian Conservation Corps were employed in a single year to eradicate *Ribes* in the national forests. People also were restricted from planting susceptible species of *Ribes* near white pine forests. Because rust spores produced on white pine can infect only *Ribes,* the fungus rapidly declined with the removal of the alternate host in the eastern United States (Figure 11.11).

Individual infected pine trees are usually doomed, as the mycelium of the rust fungus continues to invade the tissues until the tree is girdled and killed. However, healthy neighboring pines are not threatened. Individual trees can be pruned and saved if only the branches are infected, although this is not possible in a vast forest. Since 1966, individual states have been responsible for enforcement of *Ribes* quarantines. They decide where *Ribes*

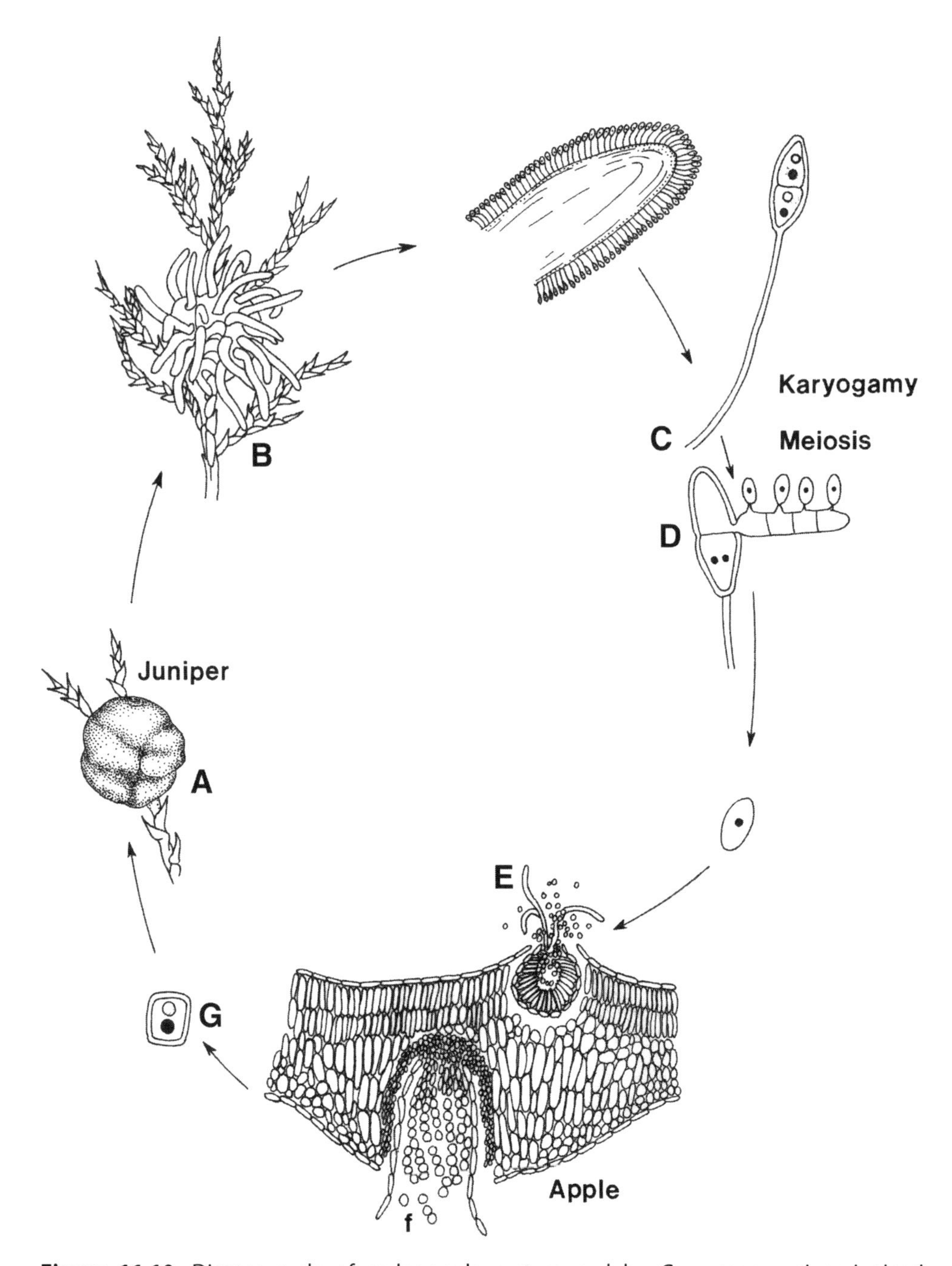

Figure 11.10. Disease cycle of cedar-apple rust caused by *Gymnosporangium juniperi-virginianae.* Gall on juniper (**A**). In spring, orange gelatinous structures expand from the gall (**B**) and are covered with teliospores (**C**). Karyogamy and meiosis occur in the teliospore (**D**), resulting in production of four haploid basidiospores, which infect apple and related plants. Basidiospore infections result in the formation of pycnia (**E**). If a dikaryotic mycelium is formed, aeciospores are produced from aecia, generally on the lower leaf surface (**F**). The dikaryotic aeciospores infect junipers (**G**). There is no urediniospore (repeating) stage.

may be safely planted without threatening white pines. In addition, rust-immune *Ribes* are now widely available from nurseries for gardeners and jelly makers.

Ribes eradication has not been as successful in the western United States, particularly in Idaho and western Montana. In 1966, replanting of timber stands of the western white pine, *Pinus monticola,* was temporarily halted because of white pine blister rust. Today, seed of blister rust-resistant western white pine is available. Resistant trees are planted as part of integrated pest management (IPM) programs, which continue to eradicate *Ribes,* prune infected branches of young trees, and avoid rust hazard zones, where environmental factors favor disease. After more than 50 years of research to develop this integrated approach, the western white pine is being grown as a viable timber species once again.

Figure 11.11. White pine blister rust (*Cronartium ribicola*). Production of aeciospores that must infect *Ribes* species (alternate hosts).

Other tree rusts are of economic importance. The fusiform rust, caused by *C. quercuum* f. sp. *fusiforme,* is a continuing problem in the southern United States, because many southern pines are susceptible and the alternate host is oak, which is very common in the South. In this case, it is not possible to consider rust management through the removal of the alternate host, although oaks near commercial pine plantations are sometimes removed. Fungicide protection and pruning can help manage infection of commercial stands. Genetic resistance is a future goal, but the long generation time of trees makes breeding them for resistance a lengthy process. As with the rust-resistant pines, the durability of resistance is also of concern, considering the many races produced by rust fungi, especially in long-lived perennials such as trees.

As discussed in Chapter 3, a similar problem plagues coffee growers, who wish to find durable resistance to coffee rust within the narrow gene pool that exists in currently available cultivars. With coffee rust, *Hemileia vastatrix,* there is the added problem of having the repeating stage occur on the host of economic importance. Even without the presence of an alternate host, which has not yet been discovered, the urediniospores continually infect coffee trees, unless the trees are sprayed repeatedly with fungicides.

Rust Fungi: Managers and Management

Several rust fungi have been used as biological controls for unwanted plants. Fungi applied in this way are sometimes called mycoherbicides. The urediniospores of rust fungi are easy to collect, store, and disperse on plants. Then, they multiply quickly to create an epidemic in the pest plant population.

A successful example comes from Chile, where European blackberry plants had become an invasive weed in pastures. Urediniospores of the rust fungus *Phragmidium violaceum* were collected from European blackberry plants and spread by helicopters. Rust infection reduced the competitiveness of the blackberries and slowed their invasion. Rust has now been used for decades to help combat invasive blackberries in Australia, Chile, and New Zealand.

The same fungus was accidentally introduced into Oregon in 2005. It infected Himalaya blackberries, which are considered a noxious weed. Defoliation by the rust fungus allows seeds of other species to germinate under the blackberry plants and also weakens the blackberries, so they are not as competitive with native species. Unfortunately, this introduced pathogen also can cause disease in commercial blackberries.

In general, rust fungi are particularly attractive as biological control agents, because they tend to be very host specific and urediniospores typically produce destructive polycyclic epidemics. The use of rust fungi is a long-term management approach in circumstances in which closely related desirable plants are not endangered by deployment of the rust fungi.

The management of rusts on desirable plants varies with the values of the plants. For instance, rust diseases of cereals and many vegetables are managed primarily with genetic resistance. Fungicides are used on some high-value ornamental plants and to prevent rusts for which genetic resistance is not yet available, such as coffee rust and soybean rust. Removal of alternate hosts of heteroecious rust fungi is important in reducing genetic variation that can overcome genetic resistance. In home gardens and other small plantings, minimizing leaf wetness during the growing season and removing infested debris at the end of the season may be sufficient to keep rusts at acceptable levels.

A possible biological control for rust diseases is the use of fungi that parasitize the rust fungi. Several such fungi, called hyperparasites, have been identified. So far, there have been no commercial applications of hyperparasites to manage rust fungi, but the research continues.

Tree diseases, including rusts, are some of the most difficult to manage. In addition, the loss of a single plant can be aesthetically and emotionally devastating. Replacement of a mature tree can take 50 years or more, so its loss is felt for a long time. The long generation time of most trees makes them, as species, particularly vulnerable to severe damage, as they are unable to rapidly change genetically to respond to the danger of an introduced pathogen. Witness the loss of the chestnut and elm trees in Europe and North America, along with the many native pines in Japan. Introduced insect pests—such as gypsy moth, Japanese beetle, and emerald ash borer—cause continuing damage.

Trees are at once particularly vulnerable and amazingly resilient to attack by pathogens and pests in their environment. The next chapter examines some other important tree diseases and how trees, the organisms with the longest life spans, protect themselves from invasion.

CHAPTER 12

Diseases of the Largest Plants: Trees

Trees live longer than any other form of life. The oldest individual tree in the world is a bristlecone pine named Methuselah in the White Mountains of California. It is nearly 5,000 years old. Some clonal trees are much older, including a spruce in Sweden and some creosote and aspen colonies. The largest tree in the world is the General Sherman sequoia in California, whose weight is estimated at 2.7 million pounds (1.2 million kilograms). In many places, trees serve as memorials to historic events long after the people who participated in the events have died.

Trees remain standing in their fixed locations, unable to move, while bacteria, fungi, insects, lightning, and people and other animals feed on or invade their tissues. Yet some trees are able to withstand this dangerous onslaught and continue to grow for hundreds or thousands of years.

The survival of trees in a hostile environment is due mainly to two key features: their protective woody bark and their mechanism for continuous growth despite being damaged. To understand how trees protect themselves, we must first examine the transition from the soft and relatively vulnerable green stem of a seedling to the woody, bark-enclosed twigs, branches, and trunk of a mature tree.

Transition to Bark and Wood: How to Live Forever

The young stem of a woody seedling (such as a tree) is composed of the same kinds of cells as those in the stem of an herbaceous plant (such as a tomato). The outer seedling's epidermal layer encloses a core of living cells called parenchyma. The vascular tissue is arranged in a cylinder within the parenchyma, with xylem (water- and mineral-conducting) cells toward the inside of the stem and phloem (food-conducting) cells toward the outside of the stem. Often, there also are bundles of thick-walled supporting fibers.

In an herbaceous plant, the meristems (active growth areas) in buds and at the tips of roots and stems are responsible for growth. There is a limit, however, to how tall an herbaceous plant can grow without falling over. To become a tree or other woody perennial, it must have bark and lateral (outward) growth to form a supporting trunk.

As the stem of a woody plant matures, two new meristems—each called a cambium—begin to function (Figure 12.1). Instead of being collections of cells at the tips of plants, these meristems are thin layers of cells around the circumferences of the stems. The first meristem is the vascular cambium, which provides new xylem and phloem cells. The second meristem is the cork cambium, which makes cells for the developing corky bark. Sometimes, the cork cambium develops gradually, in patches, producing a bark layer on some parts of the stem while leaving other parts temporarily green. Eventually, the entire stem is covered with bark. The cork cambium continues to produce cork cells to replace bark that is sloughed away by weather and wounds. Similar changes occur belowground, as the roots develop bark and thicken.

The tissues formed by the vascular cambium are called secondary xylem and secondary phloem. The layer of active secondary phloem remains relatively thin as the older phloem becomes compressed by the expanding trunk. Secondary xylem layers are commonly referred to as wood. These layers are produced each year, and unlike the secondary phloem, they are preserved for the life of the tree.

In many trees, the cells of the secondary xylem produced in the spring are larger than those produced in the summer. The resulting seasonal variation in cell size creates a pattern of annual rings, in which each year's growth can be discerned. You may have counted the annual rings visible on a stump to determine the age of a tree. In areas where rainfall and temperature are uniform and less seasonal, as in parts of the humid tropics, annual rings may not be as apparent. However, in many climates, the annual rings record the growing conditions in each year of the tree's life. Historical studies of drought periods and other environmental variations have been made by studying the annual rings of ancient trees that have fallen or been cut down and by examining sample cores taken from living trees.

The chief functions of the secondary xylem are to conduct water and minerals and to provide support. Secondary xylem is composed of several types of cells. A tracheid is an elongated, nonliving, conducting cell that consists only of cell walls. Besides having a primary cell wall, a tracheid has a secondary cell wall composed of lignin, which adds support. The tracheid wall contains numerous pits, where no secondary cell wall is present. Water and minerals pass from one tracheid to the next through the primary cell walls, where the ends of the tracheids overlap. Tracheids are the only conducting tissue in conifers (such as pines), and they also are found in hardwood trees (such as oaks and maples).

Hardwood trees also produce large conducting tubes called vessels. Like tracheids, vessels are nonliving and have secondary cell walls of lignin, but the ends of the cells dissolve. The connecting vessel cells essentially form straws, which allow water to flow more easily than occurs with tracheids. Hardwood trees also contain more fibers—a third kind of

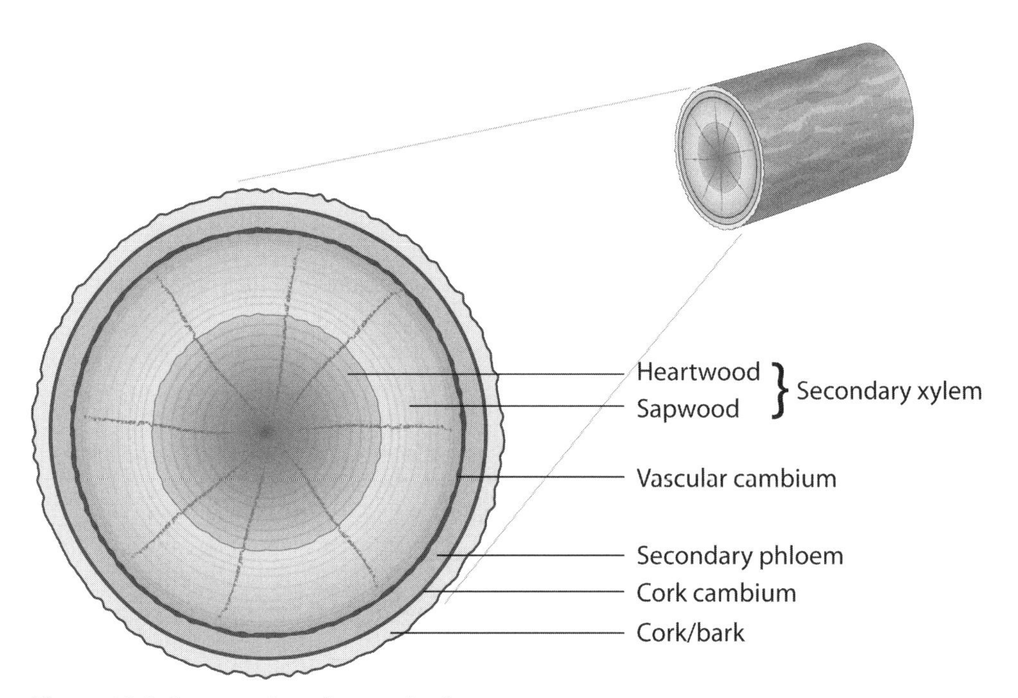

Figure 12.1. Cross section of a woody plant.

nonliving cell that adds support to the wood. This explains why conifers are referred to as softwoods. Most of the mass of a tree is composed of these nonliving cells. Living parenchyma cells form in thin radial and vertical layers throughout the nonliving xylem. The radial parenchyma cells may be visible as rays in hardwood cross sections. These living cells play a protective role for trees, which is explained later in this chapter.

In a cross section, the secondary xylem of an older tree is often composed of a dark-colored inner cylinder known as heartwood and a light-colored outer layer known as sapwood. The sapwood remains approximately the same thickness throughout the life of the tree and actively carries water and minerals. Each year, new sapwood is produced from the vascular cambium, and the oldest part of the sapwood becomes part of the heartwood. The heartwood is often dark, because it has undergone chemical changes that help protect it from decay and increase its strength. Many trees that are entirely hollow, due to decay of the heartwood, continue to function and grow, because the sapwood conducts water and minerals for the tree.

Unlike the secondary xylem, which consists mainly of nonliving cells, the secondary phloem is composed of living cells. The sieve cells transport the products of photosynthesis from the leaves to the rest of the plant. This relatively narrow band, which lies just under the bark, is vital to the health of the tree because it carries food to the roots. The phloem is vulnerable to damage when cuts are made in tree bark by accidents, animals, and mischievous humans.

Peeling bark from a twig reveals a wet, slippery layer, which is the vascular cambium. In fact, peeling away the bark removes the cork (bark), cork cambium, secondary phloem, and vascular cambium. If this is done to the trunk of a tree, the tree will die. The early European colonists in North America, who needed to plant crops quickly, peeled bark from trees in forests, causing them to shed their leaves. The colonists then cut down the dead trees when they had more time.

The protective part of bark is composed of cork cells produced by the cork cambium. Cork is nonliving tissue that helps make the tree trunk impervious to water loss and helps prevent physical damage. The continual production of cork cells is necessary, since bark is constantly subject to weathering and other damage. Commercial cork is produced from the extra-thick bark of the cork oak, *Quercus suber*. The cork industry is centered in Portugal, where this tree is native. The cork layer has to be peeled off very carefully with a sharp knife, so the cork cambium is not damaged. After 8 to 10 years, a new layer of cork can be harvested. The cork layer in most trees is much thinner.

Unlike the circulatory system in people, the xylem and phloem do not carry oxygen to plant tissues. Air exchange in a tree trunk is accomplished by lens-shaped structures in the bark called lenticels. They are composed of loosely packed living cells that allow oxygen to diffuse in and carbon dioxide to diffuse out. This diffusion process is required for the living cells of the cork cambium, phloem, vascular cambium, and parenchyma cells in the wood. Lenticels are particularly visible in thin-barked trees, such as birch and cherry. Pathogens and insect pests sometimes invade trees through lenticels.

Tree Diseases: What Parts Are Vulnerable?

Trees can live almost indefinitely. What kinds of diseases threaten their survival?

The leaves are the most vulnerable aboveground part of a tree. In contrast, the bark-covered, woody roots, trunk, and twigs are generally well protected from invasion by microorganisms and insect pests. Deciduous trees lose their leaves and replace them with new ones each year, shedding not only leaf tissue but many insects and pathogens as well.

A recommended sanitary practice is to clean up and compost, burn, or dispose of diseased leaves during autumn. Doing so helps prevent fungal spores produced on old leaves from infecting new leaves the following spring. Blighted leaves are common on ash, oak, and sycamore in wet springs. Anthracnose fungi sometimes invade woody twigs from infected leaves, especially in stressed trees or during particularly wet years, but generally, these diseases do not cause severe damage.

Besides the leaves, the xylem, phloem, and roots of trees are subject to specific diseases. This chapter describes some famous examples of diseases, but some are already familiar to you. We have described the devastation of *Phytophthora* pathogens, which commonly invade tree roots. Sudden oak death, caused by *P. ramorum,* invades all parts of the tree and can kill it through the roots or cankers in the trunk. Cankers are infections just under the bark that kill the living cells of the secondary phloem and vascular cambium. Other canker diseases that have already been described include fire blight, a bacterial disease of apples (see Chapter 3), and white pine blister rust (see Chapter 11). Some insect pests, such as the emerald ash borer, kill trees by feeding on the phloem under the bark, similar to a canker pathogen.

Once a canker pathogen, either fungal or bacterial, has girdled the entire trunk of a tree by destroying these living tissues, the tree will die. Because branch and twig cankers are sources of inoculum for further infections, it is recommended that they be pruned off before the pathogen moves into the main trunk.

In contrast to canker pathogens that kill living tissues just under the bark, a number of specialized fungi cause wood decay. Although this is an important part of the decomposition cycle, it also can cause the death of a healthy tree. We often first become aware of wood decay when fruiting bodies, such as conks, appear on the tree trunk.

As described in Chapter 3, xylem-limited bacteria and phytoplasmas can infect the active xylem and phloem of living trees, respectively. These pathogens are generally disseminated by insect vectors. Some additional specialized pathogens also cause xylem- and phloem-specific tree diseases.

Generally, indigenous pathogens cause little trouble on native trees in their countries of origin, because those trees have reached a genetic balance with their attackers over long periods of coexistence. Like plant diseases in general, the most devastating tree diseases are those caused by pathogens imported from foreign lands. Because tree populations are not likely to have natural resistance to exotic pathogens, initial disease outbreaks are often severe. Some of the most famous plant disease epidemics have involved tree diseases caused by exotic pathogens.

Dogwood anthracnose is an example of a devastating disease caused by an exotic pathogen. Unlike most anthracnose diseases, which cause primarily cosmetic problems, dogwood anthracnose has killed millions of two indigenous species: eastern flowering dogwood (*Cornus florida*) and mountain dogwood (*C. nuttallii*). Dogwoods tend to be understory trees, where moisture favors infection. Dogwood anthracnose is caused by a previously unknown fungus, *Discula destructiva,* which apparently was introduced at ports of entry in Seattle, Washington, in the West and in New York City in the East. Cankers eventually kill native dogwood trees in forests, but pruning and providing open environments can keep them alive in landscapes and along roadsides.

Dutch elm disease is a vascular wilt or xylem disease, similar to Fusarium wilt (see Chapter 9). However, it is caused by exotic pathogens, as is chestnut blight, a canker disease. The following sections discuss these two historically important tree diseases, as well as ways to keep trees healthy.

Dutch Elm Disease: Death by Drought

More than 40 million American elms (*Ulmus americana*) and 25 million English elms (*U. procera*) have been killed by the fungi that cause Dutch elm disease. Elms were popular shade trees in both Europe and North America because of their overarching branches, which shaded city streets (Figure 12.2). The impact of losing these trees is clear in old photographs of cities and college campuses, which look so different today without the elms. Elms also were planted as hedgerows and windbreaks.

Dutch elm disease is named for the country where many of the important early scientific studies of this disease were done. Two Dutch plant pathologists, Christine Buisman and Bea Schwarz, played key roles in the discovery of the cause of Dutch elm disease.

The ascomycete *Ophiostoma ulmi* was introduced into the United States in elm logs from Europe on at least three occasions. The pathogen's presence was first confirmed in Ohio in 1930, but the disease had already caused losses of elm trees in Europe, starting at the time of World War I. The fungus likely originated in Asia, because Asian elms are generally resistant to the disease. In the late 1960s, a renewed epidemic of Dutch elm disease plagued Great Britain and Europe, killing many elms that had survived the initial outbreaks. The new devastation apparently was caused by the introduction of a new species of the pathogen, *O. novo-ulmi,* imported from North America.

Dutch elm disease is a vascular wilt disease, in which the pathogen never lives independently of its host. *Ophiostoma* species produce several spore stages that help them complete their life cycles, including two types of asexual conidia. The sexual ascospores are relatively rare in the United States, because the second mating type necessary for sexual reproduction is not common.

As with other vascular wilts, the primary disease symptom is water stress. The invading fungus produces toxic substances that cause overgrowths of parenchyma cells called tyloses, which block xylem vessels. This blockage causes wilt symptoms. In addition, the fungus produces enzymes that degrade the cellulose, pectin, and other components of the vessel cell walls. The cell walls begin to break down and collapse, greatly restricting the movement of water in the xylem. This contributes to the wilt and causes a ring of discoloration observed in the active xylem of infected twigs and branches. The mycelium

Figure 12.2. American elm. **A,** This healthy tree illustrates the elm's characteristic shape. **B,** This street sign suggests that elms have been important shade trees historically, although none exist at this location today.

and budding conidia of the fungus also contribute to vessel blockage. The small oval conidia help spread the fungus throughout the xylem after it is introduced into the tree (Figure 12.3).

After the tree has died, the fungus becomes a saprophyte and grows throughout the trunk of the dead tree. A second kind of conidium is produced in a sticky mass on the tip of a stalk of fused hyphae under the bark of an elm tree that has recently died. These reproductive structures, called synnemata, are produced in tiny tunnels carved out beneath the bark by female elm bark beetles. The tunnels are maternal galleries approximately 1 inch (2 to 3 centimeters) long, in which the beetles lay their eggs. After the eggs hatch, the larvae feed under the bark, each one making its own side tunnel perpendicular to the egg tunnel. The resulting galleries of tunnels are readily visible if the bark is peeled from a tree killed by Dutch elm disease.

After a short pupation or resting stage, new adult beetles emerge through the bark and fly to a healthy elm tree, carrying the sticky conidia on their bodies. The beetles feed on the twigs and branches of a healthy elm and, in the process, deposit the conidia in the xylem of the tree—thus completing the disease cycle. Later, the female beetles return to a dying elm tree to lay their eggs under the bark.

Two species of elm bark beetles are vectors of Dutch elm disease in North America: *Hylurgopinus rufipes,* a native elm bark beetle, and *Scolytus multistriatus,* a European elm

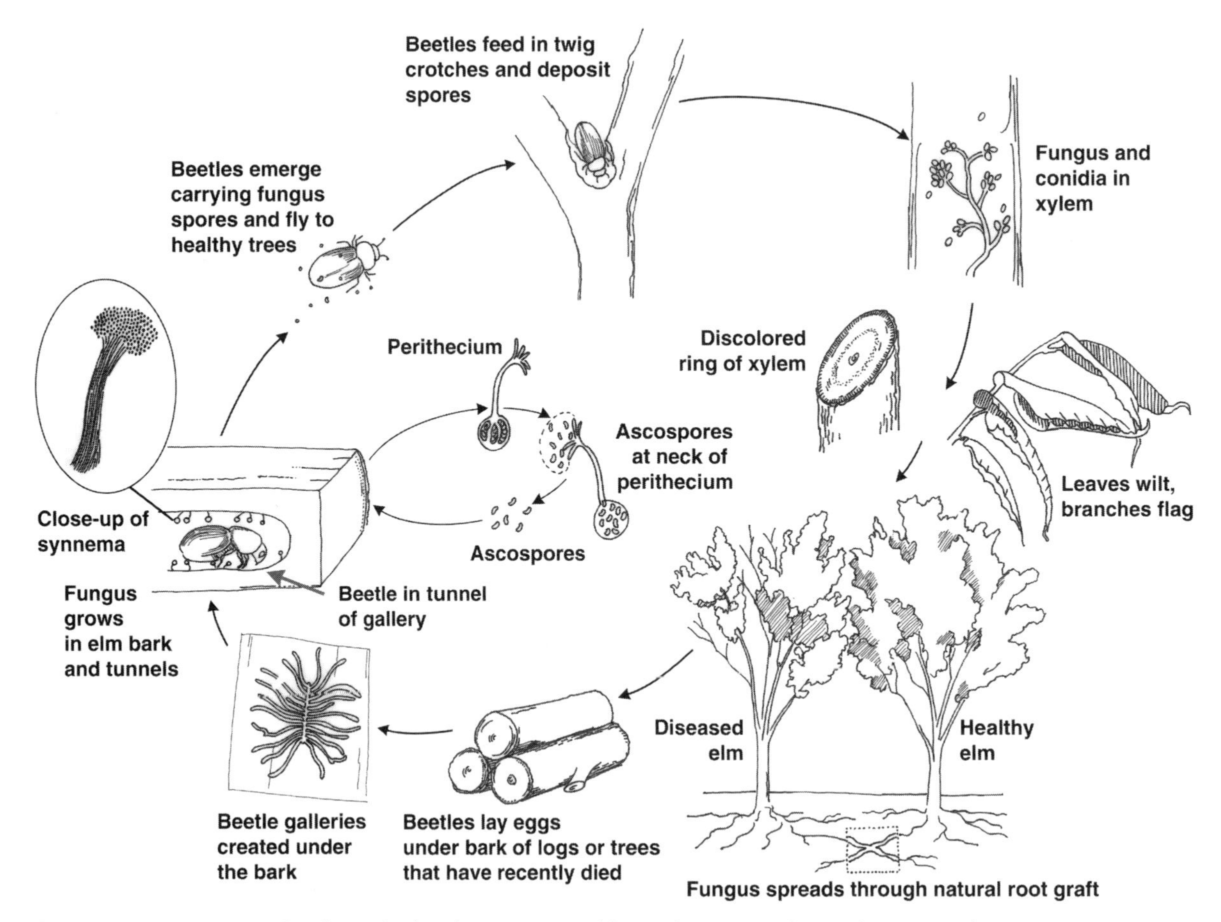

Figure 12.3. Disease cycle of Dutch elm disease, caused by *Ophiostoma ulmi* and *O. novo-ulmi.*

bark beetle inadvertently introduced from Europe. The European beetle is more numerous, because it can complete two generations per year in many U.S. states and thus is the more important vector of Dutch elm disease. The European elm bark beetle creates egg galleries that are parallel to the wood grain, whereas the galleries of the native beetle are more or less perpendicular to the grain.

Since elm bark beetles transmit the fungus to young twigs in the crown of a healthy tree, the earliest symptom of Dutch elm disease is the flagging or wilting of small branches. Because the xylem fluid moves up, the initial fungal invasion is fairly slow. Carefully inspecting elm trees and pruning out infected branches at the early stage of the disease can sometimes save a tree.

A second means of transmission is more difficult to manage. It occurred on the elm-lined streets in many towns in Europe and North America. When trees of the same species grow close together, their roots may graft together underground. Vascular connections may form in the grafts, allowing the passage of the fungus from one tree to the next. The fungus, moving up in the xylem fluid, may then colonize a tree very quickly. This means of transmission has led to the deaths of trees one after another in domino fashion along many streets.

Monoculture—whether in a wheat field or on a city street—leaves plants vulnerable to diseases. Nearly every community has an Elm Street, but today, few elms remain. Too often, they have been replaced by monocultures of ash, honey locust, maple, or oak trees, which are equally vulnerable to other diseases.

Management of Dutch elm disease is difficult. An important sanitary practice is to chip, burn, or bury dead and dying trees, thus preventing insect vectors from carrying spores to healthy trees. Also, infected logs should not be stored for firewood. After Dutch elm disease was discovered in the United States, intensive insecticide spraying was attempted to reduce the spread of the fungus by elm bark beetles. This led to publication of Rachel Carson's book *Silent Spring* in 1962 and the start of the environmental movement. On tree-lined streets, trenching and herbicides were used to break root grafts between trees to try to prevent the fungus from spreading.

Today, most remaining elm trees in the United States exist as individuals. They can be maintained if carefully watched and pruned when flagging twigs appear. Some trees receive fungicide injections as well. Unfortunately, such injections are relatively expensive and must be repeated every 1 to 3 years. Many studies have been conducted to discover microorganisms that could be injected into the tree for biological control of the disease, but so far, the results have either not been successful or have not yet been approved by the U.S. Environmental Protection Agency (EPA).

The future of American and English elms in the wild probably lies in the prolific seed production of the species. Each seed is genetically unique. As young trees grow, many become infected and die, but the introduction of an exotic pathogen has never been known to eliminate a species. Over hundreds or perhaps thousands of generations, the species will finally come into balance with its pathogens, and mature elms may once again grace the landscape.

Science Sidebar

Names for new elms

The earliest elms bred with resistance to Dutch elm disease in Europe were named for the Dutch scientists who participated in the initial studies of the disease. Unfortunately, elm trees named Buisman and Schwarz were not resistant to all strains of the fungus and also lacked the American elm's hardiness to cold.

In 1983, the U.S. Elm Research Institute released a resistant elm named the American Liberty elm, after a famous tree cut down by British troops in Boston in 1776. Elms may have been the first trees planted in the United States for landscape purposes, and that particular tree had become a rallying point for the colonists' protests against King George III. Liberty elms, as well as disease-resistant elms with other patriotic names—including Independence and Valley Forge—are available for purchase and planting.

Figure 12.4. Oak wilt disease, caused by *Ceratocystis fagacearum*. **A,** Water stress symptoms appear in the leaves of an infected oak tree. **B,** Fungus pads, which are attractive to insects, are produced under the bark of dying and dead oak trees.

Science Sidebar

Oak wilt

Oak wilt is a vascular wilt disease similar to Dutch elm disease and caused by a related fungus. It affects oak trees across most of the Midwest and eastward, and it also affects trees in Texas. Oak wilt has not yet been found west of the Rocky Mountains or outside the United States.

Red and black oaks are particularly susceptible to oak wilt disease, but all oaks can become infected by *Ceratocystis fagacearum*. Under the bark of a dying tree, the oak wilt fungus produces pads of mycelium and conidia, which are attractive to insects (Figure 12.4). Many kinds of insects are attracted by the fermenting odor, and after leaving the oak, they may move on to another oak and feed in such a way that they introduce the fungus into the xylem, commonly through bark or pruning wounds.

Because the vector relationship is not as efficient and specific as in Dutch elm disease, the spread of oak wilt has been slower. However, along tree-lined streets, root graft transmission may increase the rates of infection and death of oak trees. Management practices used for oak wilt are similar to those used for Dutch elm disease.

Tree breeders and pathologists hope to shorten that time through deliberate crossing and selection for genetic resistance. They have had some success. The earliest trees with resistance to Dutch elm disease were selected in Europe. An English elm that is apparently resistant to elm bark beetles and does not get Dutch elm disease has been propagated. In the United States, some resistant trees have been developed through crosses with Asian species, while others have been selected from populations of American elms.

The story of Dutch elm disease is yet another example of the importance of quarantines for protecting native plants from foreign pathogens, pests, and vectors. However, it is not only a historical example. Dutch elm disease is a threat today in Australia and New Zealand. Perhaps half a million susceptible elm trees provide shade and beauty on the Australian continent. The elm bark beetles have found their way to Australia, but so far, quarantines have successfully barred the Dutch elm disease fungi. New Zealand has not been so fortunate. Dutch elm disease was discovered there in late 1989 but has been restricted so far to the city of Auckland.

Some beautiful but susceptible specimens of mature elms still exist in Australia, Europe, New Zealand, and North America. If they are to survive, they will require care and protection.

Chestnut Blight: Death by Starvation

The loss of a major species of shade tree leaves whole neighborhoods with a bare and blighted look. Early in the 20th century, a major species of forest tree was nearly destroyed by a plant pathogen.

Within 50 years, 3 to 5 billion American chestnut trees were killed throughout their natural range: the Appalachian Mountains. The people of Appalachia had found many uses for the chestnut trees that dominated their forests. They used chestnut lumber for buildings, fences, and furniture. They used the chestnuts as food for both themselves and their animals. When they needed additional income, they sold the lumber to produce telephone poles, the nuts for city folk to enjoy, and the bark (rich in dark tannins) to East Coast tanneries. After chestnut blight ravaged the Appalachian forests, many of the people dwelling there became impoverished.

Chestnut blight, along with citrus canker and white pine blister rust, was one of the diseases that triggered the first U.S. quarantine legislation in 1912. Before the blight struck, the American chestnut, *Castanea dentata,* was a major tree species throughout the eastern forests, from Maine to Georgia. In the early 1900s, perhaps as many as every fourth tree in the Appalachian forests was a chestnut (Figure 12.5). By 1923, the disease had invaded 80% of the chestnut's range, and by the 1950s, 80% of the chestnut trees had died.

Chestnut blight is caused by the ascomycete *Cryphonectria parasitica.* Unlike Dutch elm disease, which is a vascular wilt, chestnut blight is a canker disease. The fungus invades the bark of a twig through small wounds, such as insect feeding sites, and it begins to invade the vulnerable secondary phloem and vascular cambium of the growing twig. Slowly, the mycelium spreads outward, until eventually the entire twig is girdled and dies. As the fungus continues growing, it may reach the main trunk, and the mature tree may eventually be killed. The American chestnut population had no resistance to this pathogen, which was apparently introduced on Asian chestnut seedlings brought to the New York Botanical Garden by a collector. As a result, many trees died within 1 year of infection.

Management of chestnut blight is theoretically possible in a single tree, if infected twigs and branches are pruned

Figure 12.5. Chestnut trees. **Top,** chestnut trees in Fairmount Park, Philadelphia, in 1878, with people harvesting the nuts using ladders. **Bottom,** forest chestnuts in the Appalachian Mountains before the blight.

and destroyed. Fungicide applications are useless, and of course, pruning trees in forests is not practical. The fungus produces sticky conidia and ascospores, which are carried to American chestnut trees by birds, insects, splashing rain, and wind. The rapid and devastating spread of the disease caused tremendous economic and ecological disruption throughout the Appalachian forests.

The tragedy did not end in North America but spread to Europe, probably by the export of chestnut wood in about 1917. Many European chestnut groves were destroyed following the initial invasion. However, chestnut trees have a growth habit that keeps the species alive despite the presence of the virulent fungus. The tree's extensive root system stores enough food reserves to send up new shoots, or suckers, from the base of the dead trunk. Year after year, these young shoots allow renewed growth, but eventually, they too succumb to the fungus, once it has invaded their tissues. Some young trees are able to grow to a sufficient size to allow fruit production, and new chestnuts sprout. There is hope that, as with the elm, the chestnut will be able to coexist with *C. parasitica* as a minor parasite, although few of us will likely live long enough to see this happen.

The continued production of suckers from the base of the dying chestnut trees inspired some optimists to think that somehow the trees could overcome the effects of the deadly parasite. Robert Frost, the famous American poet, even considered this possibility in his poem "Evil Tendencies Cancel," published in 1936:

Will the blight end the chestnut?
The farmers rather guess not.
It keeps smoldering at the roots
And sending up new shoots
Till another parasite
Shall come to end the blight.

There is now scientific evidence that this scenario actually happens. In Italy in the 1950s, some orchard chestnuts were found to be surviving despite the presence of extensive cankers. Studies of the *C. parasitica* isolates from these cankers suggested that new strains of the fungus had arisen that were much weaker than the common strains. Such strains have been designated hypovirulent. Not only are they weaker, but when they come into contact with virulent strains, the hyphae may fuse and cause the virulent strain to become hypovirulent. The weakened state is due to a virus in the hypovirulent strains that causes a disease of the fungal pathogen.

The existence of hypovirulent strains has provided a practical biological control to "cure" existing cankers. In Europe, the application of hypovirulent strains of *C. parasitica* to active chestnut blight cankers in orchards is now used to stop canker activity. In the United States, there is hope that this naturally occurring biological control may somehow be used to help reestablish the chestnut in eastern forests.

The practical application of this procedure faces some complications, however. The hypovirulent strains usually grow more slowly and produce fewer spores, which reduces their ability to spread in nature. The procedure is helpful in stopping the growth of an active canker, but so far, no way has been found to use it to prevent the disease in the first place. Also, the pathogen has numerous genetic strains. The hypovirulence virus spreads only if the strains are genetically compatible, so the hyphae can fuse. In the United States, cankers may contain 10 or more different genetic strains. This means that to achieve biological control, hypovirulent strains that match all of these virulent strains must be found.

Scientists are optimistic that this method may eventually provide useful results. Hypovirulence has been studied in other pathogenic fungi as well. These studies may lead to a better understanding of one of the factors that helps plants and pathogens come into genetic balance over time. In the meantime, The American Chestnut Foundation (TACF) has been back-crossing trees for many generations to give American chestnut trees the genetic resistance of Asian species. TACF hopes to reestablish the chestnut as a forest tree in the Appalachian Mountains.

Other Canker Diseases of Trees: Native and Exotic Pathogens

Tree species remain most vulnerable to destruction by introduced pathogens because of the long time needed to replace mature trees. The increase in global trade is a major factor in the danger of introducing new pathogens and their vectors. Potential economic and ecological losses make the protection of our tree species a high priority.

The American beech (*Fagus sylvatica*), a climax tree species of eastern forests, is being destroyed by an introduced canker fungus, *Nectria coccinea* f. sp. *faginata,* and a native fungus, *N. galligena.* By themselves, the fungi cannot penetrate bark, but they can enter through small wounds made by the beech scale, *Cryptococcus fagisuga,* a tiny sucking insect. As in chestnut blight, the invading fungus kills the vascular cambium. The future of an infected tree depends on its age and growing conditions. Old trees seem more susceptible to rapid killing, whereas young trees, although more likely to survive, become defective as timber trees. Following infection, the bright red fruiting bodies of these ascomycetes erupt through the bark of an infected tree. The fungi produce both conidia and ascospores, which can be disseminated by air, water, and vectors such as birds, insects, and small mammals.

Management of the beech canker disease in forest situations is very difficult and lies mainly in forestry management practices that provide greater species diversity and the culling of infected trees. Beech trees in yards and parks can be protected by pruning infected branches and reducing scale infestations. Scale insects are usually well protected by a waxy coating, but insecticides can be used to kill them in their crawler stage. Scrubbing of beech bark with a detergent also will help remove the scales.

The relatively sedentary life-style of the scales reduces the rate at which this canker disease spreads. It made slow progress from the site of its initial introduction in Nova Scotia, Canada, on European beeches in 1890 until its arrival in the Adirondack Mountains of New York in about 1978. The native range of the American beech includes all states east of the Mississippi River. As of 2005, beech bark disease had been confirmed in several of these states, suggesting an eventual ecological impact similar to that of chestnut blight.

Black walnut (*Juglans nigra*) and butternut (*J. cinerea*) are valued trees in the eastern forests of the United States. In the western forests, a new fungus was discovered in the early 2000s that causes so-called thousand cankers disease of black walnut. This pathogen was later discovered in Tennessee in 2010. In this disease, small cankers form wherever the walnut twig beetle provides an opening in the bark. This new pathogen may become a threat to the black walnut throughout its native range.

Butternut canker disease is apparently caused by an exotic pathogen that was named in the 1980s. The U.S. Forest Service has begun a project to find butternut trees that are resistant or tolerant to the disease to create a seed nursery for restoration of this species.

Numerous other canker fungi threaten woody plants, including many ornamental trees and shrubs. In some cases, native fungi are an important threat to imported plants. For example, Cytospora canker has caused extensive landscape losses of the Colorado blue spruce, a popular import, in the eastern United States. The Colorado blue spruce is adapted to the dry, cold winter of the western Rocky Mountains and is therefore much more prone to infection by canker fungi in the moist climate of the East. Many canker diseases are dependent on host plants being concentrated in a small area. For instance, black knot of *Prunus* species can be a serious canker disease in cherry and plum orchards, but it is usually a minor problem on wild cherry trees growing in genetically diverse and species-diverse natural ecosystems.

The damage that a canker fungus causes often depends on the vigor of the tree. A healthy tree may be able to produce callus and new vascular cambium at a rate that will keep the canker fungus from girdling a branch or the trunk. In such cases, a canker may show annual rings of callus growth, as the tree attempts to contain the growth of the invading fungus. Such a biological defense depends on the vigor of the host, so a tree stressed by other environmental factors, such as drought and air pollution, may not be successful in containing the pathogen. Many canker-forming fungi are "opportunists" and can attack only woody plants stressed by drought, extremes in temperature, other diseases, or a reduced root area.

Woody plants in landscapes and orchards should be pruned to remove canker infections while they are still limited to small branches. Since most canker fungi enter woody plants through wounds, dormant-season pruning is safest, because the air is dry and fungi are not active. Wounds from pruning are not likely to be invaded by other canker and decay fungi and bacteria from late fall to late winter.

Wood Decay Fungi: Crumbling Wood

So far, we have looked at diseases that impede water transport in the active xylem (vascular wilts) and diseases that kill the living layers just beneath the bark (cankers). A specialized group of fungi is capable of causing wood decay, and many of them do this by feeding on the lignin in the secondary walls of xylem and fibers. Lignin gives wood its strength and weight.

Most of these fungal pathogens are basidiomycetes, and many produce relatively large conks and other fruiting bodies commonly seen on the trunks and branches of infected trees. To penetrate the bark-protected woody tissues, wood decay fungi require wounds that expose the wood. Such wounds may be caused by animals, including birds and insects; weather conditions, such as ice, lightning, and wind; and many human abuses, including those from lawn mowers and automobiles.

Wood decay fungi exist in different genetic strains, similar to those of other fungi, such as the chestnut blight fungus. When their hyphae meet as they grow through the wood, black to brown zone lines form, as they wall themselves off from the incompatible strains. Woodworkers use the term spalted to describe wood with staining and zone lines caused by these fungi. They look for pieces of wood that have these unusual patterns but also structural integrity, and they carve or turn the wood into beautiful bowls and other decorative pieces.

The fruiting bodies of wood decay fungi exist in an amazing array of sizes, shapes, and colors. These are the fungi that people probably notice most often when walking in the woods. Fruiting bodies may be fleshy or dry and annual or perennial (adding new layers each year). Many have various patterns of gills, pores, or wavy layers, which increase the spore-producing area on the surface of the structure.

The basidiospores are forcibly discharged in tremendous numbers, to be carried on the wind to new victims. Most of the spores perish, but a few find their way into new trees. Saprophytic basidiomycetes, responsible for the decay of dead trees, produce similar fruiting bodies. Unless the fungus is identified, it is not usually possible to know whether a fruiting body on the trunk of a tree is the pathogen that killed it or just a saprophyte using the dead wood for food.

In either case, removing the fruiting body will not help the tree, because the active vegetative mycelium is what invades the wood. A large reproductive structure simply reflects the massive amount of mycelium that is inside the tree decaying the wood, perhaps extending 5–10 feet (1–3 meters) above or below the fruiting body of the fungus. It is common and not alarming to see fruiting bodies on branches, but once they form on the main trunk of a tree, it is doomed.

Tree Defenses: How to Stop an Invasion

Trees that succumb to invasion by wood decay fungi have failed to successfully fight the invasion. One of the secrets of long-lived trees is their ability to prevent invasion by bacteria, fungi, and other microorganisms that might cause decay or other diseases.

Rather than fight each possible enemy individually, many trees have a mechanism for isolating wounded woody tissue. This mechanism is called compartmentalization. It is a biological process that involves the isolation of wounded tissue, the production of antimicrobial compounds by the tree to fight invasion, and a change in the pattern of cell production that allows the tree to continue growing despite the presence of the wound. The type and pattern of cells in woody tissue, described earlier in the chapter, help the tree compartmentalize and "wall off" the invader.

When a wound exposes internal woody tissue to the environment, the tree immediately blocks off the area that has been exposed. Woody tissues consist of numerous small compartments that are bounded by living parenchyma cells and the vascular cambium. Living parenchyma cells allow the tree to respond to wounds by changing biochemical pathways and producing dark-colored antimicrobial substances—especially phenolic compounds, such as tannins. These compounds also help strengthen the cells. The tracheids and vessels in wounded tissue become plugged by tyloses, air bubbles, and other substances that help prevent microbial invasion.

A cross section of wounded wood usually includes a small area of decayed and invaded wood bordered by a darker layer that compartmentalizes the wounded area to prevent further invasion. The vascular cambium produces more parenchyma cells in the area of the wound, which allows a stronger biochemical strengthening of the compartment borders.

The final response of a tree to wounding is to keep growing. Unlike animals, which have a healing process that eliminates dead and dying tissue, trees maintain the wounded tissue and continue to grow around it. This natural process not only protects trees from invasion by microorganisms, but it also allows them to self-prune low or damaged branches. The branch stub in the main trunk becomes darker and stronger after the loss of the branch, but the dead stub tissue is compartmentalized to prevent invasion of the main trunk by microorganisms. This phenomenon is easily seen in knotty pine lumber.

An understanding of compartmentalization in woody plants has changed some tree maintenance practices. For instance, some tree surgeons no longer clean out decayed areas of hard, darkened wood but leave them as protective layers. In addition, tree surgeons recommend cutting a branch without damaging the branch collar, rather than flush with the trunk (Figure 12.6). This allows the tree to compartmentalize the branch stub more successfully. Also, many tree specialists no longer paint the surface of a tree wound after pruning, because the tree's own internal compartmentalization processes protect it from invasion.

At this point, it should be clear that a tree that reacts quickly to a wound may be able to isolate the tissue before invading pathogens can cause further damage. Several factors determine the success with which a tree can compartmentalize and prevent infection. The

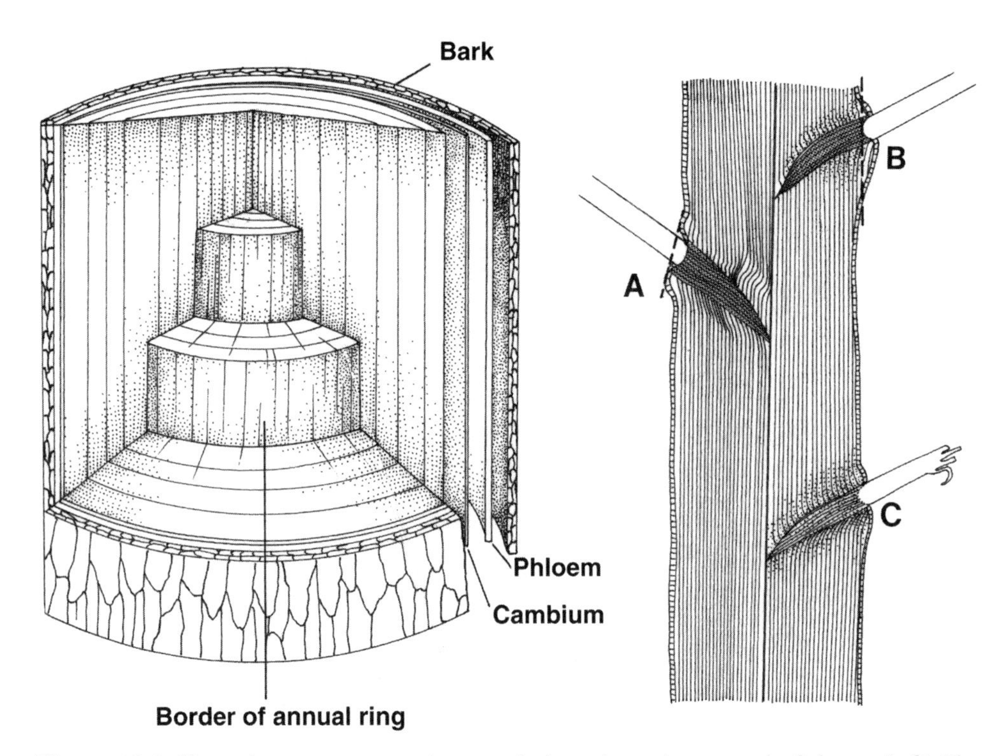

Figure 12.6. Natural compartments in trees help reduce the spread of decay. **Left,** The compartments in a mature tree. **Right,** Branch stubs may or may not cause further decay in a tree trunk. A properly pruned branch (**A**) does not disturb the collar, allowing the branch stub to compartmentalize. Improper pruning (**B**) cuts open large areas of the trunk to potential decay. Broken branches (**C**) may encourage the growth of decay fungi, which can overcome compartmentalization of the branch stub.

depth, location, and size of the wound are physical factors that affect successful compartmentalization. (This explains why the damage done by ice storms, which often tear off large branches, can be so deadly to trees.) In addition, because the compartmentalization process involves living cells, the overall health and vigor of the tree affects the rate and strength of the chemical and biological reactions of the process. Trees subjected to other stresses—such as air pollution, soil compaction, and drought—are less likely to compartmentalize successfully. Besides avoiding wounds and using proper pruning techniques, the best way to ensure the long life of a tree is to keep it healthy and vigorous through fertilization, watering, spacing, and site selection.

Mushrooms and Trees: Pathogens and Friends

It is relatively common to find mushrooms beneath many kinds of trees. Some of these mushrooms are pathogens, and some are beneficial to the tree.

Armillaria species are important pathogens that cause root rot of many tree species (Figure 12.7). They are aggressive parasites of conifers in the western United States and parasites of stressed hardwoods in the eastern United States. During autumn, clumps of the spore-producing mushrooms can often be found at the bases of infected trees. Basidiospores infect tree stumps, and then the fungus spreads to neighboring healthy trees via stringlike masses of mycelium called rhizomorphs, resulting in the common name "shoestring fungus."

In 1992, scientists discovered that a single *Armillaria* mycelium can spread from tree to tree over hundreds of years. One *Armillaria* mycelium, discovered in Crystal Falls, Michigan, was estimated to weigh 11 tons (10 metric tons) and covered 37 acres (15 hectares). It was dubbed the "humongous fungus" and was probably more than 1,500 years old. The remarkable size of the fungus prompted some misleading media accounts of the discovery of a "giant mushroom," rather than a widely distributed mycelium. Since then, even larger *Armillaria* individuals have been found in Oregon. Crystal Falls continues to celebrate its famous fungus with a summer festival, at which townspeople claim to make the biggest mushroom pizza in the world.

A successful biological control for another important root pathogen of conifers, *Heterobasidion annosum,* uses the saprophytic fungus *Phlebiopsis gigantea.* The basidiospores of the biocontrol agent can be painted on stumps just after trees are felled. In some cases, the basidiospores have even been mixed with chain saw oil to facilitate inoculation when the tree is cut. This allows *P. gigantea* to colonize the stump first and greatly reduces the chance of infection by the pathogen. *P. gigantea* does not spread or cause root rot in healthy trees.

Figure 12.7. Root rot caused by *Armillaria mellea*, the "humongous fungus." **A,** mycelial fans beneath the bark of an infected tree. **B,** mushrooms produced at the base of an infected tree.

Science Sidebar

Glowing in the dark

Armillaria species are capable of bioluminescence. On a moonless night, an infected tree emits the eerie green glow of the mycelium in infected wood and mycelial fans under the bark, which is known as "foxfire." Before the invention of flashlights, people used chunks of infected wood to light the way on dark nights.

Other types of mushrooms are commonly found in forests. Some are saprophytes, which are important for the decay of dead and dying trees, but many have an intimate, mutualistic relationship with a nearby tree. The word mutualistic means that both the tree and the fungus benefit from the relationship. The fungus obtains nutrients from the tree, and the tree receives protection from root pathogens, as well as increased surface area for absorption of water and certain minerals—especially phosphorus.

Such fungi are called mycorrhizae. Nearly all plants, herbaceous and woody, have mycorrhizae associated with their roots, either internal or external to the plant cells. Some, like orchids and maples, host endomycorrhizae that grow into the roots' cortical cells. Many trees, such as pines and oaks, host ectomycorrhizae that form a thick layer of mycelium outside the root and grow around the cortical cells (Figure 12.8). This mycelial network is called the Hartig net.

Early attempts to plant trees in treeless regions—such as the U.S. Great Plains, southern Africa, and parts of Australia—were not very successful, because the proper mycorrhizal fungi were missing. Many tree seedlings died when transplanted outdoors. Modern forest plantings include inoculation of tree seedlings and their growing media with mycorrhizae.

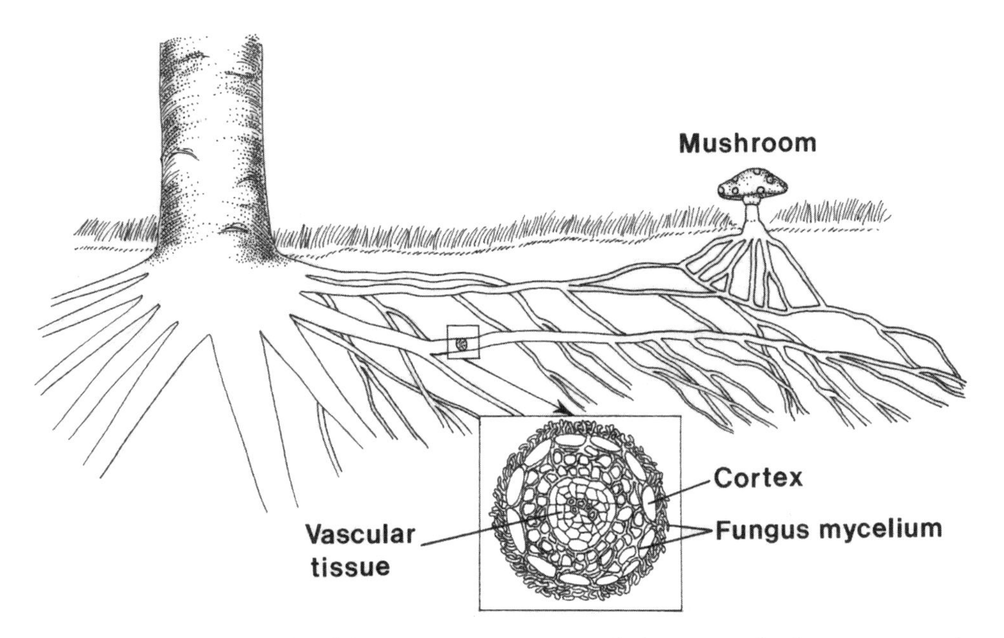

Figure 12.8. Ectomycorrhizal fungus in association with the roots of a forest tree. Mushrooms of the fungus may be found beneath the host tree. A cross section of an infected root reveals the netlike layer of mycelium on the root surface and hyphae between the cortical cells.

Mycophiles (mushroom lovers) and mycophagists (mushroom eaters) quickly learn to search for their favorite fungi under the appropriate tree hosts. Flushes of mushrooms are especially profuse under dead and dying trees, as the mycorrhizal fungi move on to new host trees via airborne basidiospores.

Protecting Our Urban Forests: Life on the Streets

Can you imagine how difficult city life must be for our trees, especially those that line the streets and sidewalks? The aboveground parts of a tree must be matched by a root system with the capacity to carry water and minerals to the crown. To absorb enough carbon dioxide for photosynthesis, trees evaporate hundreds of gallons of water every day through their stomata. According to the U.S. Department of Agriculture (USDA), the net cooling effect of a young, healthy tree is equivalent to 10 room-size air conditioners operating 20 hours a day.

Often, city trees have little soil around them. People tend to take shortcuts across their roots, compacting the soil and eliminating the air pores that keep roots alive. Many city trees are subjected daily to vehicle exhaust and other pollutants and, in some areas, salt used for snow removal on streets and sidewalks. It is not uncommon to see city trees with entire sections of bark and phloem missing, where cars have repeatedly backed into them. Many trees have wounds from lawn maintenance equipment, which make them compartmentalize large sections of wood, if they survive at all.

If people would see trees as living organisms that need their protection, many of these magnificent plants would survive to provide shade and cooling effects for many more years. Replacement trees often require decades to mature. Citizens on planning boards can encourage funding for tree maintenance and encourage species diversity, so the impact of tree losses can be minimized.

Mistletoes: Higher Plants That Infect Trees

In the days before the computer and television, people spent more time outdoors, and some of them noticed and were intrigued by unusual plants. Sometimes, these plants were associated with trees, not just on them but with living connections to their twigs and branches.

One example is the mistletoes, which have fascinated people for centuries. Their ability to stay green in winter and their seemingly magical ability to live without roots, never touching the ground, led ancient people to believe they were divine gifts—perhaps created by thunderbolts.

Many trees host various species of parasitic mistletoes. True mistletoes are large evergreen plants that invade the xylem of hardwood trees, absorbing water and minerals through specialized structures called haustoria, rather than through roots in the soil. (Recall that the word haustoria also is used to describe the hyphal structures of some pathogenic fungi, which are used to absorb nutrients from host plants.) In North America, true mistletoes (*Phoradendron* species) exist mostly in frost-free areas of California and the southern states (Figure 12.9). The name *Phoradendron* comes from

Science Sidebar

Mistletoe rituals

Mistletoes had an important place in many ancient religious rituals among tree worshippers of western Europe. Mistletoe was an important part of the summer solstice rituals of the Druids, and it also had a role in the mythologies of the Celts, Greeks, Norse, and Romans.

Druid priests cut the mistletoe with a golden sickle onto a white cloth to prevent the plant from touching the ground. Mistletoe harvest also was part of Norse ceremonies involving Balder, the son of Odin. Because of its role as a parasite of the all-powerful oak, mistletoe became a symbolic source of protective and medicinal powers. The burning of oak logs, human sacrifice, and the harvest of mistletoe were intertwined in solstice ceremonies. Beliefs about the plant's mystical properties led to the more recent custom of kissing under the mistletoe at Christmas.

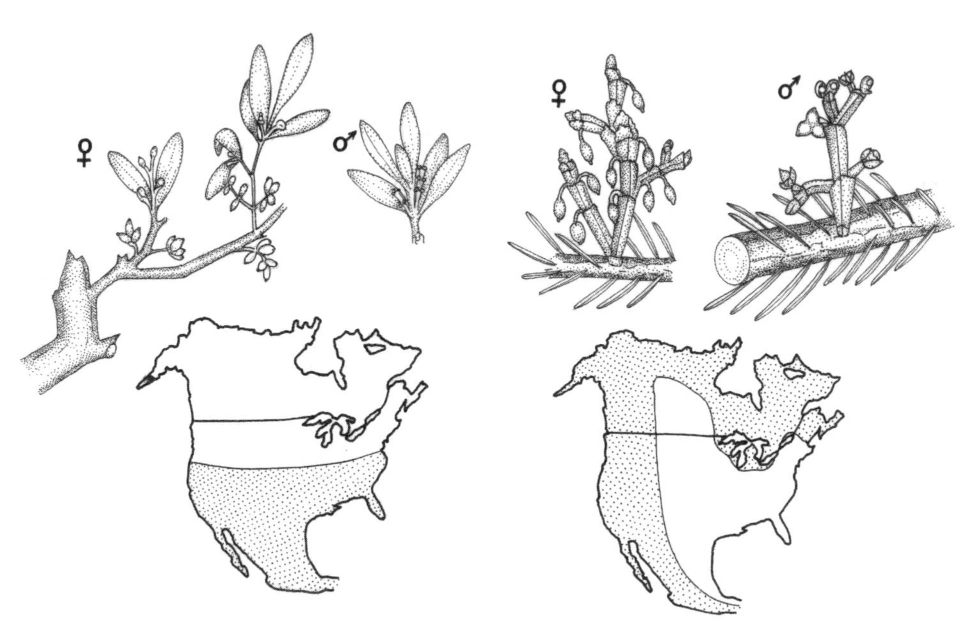

Figure 12.9. Mistletoes, dioecious plant parasites. **Left,** True mistletoes, *Phoradendron* species, infect hardwood trees and are found in the southern half of the United States. **Right,** Dwarf mistletoes, *Arceuthobium* species, infect conifers and are found in the range indicated on the map—primarily the western states, the Great Lakes region, and northern New England.

Science Sidebar

Other parasitic plants

It is amazing that more than 2,500 species of flowering plants, found in many botanical families, parasitize other plants. While some, like the mistletoes, can be economically important pathogens, most have no economic importance but are interesting curiosities.

In North America, beech-drops (*Epifagus virginiana*) and squaw-root (*Conopholis americana*) are common parasites of beech and oaks, respectively. Hikers may see them growing beneath these species in the woods.

The largest flower in the world is that of the parasitic plant *Rafflesia arnoldii,* which grows in only a few areas of Indonesia and Malaysia. The dark-red, mottled flowers can be more than 3 feet (1 meter) across and weigh 15 pounds (6 kilograms). In addition, the flowers smell like rotten flesh to attract flies, which assist in pollination. Like the mistletoes, these plants are dioecious.

Some parasitic plants cause problems in agronomic crops. These pathogens can be considered the ultimate weeds, competing directly with the host plants for food, minerals, and water through their haustoria. Dodders (*Cuscuta* species) are pathogens of many crops, including alfalfa, clover, and cranberries. The tangled, yellow strands of dodder reduce the growth of host plants and can make harvest extremely difficult (Figure 12.10). Common names include strangleweed, pull-down, and devil's hair.

Witchweed (*Striga* species) is a major agricultural pest in many parts of Africa and Asia. This plant gets its name because it is a root pathogen, stealing minerals and water from its host. The aboveground green shoots produce beautiful yellow and red flowers, which deceptively mask the deadly parasitism occurring below the ground. Witchweed infects important food crops in the grass family, including corn, rice, sorghum, and sugarcane (Figure 12.11). It is a particularly difficult parasite to manage, because a single plant can produce more than 50,000 tiny seeds that can remain viable in soil for up to 20 years. A badly infested field can lose 90% of the crop to witchweed. Heavy infestations have led to abandonment of farmlands in areas of the world such as Africa and Asia, which cannot afford to lose food.

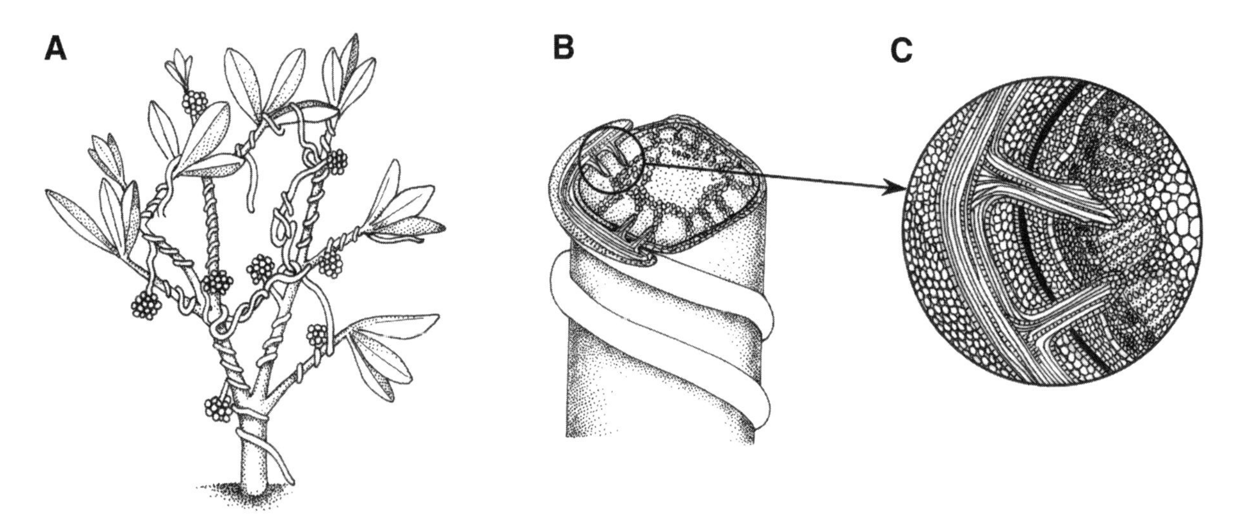

Figure 12.10. A, dodder on a host plant. Dodder winds around the host stem as it grows (**B**), invading with haustoria to absorb water and nutrients from the vascular tissue of the host (**C**).

the Greek words for "tree" and "thief." In Europe, true mistletoes are classified in the genus *Viscum.* Mistletoes, like some other flowering plants—such as ginkgo and marijuana—exist as separate male and female plants and are termed dioecious, which means "two houses." Female mistletoe plants produce white berries that attract birds, which aid in seed dispersal.

Another type of mistletoe is an important parasite of conifers. The dwarf mistletoes (*Arceuthobium* species) invade both the xylem and phloem of host trees, obtaining food, water, and minerals. They produce tiny sticky seeds that are shot from mature fruits, sometimes landing on the needles and twigs of host trees. The seeds germinate, penetrate, and grow into the woody tissues. Conifers heavily infected by dwarf mistletoes become distorted and stunted, producing poor-quality timber. Infected trees often develop clusters of branches known as witches' brooms.

Natural fire cycles reduce many pathogens and insect pests, but fire control is required, as people have built houses in forested lands. Modern forestry practices have increased the problems caused by dwarf mistletoes. Clear cutting of timber lands, which causes soil erosion, has been replaced by selective cutting of mature trees and replanting over a period of time. Unfortunately, pathogens and pests can flourish when both mature trees and new seedlings grow in the same area.

In this chapter, you learned about both the largest plants—trees—and the largest plant pathogens—fungi in the genus *Armillaria.* In the following chapter, we investigate the smallest known pathogens—the viruses and viroids.

Figure 12.11. Witchweed (*Striga* species), a parasite of host roots. **A,** witchweed at the base of a corn plant. **B,** corn growing poorly due to witchweed infection. Witchweed plants can be seen between the corn rows.

CHAPTER 13

The Smallest Pathogens: Viruses and Viroids

The pathogens discussed in previous chapters—fungi, bacteria, and nematodes—are all relatively easy to see, although many require magnification. In addition, these pathogens often induce characteristic symptoms that aid in disease diagnosis.

In contrast, viruses are too small to see with the naked eye or the light microscope. Diagnosis is complicated, because many viruses induce symptoms (including stunting, yellowing, distortions, and reduced crop quality and yield) that also can be caused by a variety of other factors (such as poor soil fertility, soilborne pathogens, herbicide residues in soil, and even air pollution).

Plant viruses have a long scientific history. Many of the early breakthroughs in virology were made with plant viruses. Today, more than 900 plant viruses have been described. The other pathogens discussed in this chapter—the viroids—were first discovered in plants in the 1960s. Viruses and viroids differ from the previously discussed pathogens in how they cause diseases in plants and in some of the approaches used to manage the diseases they cause.

Viruses as Plant Pathogens: Early Scientific and Economic Impacts

Viruses are small, relatively simple pathogens. Most plant viruses are composed of only two components: a nucleic acid genome, which provides the genetic information, and a protein coat, which protects the genome and also has other important functions—for example, in virus transmission. The nucleic acid may be single-stranded (ss) or double-stranded (ds) ribonucleic acid (RNA) or deoxyribonucleic acid (DNA), although the vast majority of plant viruses have a genome of ssRNA.

In 1886, Adolf Mayer, a German scientist, first identified mosaic as a disease in tobacco. He observed a random pattern of light- and dark-green color in leaf tissue (Figure 13.1). However, he considered the causal agent to be a bacterium.

Figure 13.1. Symptoms of tobacco mosaic disease.

The first indication that a new kind of pathogen was involved came from experiments by Russian scientist Dmitrii Ivanowski in 1892. He passed sap from an infected tobacco plant through a filter that would remove bacterial cells, and the sap remained infective. This demonstrated that whatever caused tobacco mosaic disease was smaller than bacteria. However, Ivanowski doubted the validity of his results, and it was not until 1898 that Martinus W. Beijerinck, a Dutch scientist, proved that this infectious sap contained a pathogen that multiplied. Beijerinck called the causal agent a "living, contagious fluid," or virus, to differentiate it from pathogens composed of cells, such as bacteria and fungi.

Further understanding of the structure and function of viruses had to await the development of new technologies in the 20th century, which allowed chemical analysis and visualization of viruses. In fact, many of these new techniques were invented as tools to unravel the mysteries of these pathogens.

Diseases caused by viruses were particularly frustrating to scientists and physicians. Unlike diseases caused by fungi and bacteria, virus diseases had no visible pathogens, and common culturing techniques failed to isolate causal agents. In addition, virus diseases of animals and humans were not cured by antibiotics, the "miracle drugs" that were so successful against bacterial pathogens. Although humans and animals might recover from virus diseases through the production of antibodies by their immune systems, virus-infected plants had to be destroyed.

Virus diseases of plants had been recognized for many centuries, even though no one knew anything about their causal agents. For example, virus-infected flower bulbs caused economic mayhem in western Europe in the 1630s. About 80 years earlier, a Flemish diplomat, Ogier Ghiselin de Busbecq, had introduced tulips to Europe after discovering them as wildflowers in Turkey. Tulips became a craze, and an unstable speculative market grew, as people tried to make their fortunes on the new flowers. The most valuable bulbs were those that produced flowers with streaks and so-called flames and feathers of contrasting colors. Known as "bizarres," these bulbs were particularly valuable because

Figure 13.2. A painting by Dutch artist Jakob Marrel (1613/4–1681) that includes "bizarre" tulips, with color breaking due to virus infection.

the seed produced by the flowers usually grew only into plants with normal, solid-color flowers.

Today, it is known that such color breaking is a common symptom in virus-infected plants. These plants may transmit a virus to their seed but usually at a low rate—perhaps 1–10%. Color breaking contributed to the value of bizarre tulip bulbs in the Dutch markets, one of which was traded for the exorbitant price of $8,395. The market finally crashed, leaving many dealers bankrupt, but the Netherlands remains the producer of 95% of the world's tulip bulbs.

During the period of so-called tulipomania, these flower symptoms were portrayed in paintings by many Dutch and Flemish artists (Figure 13.2). Even today, one popular tulip cultivar is known as the Rembrandt. Modern tulips are available with genetic mutations that consistently give beautiful color patterns. Mutations were needed to accomplish this, because virus-infected plants eventually become stunted and produce small and weakened bulbs.

As described in Chapter 1, potato growers saw the same "running out" in virus-infected potato tubers. By periodically harvesting the seeds produced by sexual reproduction, they could obtain virus-free plants and restore the yield, which had been greatly reduced by virus infection.

Structure and Replication of Viruses: Simply Effective

The exact nature of viruses remained a mystery for many decades. That changed in the 1930s, with the introduction of the electron microscope and staining techniques that allow detailed visualization of viruses.

Viruses were found to be approximately 10 times shorter in length than bacteria. They are measured in nanometers (1 nanometer = 10^{-9} meters). The first plant virus visualized was the causal agent of tobacco mosaic disease, *Tobacco mosaic virus,* which is a rod 15 nanometers wide and 300 nanometers long. Plant viruses may be short to long rigid or flexuous rods, spherical (actually polyhedral), or shaped somewhat like small, rod-shaped bacteria (Figure 13.3).

In 1946, Wendell M. Stanley, an American scientist, received a Nobel Prize for the first crystallization of a virus, which was, not surprisingly, *Tobacco mosaic virus*. For the first time, a relatively large amount of purified virus was available for analysis. Shortly thereafter, Frederick Bawden and Norman W. Pirie, in Great Britain, demonstrated that the virus was a nucleoprotein. Later work showed that the nucleic acid was necessary for virus infection, and the protein served a protective role. Of course, further understanding of viruses depended on publication of the structure of DNA by James D. Watson and Francis Crick in 1953, as discussed in Chapter 6.

Plant viruses have not been given Latin binomials, but each virus name is written in italics and the first letter of the first word is capitalized. Viruses are often named for one of their host plants and the symptoms they cause, and virus names are abbreviated using initials. For example, *Tobacco mosaic virus* causes mosaic symptoms in tobacco plants and is commonly shortened to TMV. Note that the disease name,

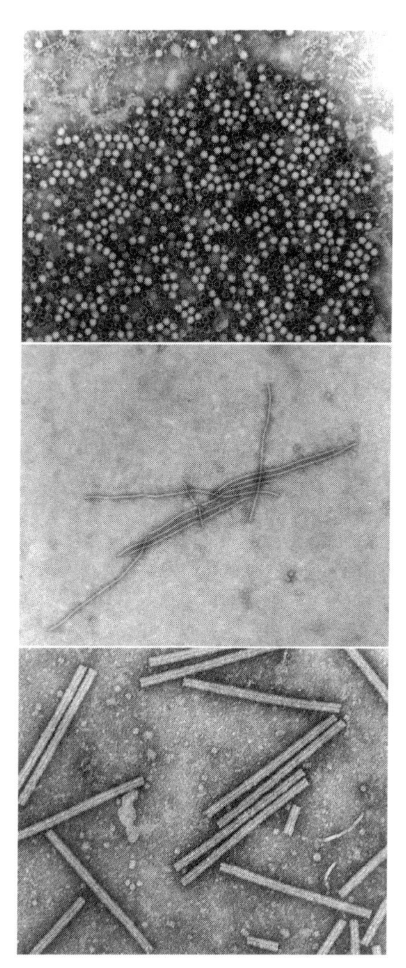

Figure 13.3. Common shapes of plant-pathogenic viruses. **Top,** spherical (*Cowpea mosaic virus*). **Center,** flexuous rods (*Maize dwarf mosaic virus*). **Bottom,** rigid rods (*Tobacco mosaic virus*) (electron micrographs).

tobacco mosaic, is not capitalized or italicized. Older classification systems were based on properties such as virus particle structure and means of transmission, but now, viruses are placed in genera and families based on the structures of their genomes and how they reproduce, or replicate.

The genome of a plant virus can be composed of either DNA or RNA, unlike all other groups of plant pathogens, which have DNA genomes. Many viruses consist of one type of virion (virus particle), which contains the complete genome of the virus. However, the genome in many other plant viruses consists of several different parts. In such a virus, all parts of the genome must be present in a plant cell for the virus to function. The need to infect the cell with several different pieces of nucleic acid does not necessarily reduce transmission efficiency; many easily transmitted plant viruses are of this type. One example is *Cucumber mosaic virus* (CMV), a common cause of mosaic symptoms in cucumbers and squash.

Chapter 6 explained that genetic information in DNA is transcribed to the messenger RNA (mRNA), which moves from the nucleus into the cytoplasm. The genetic code of the mRNA is then translated through triplet codons. Transfer RNAs (tRNA) carry amino acids to the mRNA, and with the help of the ribosomes, a protein is eventually produced. These same mechanisms are used in the replication of plant viruses.

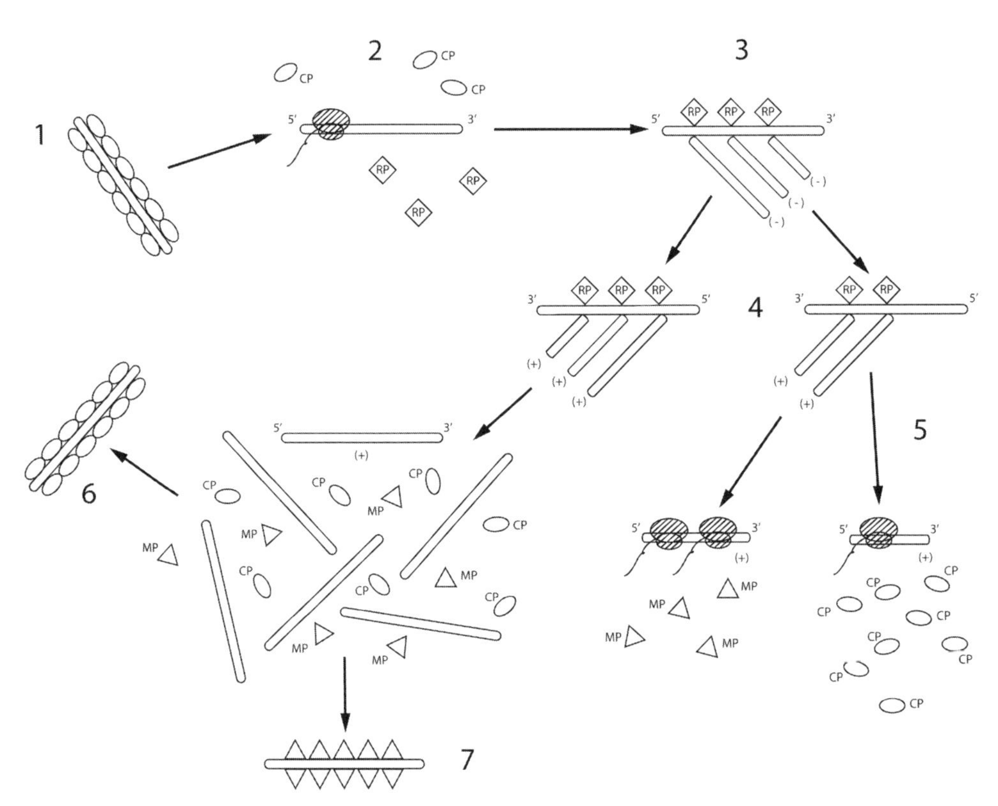

Figure 13.4. Replication cycle of *Tobacco mosaic virus* (TMV) (**1**). As the coat protein (CP) is stripped away, host ribosomes translate the replicase-associated proteins (RP) from the RNA (**2**). The replicase proteins (**3**) produce a negative-sense (– sense) RNA template, which will be used to make new virus RNA (**4**). The proteins also generate two other RNAs (subgenomic RNA, or sgRNA) (**5**), which can be translated into the movement protein (MP) and CP. The new virus RNA is encapsidated by the CP to form new TMV particles (**6**) or wrapped in MP (**7**), which allows it to enter a new cell and continue replication.

When a plant virus enters the plant cell, the virion disassembles so that the nucleic acid is exposed. In most plant viruses, the nucleic acid is an ssRNA that serves as an mRNA, using the same genetic code that functions in all living organisms. Part of the virus genome codes for an RNA polymerase or replicase—an enzyme necessary for replication, or multiplication, of the virus RNA. A mirror-image copy of RNA is made from the original virus RNA, which then serves as a template for producing many copies of the virus RNA. After the original ssRNA is replicated, another section of the RNA codes for coat protein. The coat protein then coats, or encapsidates, the RNA copies, and new virions are formed.

Remember that this entire replication cycle takes place within the living cell of a plant host (Figure 13.4). Thus, every virus is an obligate parasite. All the materials that the virus needs to produce new RNA and protein come from the host cell.

The entire genomes of many plant viruses have now been sequenced, and they typically code for fewer than ten proteins. For instance, the TMV genome contains 6,400 nucleotides that encode only four proteins. Two of the proteins are replicase enzymes necessary for replication of the virus RNA. The third is a movement protein, which facilitates spread of the virus in an infected plant, and the fourth is the coat protein.

Much current research is focused on trying to discover how viruses cause diseases in plants. Although viruses do not produce disruptive enzymes or toxins, they do usurp substantial amounts of a plant's resources, including nucleotides and amino acids, as well as the use of host ribosomes and tRNAs. Recent research has found that molecules produced during virus replication, including RNAs and proteins, disrupt specific physiological processes in plants, leading to the visible symptoms associated with virus infections.

Virus Transmission: Hitching a Ride

Because viruses are very small and unable to grow or move on their own, they typically enter plant cells through wounds created by other living or nonliving agents. There are a few exceptions, such as viruses that enter ovules during fertilization by virus-infected pollen grains and viruses that pass through root grafts between neighboring plants. Many agricultural practices result in the sap transmission of viruses when wounds are produced during activities such as cultivation, grafting, mowing, and pruning. A small number of viruses also may be transmitted when infected plants rub against neighboring plants during windy weather and storms. As discussed later in this section, mechanical or sap transmission is an important tool for plant virologists in the laboratory.

Figure 13.5. Important insect vectors of plant viruses: aphid (**Top**) and whitefly (**Bottom**).

In nature, various vectors are responsible for virus transmission, the most important of which are arthropods—especially insects with piercing/sucking mouthparts called stylets. The insects use their stylets to pierce plant cells and to suck up the contents. In doing so, they inadvertently ingest viruses that are present in the plant cells, which they may then carry in their stylets or bodies as they move on to healthy plants. The most important plant virus vectors are aphids in temperate regions and whiteflies in tropical regions (Figure 13.5).

The relationships among vectors, viruses, and host plants are complex. Just as a particular virus can infect only certain host plants (its host range), a particular virus can be transmitted only by certain insect species—a specificity associated with the virus coat protein.

For example, a single aphid can feed on a plant infected with two different viruses and transmit one virus but not the other.

Early studies found that the transmission patterns of different viruses were not all the same, and categories of insect transmission were created. One category is called nonpersistent (or styletborne). In nonpersistent transmission, the insect becomes capable of transmitting a virus after only a few seconds of probing its stylet into an infected plant, but it remains viruliferous (capable of transmitting the virus) for only a short time—often less than a day.

Another category, called persistent (or circulative) transmission, involves a longer feeding period to acquire the virus and usually a latent period before the insect becomes able to transmit the virus. During the latent period, the virus moves through the insect's digestive system, circulates in the blood, and eventually passes into a salivary gland. The insect can then excrete the virus, along with salivary secretions, into plants as it feeds. The insect may be able to transmit the virus for days or even weeks. In a few cases, the virus even infects the insect and replicates within it, which can greatly increase the transmission potential. There also are patterns of insect transmission that do not fall neatly into the persistent and nonpersistent categories.

In addition to aphids and whiteflies, other insects transmit plant viruses, including leafhoppers, mealybugs, and beetles. Viruses also are transmitted via infected seed and pollen; through human activities, such as vegetative propagation and grafting; and by other organisms, including both primitive fungi and ectoparasitic, migratory nematodes.

Virus Movement in Plants: The Short and the Long of It

Once the virus is in the host plant, the spread of infection throughout the plant may involve the movement of virus particles or the movement of the uncoated virus genome associated with one or more proteins, including the movement protein. Movement over short distances, from cell to cell, takes place through cytoplasmic channels that connect living plant cells, known as plasmodesmata. These channels must be modified by the virus for it or its genome to fit through them. Spread of infectious particles from one cell to another is a relatively slow process, often measured in micrometers per hour.

When the infectious particles reach the vascular tissue, long-distance movement throughout the plant is possible. Because viruses require living cells for replication, most infectious particles of viruses move in the living cells of the phloem, rather than in the nonliving xylem vessels. Viruses follow the same pattern of movement in the translocation stream as do the products of photosynthesis. Early experiments demonstrated that virus infection typically spreads first to roots, then to young shoot tissues, and gradually to older leaves. Compared to cell-to-cell movement, long-distance movement can be quite rapid, measured in centimeters per hour.

In nature, viruses usually infect their hosts systemically. They can be found in almost all of the cells of a small herbaceous plant only 30 days after infection, whereas systemic infection of a large woody plant may take years. Such a long time lag can have important economic implications. Many woody ornamentals and fruit trees are composed of grafted root stocks and scions (shoots). If the plant tissue on one side of the graft union is infected with a virus, considerable time may pass before the virus reaches the other side. If the healthy part reacts strongly to the invasion by the virus, a hypersensitive reaction may occur, resulting in a layer of dead cells at the graft union. Virus infection is the most common cause of graft failure, which results in the death of the tree.

Virus Detection: The Classic Methods

Diseases of plants caused by viruses were often overlooked in the past and are sometimes still overlooked today. Barley yellow dwarf, a disease of cereals and grasses caused by two closely related viruses, is a good example. In 1951, barley fields in California turned brilliant yellow within a single week. This led scientists to closely examine the problem, which was eventually diagnosed as viral in nature.

Looking back in the scientific literature, it is clear this disease had plagued various grain crops—including barley, oats, and wheat—for many decades. Even though viruses had been known to cause plant diseases since the late 1800s, the symptoms and means of spread of these cereal diseases were unlike those of the previously described virus diseases. For these reasons, the epidemics had been attributed to cold, wet weather or bacterial infection. Today, the viruses that cause barley yellow dwarf are known to infect more than 150 grass species worldwide, including all of the most important cereal crops, yet the losses caused by this disease still often go undetected.

In some cases, viruses cause distinct and obvious symptoms in plants, including mosaics, ringspots, distortions, and changes in flower color. Mosaic and ringspot patterns are two symptoms that should immediately cause the observer to think that a virus may be involved. Many viruses produce more subtle symptoms, such as yellowing and stunting, which can be confused with many other plant problems. Other viruses produce no obvious symptoms, even though they can cause significant yield losses.

It is critical to detect viruses in plants before the plants are used for breeding, seed production, propagation, or agricultural production. Because not all plants produce obvious symptoms of virus infection, other means of detection are necessary. One of the first detection systems developed involves the use of so-called indicator plants. Scientists discovered through trial and error that when sap from a test plant is rubbed on the leaves of an appropriate indicator plant, distinct symptoms appear within 1 to 2 weeks, if a virus is present. In most cases, the symptoms consist of small necrotic spots called local lesions (Figure 13.6).

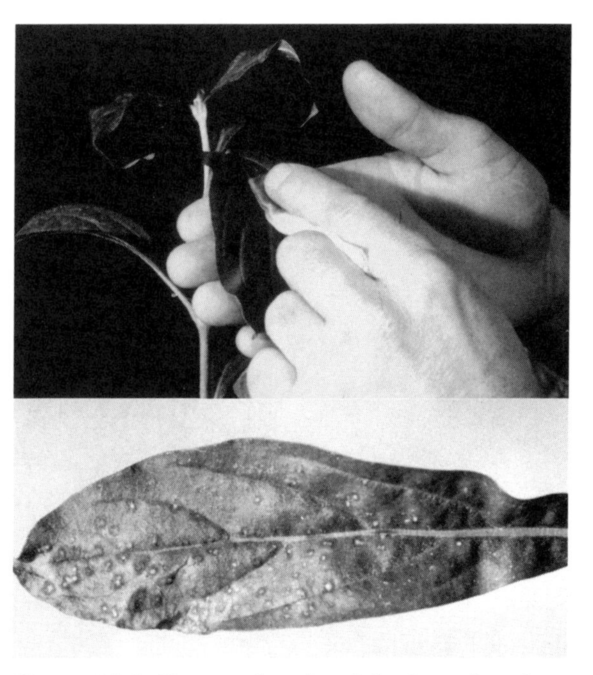

Figure 13.6. Bioassay for virus infection of a plant. **Top,** Sap from a test plant is rubbed on a leaf of an indicator plant, which shows distinct symptoms if the virus is present. The plant's leaves are usually dusted with a fine abrasive powder, so that small wounds are produced when the sap is applied. **Bottom,** About 1 week later, small necrotic local lesions indicate a positive test for the presence of the virus.

Sap transmission of many viruses is accomplished by gently rubbing sap from an infected plant onto the leaves of a healthy plant. Usually, the leaves are lightly dusted with an abrasive powder, so that the rubbing creates tiny wounds for virus entry. However, some viruses cannot be transmitted mechanically and must be transmitted by a vector or through a living plant connection, such as by grafting a piece of the test plant to an indicator plant. Indicator plants are used commercially and in many research laboratories.

The disadvantages of this detection method include many of the factors associated with any biological system. Those factors include symptom variation caused by environmental conditions, low virus titer (or concentration), misses, cross-contamination during the inoculation of a large number of test plants, and, of course, the time and space required to grow the indicator plants. Perhaps the most important disadvantage of this detection method is the time spent waiting for the expression of symptoms. For that reason, indicator host assays have largely been replaced by assays based on quick, direct detection of virus protein or nucleic acid in infected plant tissue.

Virus Detection: The Modern Methods

Scientists have long searched for methods of virus detection that are specific, fast, and well suited to the large quantity of tests necessary to support creating virus-free plants in commercial plant production. Some new methods make use of serology, relying on the immune systems of animals.

When a foreign protein (an antigen), such as the proteins associated with viruses, invades an animal, the animal produces antibodies against it. Antibodies are themselves proteins that physically fit onto the foreign protein to help prevent an invading virus from functioning and to expedite its removal from the body. If the antibodies produced by the immune system successfully inactivate the virus and stop its replication, the individual may recover from the disease. Antibodies are specific for each type of virus, and each time infection by a new virus occurs, new antibodies must be produced. Because the viruses that cause the common cold change frequently, many people get colds each year and must wait for their bodies to produce new antibodies before they can recover.

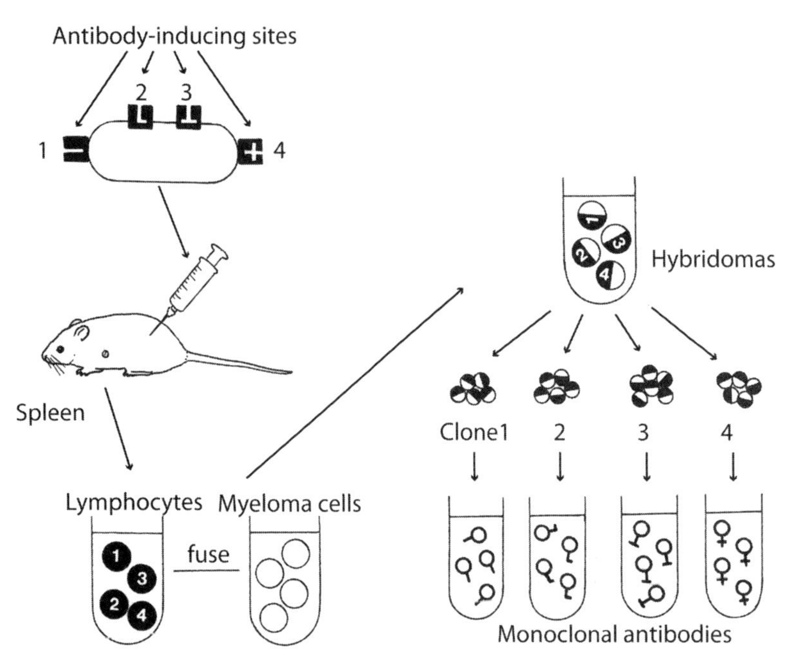

Figure 13.7. Steps in the process of generating monoclonal antibodies. Antibody-inducing sites on viruses injected into a warm-blooded animal cause lymphocytes in the spleen to produce specific antibodies. Each lymphocyte can be fused with a myeloma cell to produce a set of hybridomas with indefinite life. Hybridomas can be cloned, stored frozen, and later thawed and grown to produce antibodies that react with specific sites on the virus. Monoclonal antibodies can be produced for detection of a wide variety of proteins.

To obtain antibodies for serological tests of plant tissue, scientists first inject purified plant viruses into animals—usually rabbits. Even though plant viruses cannot infect the animals, the animals' immune systems recognize the plant virus proteins as foreign and produce antibodies against them. Several weeks later, blood is drawn from the animals, and the antibodies are separated from the blood cells and serum (blood fluid). These antibodies specifically react with the virus used to induce their production when it is present in a test plant, allowing the virus to be detected. Early serological tests involved visualizing the antibody/virus reaction as a cloudy aggregation. Modern tests amplify the reaction, so that a much lower level of virus can be detected.

Antibodies produced in this way are a mixture, for two reasons. Each virus triggers the production of several different antibodies, and the original purified virus preparation may contain small amounts of plant protein contaminants. These antibody mixtures are called polyclonal antibodies. More specific antibodies, called monoclonal antibodies, were first produced by Georges Köhler, César Milstein, and Niels Kaj Jerne in 1975, a discovery for which they received a Nobel Prize in 1984.

The first step in the production of monoclonal antibodies involves injecting the purified virus into an animal—usually a mouse or rat (Figure 13.7). Unlike plant cells, animal cells cannot be grown in culture indefinitely, so antibody-producing cells cannot be cultured for long periods. The monoclonal antibody technique circumvents this problem by fusing an antibody-producing cell from the injected animal's spleen with a lymphoma cell (cancer cell), which divides continuously, to produce a new cell called a hybridoma. Pure cultures of cells from a single hybridoma can be maintained indefinitely, so that each culture produces a pure supply of a single kind of antibody—thus, the term monoclonal antibody. The end product is a relatively inexpensive source of very specific antibodies, which eliminates the need for maintenance and repeated bleeding of living animals.

Because a continuous supply of monoclonal antibodies can be produced, it is now possible to use immunoassays to detect more complex pathogens and nearly any protein. If an identifying protein can be found that is specific to a bacterium, fungus, or nematode, monoclonal antibodies can be produced to rapidly detect and identify its presence in plant tissue, without using complex and time-consuming isolation techniques. Commercial immunoassay kits that use monoclonal (with a single source) and/or polyclonal antibodies (with mixed sources) are available for detection of many common plant viruses and for detection of other types of pathogens in flowers, turfgrasses, vegetables, and woody plants. Biotechnology companies have developed products that use monoclonal antibodies to detect other important compounds, as well, such as various mycotoxins and pesticides. Many of these kits use a detection method known as an enzyme-linked immunosorbent assay (ELISA).

In ELISA, polyclonal antibodies or monoclonal antibodies are linked to an enzyme. That enzyme causes a visible color change when a particular substrate is added, indicating that the antibody has reacted with the correct antigen (for instance, a virus). ELISA can detect very low virus titers—even nanogram amounts of virus—and virus concentration can be quantified by measuring the color intensity.

ELISA allows rapid testing of numerous plants using only small sap samples. Many ELISA protocols are completed in less than a day, providing rapid and specific diagnosis. So-called strip ELISA tests have been developed to detect economically important viruses

Science Sidebar

DAS-ELISA

Several different types of ELISA have been developed. One type commonly used for plant virus detection is the double-antibody sandwich, or DAS-ELISA, so called because the virus is "sandwiched" between two antibodies.

In DAS-ELISA, the first antibody is adsorbed to a small cylindrical well in a plastic ELISA plate (Figure 13.8). Excess antibody and other materials are washed away. In the second step, sap from the test plant is added. If the virus used to create the antibody is present, it will be captured by the antibody attached to the wall of the well and held in the well while other material is washed away. Other viruses and proteins will not be captured. In the third step, an antibody to which a special enzyme has been attached is added. If the expected virus has been captured, the enzyme-linked antibody also attaches to the virus, completing the "sandwich," and it remains in the well while all excess material is washed away. In the final step, a substrate is added that changes color if the enzyme is present.

If the virus were not present in the plant sap, the enzyme-linked antibody would have been washed away and lost. If color appears after the substrate is added, then the enzyme and virus must be present.

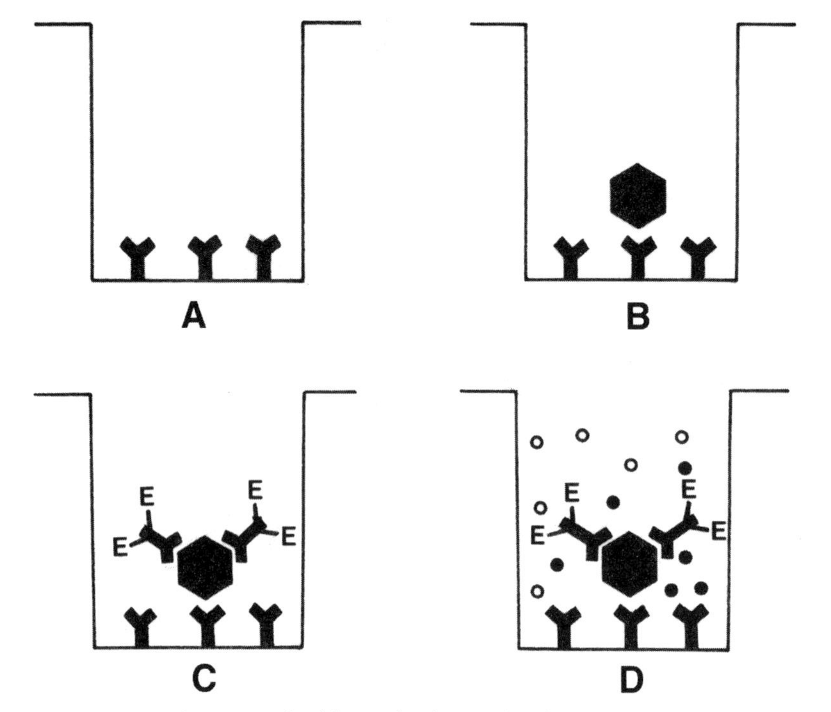

Figure 13.8. Steps in double-antibody sandwich enzyme-linked immunosorbent assay (DAS-ELISA). **A,** Specific antibody is adsorbed to the plate. **B,** The test sample containing the virus is added. **C,** The enzyme-labeled (E) specific antibody is added. **D,** The enzyme substrate (circles) is added and cleaved to give the color (dark circles). Each step is incubated in the plate for a specific period of time and then washed out of the wells.

of valuable greenhouse crops. Antibodies are adsorbed (adhered in a thin layer) onto a plastic strip that is dipped into leaf sap from a plant to be tested. Scientists are certain to devise even more significant improvements in serological diagnosis and detection of plant viruses in the future.

It is interesting to note that the discovery of immunoassays—in particular, the ELISA technique—has revolutionized many aspects of medicine. Immunoassays are used for a number of human pathogens, allowing medical laboratories to quickly and accurately provide diagnostic information. Previously, the diagnosis of a bacterial disease required several days, as cultures had to be grown on selective media, but ELISA tests can be completed in minutes or hours. Rapid pregnancy tests are based on monoclonal antibody immunoassays that detect a hormone in the urine of a pregnant woman. In the case of acquired immune deficiency syndrome (AIDS), the human immunodeficiency virus (HIV) is difficult to detect, so HIV tests detect the antibodies produced in people who have been exposed to the virus. Thus, this test is done using antibodies produced against other antibodies.

RNA-Based Virus Detection: Using the Other Component

The serological tests just described are based on detection of the protein component of a plant virus—its "coat." Other modern detection methods are based on detection of the virus nucleic acid. These techniques may be more sensitive than ELISA, but they also may require more equipment and expertise and may therefore be more expensive.

Since the 1980s, a technique called polymerase chain reaction (PCR) has become widely used for plant virus detection because of its relative simplicity and great sensitivity (Figure 13.9). Kary Mullis developed PCR in 1983 and received a Nobel Prize in 1993 for his work on this technique.

PCR allows the amplification and hence detection of a specific DNA in a mixture of DNA molecules—for example, a virus genome in plant sap. However, since most plant virus genomes are RNA, not DNA, the genome must first be reverse transcribed to produce a DNA copy suitable for PCR. This procedure is called reverse transcription PCR (RT-PCR). After production of the copy DNA (cDNA), the two strands of the cDNA molecule are copied many times by being separated by heat and then bound to small, specific pieces of DNA, called primers, which allow DNA polymerase to bind and copy the molecule.

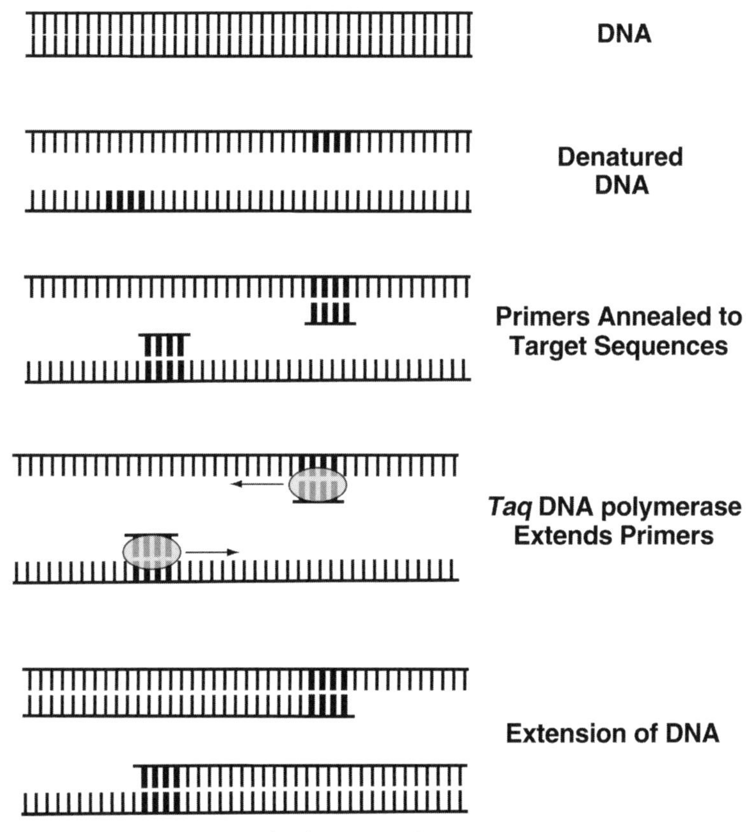

Figure 13.9. First steps of polymerase chain reaction (PCR).

After many cycles of heat denaturing and DNA synthesis, the amount of DNA can be exponentially multiplied more than a million-fold into an amount that is readily detectable. PCR has the same advantages of ELISA in specificity and rapidity, with some protocols requiring less than a day.

Science Sidebar

Other uses for PCR

Like ELISA, PCR also is used in human medicine for detecting and diagnosing both infectious and hereditary diseases. In addition, PCR is used for identifying so-called genetic fingerprints, or profiles used in paternity testing and investigations of crimes such as rape.

DNA profiles are used by forensic scientists to help them identify individuals based on their DNA. This is possible because although 99.9% of human DNA sequences are the same in everyone, the remaining 0.1% is sufficient to differentiate individuals. DNA from a criminal suspect can be multiplied in PCR and the unique sequences compared with those in samples found at a crime scene. In the United States, the FBI uses a set of 13 such assays for identification of suspects in criminal cases.

Management of Virus Diseases: Prevention, Not Cure

Plant virus diseases cause tremendous losses each year, and their management is difficult. Because there is no cure for virus infections, infected plants must be destroyed. Prevention is thus the most effective means of virus management.

In some cases, management of the vector can prevent virus transmission. If an important virus is transmitted by an insect vector in a persistent manner, an insecticide is sometimes applied to a high-value crop to prevent infection. Unfortunately, many viruses are transmitted by insects in a nonpersistent manner. Most insecticides are then ineffective, because they do not kill the insects fast enough to prevent transmission. Oil sprays, which interfere with nonpersistent transmission, have been investigated but are not widely used. Reflective mulches, which reduce the landing rate of vectors, are used to reduce virus incidence in some high-value crops, such as vegetables. Soil may be fumigated before planting certain high-value crops, such as raspberries, to reduce virus transmission by nematodes and some primitive soil fungi. Plant cultivars with resistance to virus vectors also have been developed.

Other strategies used for virus disease management are familiar from their use with other types of pathogens. For example, quarantines and avoidance (planting crops where a virus or its vector is unlikely to be found) are commonly employed to manage virus diseases. Weed management also is important, especially in annual flower and vegetable production, because many viruses overwinter in common weeds.

Exclusion of viruses by the production of virus-free seed and propagative parts is a common and economically sound means of virus disease management. Most flower and vegetable seed is certified as meeting acceptable standards, so that very few, if any, of the resulting plants will be infected with viruses. The seed is carefully tested to determine the percentage of virus infection. In some cases, even a 1% virus infection in seed is unacceptable for commercial use. Commercial growers commonly walk through their fields early in the growing season looking for potentially virus-infected plants and "rogue" them out before the virus can be transmitted to healthy plants.

Seed- and pollen-borne viruses are usually found in relatively low levels in seed, but propagative parts—such as cuttings, tubers, bulbs, and corms—will all be systemically infected if produced by a virus-infected plant. Thus, commercial producers of propagative parts have elaborate schemes of virus testing and clean propagation, enabling them to make virus-free stock available to growers. This is especially important for high-value perennial plants, which require a major investment of time and money before the plants mature. Examples include many woody ornamentals, fruit trees, strawberries, and bramble fruits. Virus-free propagation of many flower crops—including carnations, chrysanthemums, and geraniums—also is necessary.

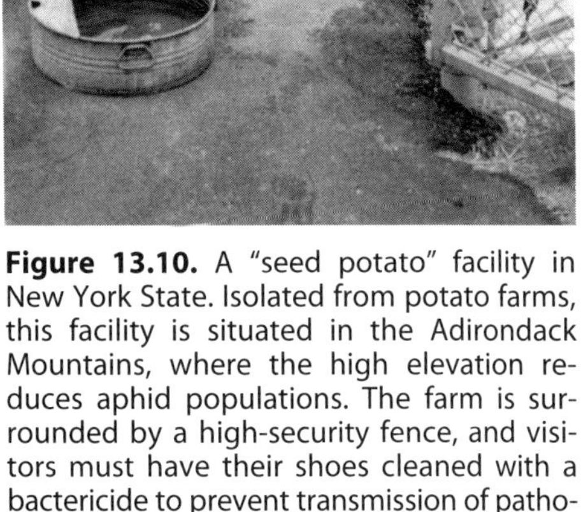

Figure 13.10. A "seed potato" facility in New York State. Isolated from potato farms, this facility is situated in the Adirondack Mountains, where the high elevation reduces aphid populations. The farm is surrounded by a high-security fence, and visitors must have their shoes cleaned with a bactericide to prevent transmission of pathogens to the site.

Production of potato seed pieces for planting begins in laboratories, where the first plants are grown in a completely protected environment. The resulting tubers from these plants are planted by foundation growers, who repeatedly inspect their fields and remove virus-infected plants (Figure 13.10). Government inspectors certify the resulting tubers as meeting acceptable levels of virus infection, based on several visual inspections and laboratory tests.

A sneak preview of the next season's results is provided by the so-called Florida test. Samples of the harvested tubers are planted during the winter months in south Florida to allow inspectors to determine how many plants are infected with viruses. This test was developed because although the mother plants may appear healthy, aphids may have transmitted a virus to the plants late in the season, and the virus may have been transported through the phloem to the developing tubers.

Commercial growers and even home gardeners find the cost of certified seed well justified. Potatoes grown from supermarket tubers will commonly all be infected with viruses and produce much lower yields than those grown from certified seed.

Propagation of Virus-Free Plants: A Clean Start

Exclusion methods are used to provide virus-free seeds and propagative parts, which minimize the chance of a virus disease epidemic. Yet in many instances, all of the plants in a population are found to be infected with viruses. Also, in breeding programs, only a single virus-infected plant of a particular genotype may exist.

The solution to these infection problems, mentioned briefly in Chapter 6, was derived from F. C. Steward's successful application of theoretical totipotency through the regeneration of carrots from mature cells grown in coconut milk. In theory, an entire plant can be regenerated from a single cell. In practice, success is most likely when a tiny sample of cells from the meristem of a plant is used as the starting material. The first virus-free plants were regenerated from meristem tips taken from virus-infected plants in 1952.

Even though studies have shown that, in general, meristems are relatively virus free, due perhaps to the lack of well-developed plasmodesmata in their cells, testing regenerated plants for virus infection is always prudent. Whether the meristem is virus free is still somewhat a matter of chance. Plants grown at high temperatures (86–104°F [30–40°C]) are more likely to have virus-free meristems, probably because virus replication is reduced.

The technology of tissue culture had to make certain advances before this means of virus-free propagation could be economically feasible. Tiny meristem tips taken from mature plants cannot photosynthesize and must be grown on a sterile nutrient medium, just as microorganisms are (Figure 13.11). The early use of coconut milk was successful, because it contained both sugars and plant hormones—now more commonly called plant growth regulators. As understanding of specific growth regulators and their functions developed, more precise control of plant development became possible.

The regeneration of whole plants from excised meristems or protoplasts (see Chapter 6) begins in tissue culture, where a mass of undifferentiated cells is grown on a nutrient medium containing sugars, nutrient salts, and a combination of growth regulators. The type and balance of growth regulators varies for each regeneration step and for each kind of plant. The undifferentiated tissue is transferred to a new medium, which stimulates shoot formation. Many shoots may form from the original small mass of cells. Individual shoots are then transferred to a third medium to induce root formation. At this point, the tiny plant may be transferred to a soil medium and grown in normal environmental conditions.

Figure 13.11. Left, Removal of the meristem tip requires sterile conditions, good light, and a steady hand. **Center,** Growth of the meristem tissue on nutrient medium containing appropriate growth regulators results in numerous shoots. **Right,** Tissue culture allows thousands of plants to be rapidly propagated under pathogen-free conditions in a relatively small area.

Science Sidebar

Plant growth regulators (Hormones)

There are several different kinds of plant growth regulators. Auxins play a role in apical dominance, in which the growth of lateral buds is inhibited as long as the auxin-producing apical bud is present. Auxins also induce cell elongation and encourage root development. Plant cuttings dipped in auxin solutions or powders available in garden stores develop roots more quickly than other cuttings. A number of plant pathogens produce auxinlike compounds and/or stimulate host plants to increase auxin production, causing growth effects—especially gall formation. These pathogens include root-knot nematodes, crown gall bacteria, and the pathogens responsible for corn smut and cedar-apple rust.

A second group of plant growth regulators is the cytokinins. They were named for their role in cell division, which also is called cytokinesis.

The gibberellins are a third important group of plant growth regulators. They were discovered because of their production by a plant pathogen. In the 1930s, the fungus *Gibberella fujikuroi* was determined to be the cause of the foolish seedling disease of rice, which caused excessive elongation of the young plants. The active compound isolated from the fungus was named gibberellin, and in 1956, plant-produced gibberellins that play a role in cell division and elongation also were discovered. Gibberellins applied to seedless grapes and other fruits result in larger fruits.

Plant growth regulators, like animal hormones, are active at low concentrations and interact with each other. Thus, it is the relative balance of growth regulators in specific plant tissues that determines the pattern of plant growth.

Regeneration of plants through tissue culture is an area of intense research activity, for two main reasons. First, pathogen-free plants can be multiplied rapidly in a completely protected environment at a cost that approaches traditional propagation costs. The value of such plants lies in their increased yield and vigor and the longer production time before replacement is necessary. In the past, propagative parts had to be grown in natural soils for several years, during which they were exposed to various virus vectors and soilborne pathogens before they were sold for commercial production. Plants are now sold directly from pathogen-free laboratories and greenhouses to commercial growers.

The second reason for intense research activity in this area is that regeneration of plants from protoplasts or single cells is necessary for successful genetic engineering. Once gene transfer has occurred, the transformed cell must be regenerated into a plant, in which each cell contains the new genetic information.

Resistance to Virus Diseases: A Mainstay

Genetic resistance to viruses, selected through traditional breeding programs that have used repeated crosses between plants, has been quite successful for many important viruses. Scientists and breeders are always concerned, however, that a new virus strain may overcome the resistance. It also is difficult to provide resistance to all the viruses that commonly infect important crop plants. In a few cases where no other effective management strategy exists, plants have been deliberately inoculated with mild strains of viruses to protect against infection by severe virus strains, making use of a phenomenon called cross-protection. However, several potential problems with this technique—including the fear that a mild strain might mutate into a severe one—make this technique a last resort in virus disease management.

The virus-resistance success story mentioned in Chapter 6 may be more easily understood now. The Ti plasmid of *A. tumefaciens* was used to transfer the coat protein gene of TMV to the nucleus of a tobacco cell. The tobacco plants regenerated from the transformed cells were able to delay disease development when infected with TMV. Dozens of different plant species with this pathogen-derived resistance to dozens of different plant viruses have now been created. Many have introduced virus coat protein genes, but other virus genes and even noncoding sequences have been used to protect plants.

It is now believed that plants are able to recognize RNAs produced from these virus sequences as nonself and trigger their degradation—a form of RNA interference. This degradation is ongoing in plants with pathogen-derived resistance, so that introduced viruses are immediately destroyed and no infection results. One commercial success story is transgenic papaya plants, protected against infection by the devastating *Papaya ringspot virus*. These plants were approved for sale in the United States in 1997 and in Japan in 2010.

Viroids: Something Even Smaller

Other types of pathogenic, replicating molecules have been discovered. Viroids are small, circular RNAs that lack coat protein. They were first discovered in the 1960s by Theodore Diener, who named them in 1971. Viroids cause a number of economically important diseases, including potato spindle tuber disease, two different viroid diseases in chrysanthemum, and cadang-cadang ("dying-dying") disease of coconut palm. More than 30 million coconut palms have died of this viroid disease in the Philippines (Figure 13.12).

Science Sidebar

Viruses for genetic engineering

There is great interest in the use of plant viruses to carry new genes into plant cells, especially in plants that are not easily infected by the crown gall bacterium (*Agrobacterium tumefaciens*) now used in genetic engineering. Scientists have developed gene delivery systems that use several different plant viruses, including *Tobacco mosaic virus* (TMV). The viruses are modified to carry beneficial genes without causing disease in the plant.

One of the potential uses of these virus gene vectors is the production of plants with genes for important pharmaceutical proteins. Using the vectors also will help achieve a better understanding of the mechanisms of virus pathogenicity.

Figure 13.12. Cadang-cadang disease of coconut palm caused by a viroid.

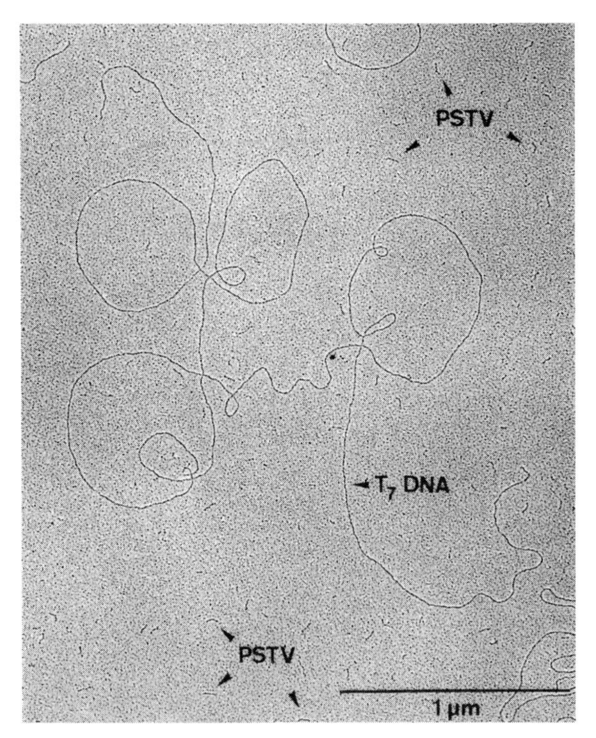

Figure 13.13. Electron micrograph of *Potato spindle tuber viroid* (PSTVd) in relation to the DNA of a bacteriophage (a virus that infects bacteria) (T7). Arrows indicate the tiny viroid RNAs.

Viroids are so small that they do not contain enough genetic information to code for a protein (Figure 13.13). Their origin may be RNA from host plants, a theory supported by the fact that viroids have mostly been found in plants that are intensively vegetatively propagated. How they cause disease is not well understood, but it likely involves interference with host protein and/or mRNA functions. No vectors are yet known, and management of these diseases is very difficult. Viroids are the only pathogens that cannot be detected by immunoassays, because they have no protein coat. They can be detected by RT-PCR, however.

Other pathogenic viruses and RNAs that require "helper" viruses for their replication also have been discovered. Some modify the disease symptoms of virus infection in plants, either increasing or decreasing their severity.

In summary, the infectious pathogens of plants range from subcellular viruses and viroids, to single-celled prokaryotic bacteria, to multicellular eukaryotic fungi and nematodes. Some plants even infect other plants. However, some diseases are not caused by infectious pathogens. These environmental stresses and the challenges we face to provide food and fiber for a growing population are the subjects of the final chapter.

CHAPTER 14

Food for a Hungry Planet

We began this book with the story of the Irish potato famine of the 1840s. Some of the factors that made this historical event particularly disastrous were directly related to the biology of *Phytophthora infestans* and the potato disease late blight. First of all, the potatoes were vegetatively propagated and therefore genetically uniform. This made them susceptible to devastation when exposed to a virulent pathogen from which they had been geographically isolated. Second, the cool, wet weather enhanced the spread of the disease.

Other factors were more sociological and economic. In 1845, Ireland's rapidly growing population depended almost entirely on a single plant species for food: the potato. Athough the famine that resulted affected many countries, it was the worst in Ireland because of political conditions that required the export of some types of food and prevented the import of food from areas where it was more abundant.

Much has changed in the world since the "Hungry Forties," but famines still occur. Many of factors that led to the suffering and deaths of the Irish in the 1840s deny good health and survival to millions of people in the world today. The statistics are disturbing. According to the United Nations World Food Programme, 25,000 people die every day from hunger-related causes—most of them children. Hunger is the world's greatest health risk. Yet in the United States, an advisory committee for the 2010 Dietary Guidelines for Americans called obesity "the single greatest threat to public health in this century."

The effects of hunger and malnutrition are well known but frequently ignored by the well-fed people of the world. The hungry people of the world endure impaired physical and mental development and general lethargy and susceptibility to disease. They are deprived of the ability to enjoy even the most basic qualities of life.

Certainly, no problem is more complex than that of how to feed Earth's human population. To address its complexities with simplistic solutions is to fail to provide an adequate solution. Many academic disciplines can contribute to a better quality of existence for all people on Earth. Historians, anthropologists, sociologists, and political scientists examine the political and social interactions that confound our ability to cooperate sufficiently for political and economic stability and self-determination. Economists contribute theories on the distribution of wealth and economic development within the constraints of available resources. Ecologists evaluate the carrying capacity of the planet's ecosystem. They attempt to determine the human population that can be adequately maintained without permanently disrupting and potentially destroying the fragile environment.

Agricultural scientists, including plant pathologists, also contribute to efforts to provide a reliable and sustainable food supply. As noted in Chapter 11, Norman Borlaug, a plant pathologist, received the Nobel Peace Prize in 1970 for his contribution to the development of "miracle" high-yielding wheat (Figure 14.1). This development also led to the media phrase "Green Revolution." The Nobel committee does not present a specific prize for contributions to agricultural science, but in honoring Borlaug, it recognized that an inadequate food supply is directly related to political unrest and war.

Figure 14.1. Norman Borlaug, winner of the 1970 Nobel Peace Prize.

World Population and Food Demands: What Is the Limit?

In 2011, the world's population was close to 7 billion. However, it did not reach its first billion until around the year 1800. A growth rate as low as 2.8% can result in a doubling of the population in only 25 years. Even with extensive food shipments and economic aid, it is hard to imagine the technological, political, and sociological problems involved in providing for a population that increases so quickly. To build schools and houses and provide jobs and medical care, in addition to food, for a population that doubles in 25 years is beyond the capability of most governments.

The world's population has been growing continuously since about 1400, when the "Black Death" ended, and it is expected to continue to grow for at least 30 to 40 more years. In the meantime, approximately 1 billion people in developing countries do not get enough to eat on a regular basis, leaving them without the energy to carry out a productive life. Another 1 billion people have food, but it is deficient in important qualities, such as protein and other specific nutrients necessary for a healthy life. The largest population of poor and hungry people lives in Africa, where the challenges include recurrent food crises, rapid population growth, war, and the acquired immune deficiency syndrome (AIDS) epidemic.

Scientists have estimated that more food will need to be produced in the next 50 years than has been produced in the previous 10,000 years. This challenge occurs at a time when crop yields are not increasing as quickly as they have in years past. In one study, cereal yields were found to have increased 3% in 1961, 2% in 1975, and 1.4% in 2000. Unusual weather conditions that reduce crop yields even a little can lead to starvation in some countries.

What are the limitations to increased food production? Is it possible to produce and distribute sufficient food for so many people? What are the political and economic implications for a government whose people must rely on food from outside their nation's borders? What is it like to live in a country that is not capable of producing enough food to feed its people?

Arable Land: Less Every Year

One way to increase agricultural production is to bring more land into cultivation. This is an area of considerable controversy because of disagreement about how much arable, or cultivable, land exists in the world.

Total cultivable land is estimated at 7 billion acres (2.8 billion hectares), of which 3 billion is now under cultivation. This seems to suggest that a lot of land is available for conversion into farmland. However, most of the easily farmed and highly productive land is already being used for intensive agriculture. In addition, much good agricultural land is lost every year.

The United States alone is losing 1.2 million acres of farmland annually. Agricultural land is being converted to building sites and used for roads in areas of urban sprawl. Much of the most valuable farmland in the United States, especially around cities and rivers, has been permanently lost, because it brought a better price as building lots for homes than for agricultural production. California, Florida, and many northeastern states are rapidly and permanently losing significant areas of potential agricultural production.

Agricultural land also is being lost in the world's poorest countries. Because population growth is high in these countries, land must be taken out of food production to provide living areas for people.

To reduce the loss of farmland in densely populated areas of the United States, land banks have been created by providing tax incentives to farmers who keep their land in production, rather than sell it for development. By 1983, government programs had put 78 million acres (31.6 hectares) into soil-conserving uses, but the growing demand for land had reduced that amount to 30 million acres (12.1 million hectares) within 2 years.

In response, the Food Security Act of 1985 implemented new programs to specifically preserve land that was highly erodible (the Sodbuster program) or that was a wetland (the Swampbuster program). Cropland erosion fell about 40% from 1982 to 1997. The loss of wetlands for agriculture also declined, from 235,000 acres (94,000 hectares) per year during 1974–1982 to 19,000 acres (7,600 hectares) per year during 1992–2002. Several different factors are responsible for these changes, including reduced tillage programs and taking land out of production. However, approximately 25% of these reductions can be attributed to linking participation in the Sodbuster and Swampbuster programs to farm program subsidies.

Other land is lost to agriculture because of severe erosion and desertification (the formation of deserts). Erosion strips away the topsoil from areas of intensive agricultural production. This topsoil, which accumulated for millions of years beneath prairies and forests, is exposed to wind and water erosion after land is tilled. Topsoil is a nonrenewable resource that can be lost during every growing season. Water erosion carries away valuable topsoil to rivers and oceans in tremendous quantities. The Mississippi River carries 300 million tons (272,000 metric tons) of soil into the Gulf of Mexico each year. Every major river system in the world carries a similar load.

Figure 14.2. Rice culture in the highlands of the Philippines.

Some technologies were developed years ago to help reduce erosion. In parts of Asia, where population density has been high for centuries, complex terrace systems help maintain topsoil for sustained production (Figure 14.2). Contour tillage, which produces the alternating strips of crops familiar in midwestern farmlands, also reduces soil erosion. This method became popular in the United States after the "Dust Bowl" days of the 1930s.

New technologies are reducing erosion even more. No-till agriculture in row crops using genetically engineered, herbicide-resistant plants makes plowing unnecessary. Besides reducing wind and water erosion, this approach reduces the rate of decay of organic matter in the soil and helps sequester carbon. It also helps retain water and minerals in the soil. As yields increase, the amount of vulnerable land that must be brought into agriculture is reduced.

The most arable land loss occurs where the population is dense and growing and/or where changing environmental conditions are pushing already vulnerable soil out of production. China and India are large population centers, and sub-Saharan Africa has a growing population living on very fragile soils.

Brazil's agricultural land is still expanding and has the greatest potential to still grow. Brazil doubled its agricultural production between 1970 and 1990 and again between 1990 and 2005, and it has become an important producer of beef, coffee, orange juice, soybeans, sugar, and ethanol. Brazil's new agricultural production land has come from both savanna and tropical rain forests.

Although traditional farmland is limited and declining, other alternatives are in development. Urban agriculture—sometimes on building rooftops—has the potential to produce fresh vegetables and fish in city neighborhoods. Another alternative is to consider using the oceans, which cover 70% of Earth, for agriculture.

Water Resources: Approaching the Limits

Agricultural production is limited by the lack of water in many areas. Irrigation is necessary for production throughout the world, including many parts of the United States. Rivers have been diverted, and water rights have been the subject of numerous court battles. Recent droughts have led to enforced water restrictions in urban areas, including bans on car washing and watering of lawns.

However, water use by individuals and cities accounts for only a small proportion of total water use. In years past, irrigation used the most water in the United States, but today, it is the second largest use at 39%. The largest use now is cooling water for electric power generation. Worldwide, the lack of water for farming is a major limiting factor in agricultural production. It has been estimated that 40% of corn losses are due to insufficient water.

The large, green circles surrounded by brown desert, visible from airplanes when flying over the western states, are evidence of how water can bring agricultural production to otherwise unproductive areas (Figure 14.3). In many cases, irrigation water is pumped from underground aquifers, in which it has been accumulating for millions of years. However, this "fossil water"—such as the Ogallala Aquifer, which stretches from the high plains of Texas to South Dakota—is not being replaced by infiltration as quickly as it is being removed. Over time, the cost of pumping water from greater and greater depths reaches a point where the costs are excessive, and the land must be abandoned for agriculture. The removal of water in some areas has been so rapid and extensive that sink holes have developed, causing land to collapse into the empty recesses left after the water has been pumped out. Improvements in irrigation techniques in dry-land agriculture may allow sustained agricultural production even as irrigation water becomes scarcer. For instance, drip irrigation greatly reduces water loss to evaporation.

Figure 14.3. Center pivot irrigation system used on corn in Minnesota. Agriculture would be severely limited in many states if irrigation water were not available.

There is considerable concern that the loss of significant tracts of the world's forests will disrupt both the carbon cycle (see Chapter 4) and the hydrological cycle, which affects rainfall rates and patterns throughout the world. The dense vegetation of the world's rain forests significantly affects temperature and precipitation through the absorption of CO_2 and loss of water by evapotranspiration. Unusual droughts and weather patterns also contribute to the loss of arable land, causing erosion and salinization (salt accumulation), where the amount of water is insufficient to elute (wash out) accumulating salts from the soil.

Cereal breeding programs are developing corn and sorghum cultivars that require less water and can produce crops in drought-prone areas, such as the Sahel region of Africa. Another goal is the development of cultivars with salt tolerance. These cultivars would allow the use of water sources with high levels of salt, which damage currently used cultivars. Salt tolerance also would be a useful trait in areas where the rising ocean level infiltrates farmlands near sea level. Irrigation water in these areas can become salty, which makes the problem even worse.

Agriculture can affect water quality. Wells in rural areas are sometimes polluted by nitrate fertilizers and pesticides, especially when these chemicals are not applied properly. Runoff of nitrogen and phosphorus fertilizers into surface water, especially from row crops and bare soils, causes so-called dead zones in oceans far from the origin of the problem. These nutrients cause algal blooms, and when the algae populations die, their decay robs the water of oxygen, causing fish and other aerobic organisms to die. This can lead to areas devoid of all aerobic aquatic life, such as the enormous dead zone found in the Gulf of Mexico, where the nutrients carried by the Mississippi River are dumped.

Soil Fertility: For Both Yield and Food Quality

The world's fastest-growing population areas are concentrated in the tropical and subtropical regions. Many of the soils in these regions are shallow and particularly vulnerable to destruction, and they can support agricultural production for only a few years. In addition, these soils are quite acidic and contain high concentrations of iron and aluminum—all of which are detrimental to the growth of food crops. More research must be done on how to increase and maintain the productivity of such soils.

Fertilizers are a mainstay of agricultural production in developed countries. The use of synthetic nitrogen fertilizers, in particular, has continued to rise since discovery of the Haber-Bosch process of nitrogen fixation in the early 20th century. This energy-intensive process is one of the biggest users of fossil fuels.

Nearly 1.1 billion tons (1 billion metric tons) of synthetic nitrogen fertilizer is used each year. No-till and reduced-till agriculture have helped lower the need for fertilizer and reduced pollution from leaching and runoff. Nitrogen, which accounts for more than 80% of the energy used in manufacturing fertilizers, can be made available to plants in several ways through biological fixation (see Chapter 4): using improved strains of rhizobial bacteria, which parasitize various leguminous plants; using free-living, nitrogen-fixing bacteria; and through genetic engineering of the nitrogen-fixing process in nonlegume plants.

The three most important macronutrients for plants are nitrogen, phosphorus, and potassium (N-P-K). Of these, potassium is the most abundant and least polluting. Phosphorus must be mined from rock reserves, mostly in Morocco and the western Sahara Desert. According to estimates, the world's limited supply of phosphorus will last another 280 years at the current usage rate. Soil testing to determine plant requirements can reduce unnecessary applications of phosphorus. Like nitrogen, excess phosphorus contributes to water pollution and algal blooms. These effects have led to bans on the use of phosphorus in detergents and lawn fertilizers in some states.

Agricultural production is directly related to the availability of land, the quality of the soil, and the availability of water and fertility. The previously noted statistics demonstrate that there is reason for great concern about each of these major factors. The judicious and sustained use of land for agriculture and the critical inputs of water and fertility have scientific, social, and political aspects that must be approached both locally and globally.

Air Pollution: An Added Stress

One of the most difficult environmental problems to diagnose and quantify is the chronic damage caused by air pollution. Industrial activities, automobiles, and the burning of fossil fuels for energy have increased the concentration of certain chemicals in the atmosphere.

Some air pollutants are released into the atmosphere and damage plants directly. They are called primary pollutants. For example, hydrogen fluoride (HF) causes marginal necrosis in sensitive plants. It is a common pollutant from ore refineries (especially those that produce aluminum) and ceramic factories. The incomplete combustion of fuels is a source of other important pollutants. They are present in exhaust from internal combustion engines and released from factories. Both nitrogen oxides (NO_x) and sulfur dioxide (SO_2) are common pollutants that cause toxic effects in plants at relatively low levels (less than 3 parts per million [ppm]).

Sulfur dioxide and nitrogen oxides cause interveinal yellowing of affected tissues and general growth suppression at low concentrations. Tissues may show bleaching and bronzing at high concentrations. An important source of sulfur dioxide is the burning of high-

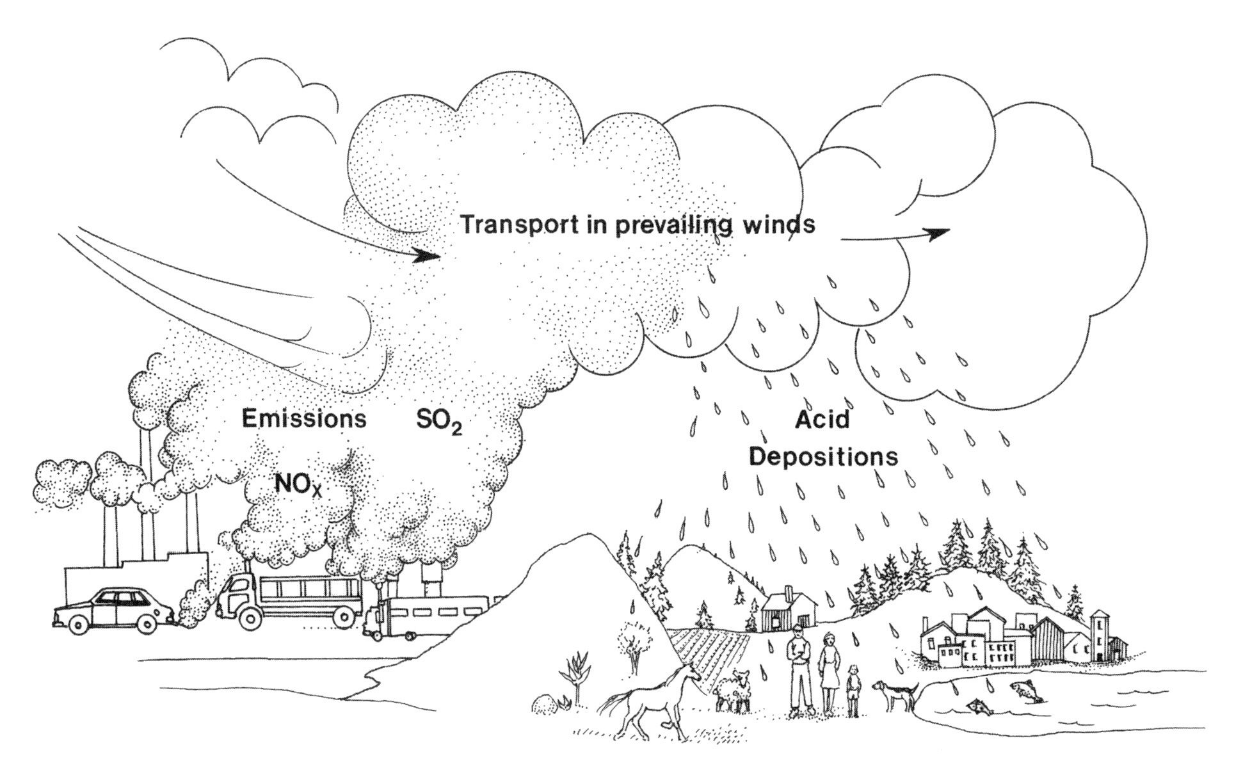

Figure 14.4. Origin of acid rain. SO_2 = sulfur dioxide; NO_x = nitrogen oxides.

sulfur coal for energy production. Sulfur dioxide and nitrogen oxides combine with water in the atmosphere to produce sulfuric acid and nitric acid, respectively. This results in acid precipitation, principally in the form of acid rain (Figure 14.4). Acid rain is considered a secondary pollutant, because it results from a chemical reaction of substances in the atmosphere.

Two other secondary pollutants originate from engine exhaust and must be chemically converted in the presence of sunlight to form chemicals toxic to plants. These pollutants are ozone (O_3) and peroxyacetyl nitrate (PAN), both of which are toxic components of smog. Because these pollutants are formed in photochemical reactions from sunlight, their levels follow a diurnal pattern, increasing during the day and declining at night. Warm temperatures also increase the rate of the reactions, so these air pollutants are produced at higher concentrations during the warm summer months.

Ozone enters leaves through their stomata and causes necrotic specks to form on the upper surfaces (Figure 14.5). High concentrations of ozone can cause premature defoliation of sensitive plants, while chronic damage from low concentrations inhibits photosynthesis. Ozone is the most destructive air pollutant and is the predominant air pollutant in the eastern and midwestern United States. Acute ozone damage has been documented in some crops in some areas, such as spinach grown in New Jersey and Long Island, New York. Yield reductions resulting from chronic ozone effects have been demonstrated in a number of other crops, as well, such as citrus, cotton, and soybeans.

Symptoms of PAN injury are sometimes called "silver leaf," because the affected lower leaf surfaces become silvery. At high concentrations, the damage may be similar to that caused by ozone. PAN is an important component of smog in highly populated areas of the western United States, especially in the region around Los Angeles.

Stress-triggered decline of forests worldwide is the result of the combined effects of environmental stresses (such as drought and pollution) and insects, pathogens, and secondary disease organisms that lead to decline and death. Acid rain is an important stress factor for forests in certain areas. Acid rain and other pollutants may cause direct damage to trees or indirect effects on the soil and water in the trees' environment. Because of the importance of trees for fuel and for sequestering carbon in the carbon cycle, understanding the impact of environmental stress on the world's forests is critical.

Air pollution affects the health of all members of an ecosystem, including humans. The severity of the effects is difficult to assess, but federal air quality standards have been established in an attempt to reduce air pollution. Each year, these standards are exceeded in areas of concentrated vehicle traffic and near certain industrial areas. Cars with catalytic converters produce cleaner automobile

Figure 14.5. Ozone (O_3) injury symptoms on tobacco. Bel W3 tobacco is particularly sensitive to ozone injury and is used as an indicator plant for the presence of ozone in the atmosphere. The damaged tobacco leaf on the right is from a plant grown outdoors in Waltham, Massachusetts, and the leaf on the left is from a healthy plant with no injury.

exhaust than other vehicles, but their gas mileage is reduced, increasing transportation costs. Despite more stringent regulations for auto exhaust, traffic continues to increase, causing higher ozone and PAN atmospheric pollution.

To reduce air pollution, should drivers be restricted to alternate-day driving? Should private automobiles be banned from cities and only public transportation be used? Should incentives be offered for driving low-polluting hybrid or electric vehicles?

Unfortunately, air pollutants do not honor political boundaries. Pollutants are detected at ever-higher concentrations even in areas far from their sources, such as Antarctica. On some days, views of the Grand Canyon, in Arizona, are blurred by air pollution produced hundreds of miles away. These toxic compounds not only detract from the natural beauty of our environment, but they also affect the health of all organisms, both directly and by contributing to the stresses that increase susceptibility to other problems.

The air pollution level has become a regular part of the weather report on television news in many areas, and warnings to reduce outdoor activities during periods of heavy pollution are commonly issued in cities during the summer. In 2010, the Environmental Protection Agency (EPA) adopted strict new standards for atmospheric SO_2 concentrations, because high levels have been linked to asthma and other respiratory problems that can lead to premature deaths of people. SO_2 is discharged into the atmosphere by coal-burning power plants and diesel-burning vehicles. According to estimates, implementing the new EPA standards will cost $1.5 billion in the next decade. However, the health benefits that will result are valued at $13–33 billion.

Food Crops: What Are Our Choices?

An important lesson learned from the Irish potato famine is that it is dangerous for a population to rely on a single plant species as its food source. Is the population of the world in the 21st century any less vulnerable than the Irish were in the 1840s?

Figure 14.6. Pile of wheat at a grain terminal on the Snake River, Washington.

Approximately 800,000 species of plants exist in the world, of which at least 3,000 are edible. Of these plants, perhaps 150 are commonly eaten. In fact, 90% of the world's food comes from 15 plant species and 8 animal species. Some of the important plant species include banana, beet, cassava (manioc), coconut, common bean, corn (maize), millet, potato, rice, sorghum, soybean, sweet potato, sugarcane, wheat, and yam.

Three species of plants dominate world crop production today: corn, rice, and wheat. Wheat, a single species, is the primary source of calories for millions of people worldwide and provides 55% of the carbohydrates and 20% of the food calories consumed every day. Only approximately 12% of the U.S. corn crop is used for direct food (such as corn meal) or indirect food (such as corn syrup). Cassava and potatoes are important root and tuber crop sources of calories for the world's human population.

This heavy reliance on only a few plant species by more than 7 billion people reflects a vulnerability similar to that of the Irish population in 1845. A large decrease in the annual yield of even one of the key crops would significantly reduce the total worldwide calories available that year. These major crops also are the major stored food of the world (Figure 14.6). In 2007, the Food and Agriculture Organization (FAO) reported that wheat stores had dropped to a level that corresponded to 12 weeks of the world's total consumption, and only 8 weeks' worth of corn was stored.

Improved food storage is of critical importance in areas of high population growth. Storage is particularly critical in tropical areas, where warm temperatures increase the rate of food degradation by pathogens and decay organisms, which also are a source of mycotoxin contamination. In addition, insect and rodent pests cause significant losses of stored food every year. Hungry months between harvests is a normal way of life for 2 billion people every year.

Many of the world's minor crops deserve more research and improvement. For instance, quinoa is an important crop that was cultivated as a sacred grain for many years by the Inca in South America. It belongs to the genus *Chenopodium* and is related to the common weed known as lambsquarters.

Amaranth is another neglected crop. It is a high-protein seed producer that was cultivated in Mexico before the arrival of the Spanish in the 16th century. It belongs to the genus *Amaranthus,* which includes the common garden weed known as pigweed. Amaranth is well adapted to the arid, stressful climates of many areas that are not particularly well suited for growing traditional cereals. Long ago, the cultivation of amaranth was banned by Spanish church officials, because the seeds were used by the native people in religious ceremonies. Today, there is a new effort to reestablish the crop, particularly for use in marginal croplands.

Although cereal and other grain-type crops are popular and easy to store, many root and tuber crops produce large amounts of carbohydrates, especially in tropical countries. African yam, cassava, sweet potatoes, taro, and others deserve more attention for their potential to help create a stable and reliable base of plant foods for the world's growing population.

Although monoculture production of major food crops will not disappear, mixed farming systems are being studied, particularly for food production in developing countries. Small plots of land planted with a number of crop species can be highly productive and reliable, because poor yields in certain crops caused by weather, disease, or pests can be compensated for by good yields in other crops. Farming mixed species probably results in fertility and pathogen and pest management benefits that are not yet well understood. Sustainable year-round production from small plots of land is crucial in tropical areas, where population and environmental pressures are the greatest.

Global Climate Change and Energy: Hotter and More Extreme

As discussed in Chapter 4, the opening of land for agriculture and the use of trees and fossil fuels for energy have increased the level of CO_2 in the atmosphere, which has led to global warming. Exceptionally high temperatures have prevailed worldwide since 1985.

Evidence for climate change comes from almost all areas of biology. Recent studies have shown that many species of trees and shrubs are moving higher in elevation or expanding their natural ranges farther toward Earth's poles. Birds, insects, and other animals are moving with them and changing migration patterns and life cycle timings. These kinds of changes will directly impact the survival of pathogens and their potential to cause plant diseases.

Global warming also is going to directly impact agricultural production because of both temperature changes and rainfall distribution and variability. In the United States, agricultural productivity will probably fall in the basins of the Colorado and Rio Grande Rivers, the Ogallala Aquifer range of the Great Plains, and the southeastern states. Agriculture will be affected the most in developing countries near the equator, which will have warmer temperatures and longer dry periods. These changes will affect what crops can be grown and how productive they will be. In areas where the climate is most conducive to agriculture, efforts are needed to ensure that crop production is as efficient as possible.

Researchers are studying ways to increase photosynthesis efficiency by improving the capture of light by the leaf canopies of crops and by broadening the wavelengths of light that can be used in photosynthesis. The chemical processes involved in photosynthesis also are under study. Biologists have long known that plants with C4 photosynthesis are more efficient at fixing carbon into biomass. Corn, sorghum, and sugarcane are examples of C4 plants. Is it possible that the less-efficient C3 plants—such as rice, soybeans, and wheat—could be engineered to perform C4 photosynthesis?

Energy demands are causing climate change by increasing CO_2 levels in the atmosphere. The need to sequester more carbon, primarily in trees, is going to increase. In many areas of the world, trees are necessary for fuel to cook food. Using alternative solar-powered stoves could reduce this demand in population-dense areas. Because trees sequester four to eight times more carbon than crops, there will be competition for land currently used for agriculture. Biofuel crops for renewable energy also will compete for land used for food crops. One alternative would be to engineer certain algae to produce oils that could be used as biofuel. They could absorb CO_2 from industrial processes and fossil fuel production and produce renewable energy without using farmland.

In addition to finding renewable forms of energy, people in industrial countries must significantly reduce their energy consumption. Relatively simple changes can have significant effects, including the use of insulation and better windows, home thermostats on timers, alternative transportation choices, and energy-efficient appliances.

In countries where obesity is a major public health threat, the energy consumption of the food industry should be considered. Many processed foods are not only more expensive than unprocessed foods, but they also contain unhealthy levels of fat, salt, and sugar. Processed foods also tend to be sold in large portions, require packaging produced by fossil fuels, and must be refrigerated or frozen, which uses even more energy. Many dairy and juice products could be placed on shelves, rather than in refrigerated cases, if they were ultrapasteurized, as they are in some parts of the world. Finally, by moving to a more plant-based diet, we would be eating "lower on the food chain" and could save much of the energy required to convert plant protein into animal protein.

Previous chapters discussed many plant diseases that have threatened food, feed, and fiber crops in the past, some of which are still significant problems today. Two important tropical crops currently threatened by diseases are cacao and cassava.

Threatening Diseases: Cacao, a Cash-Crop Example

Chocolate is derived from *Theobroma cacao,* a tree that originated in the tropical rain forests of the Western Hemisphere. The scientific name for the cacao tree means "food of the gods." A number of European countries are famous for the quality of their chocolate, but none of them is a source of the crop itself.

The first product developed from cacao seeds was a bitter drink produced by the Maya and the Aztec. The Spanish explorers—Hernán Cortés, in particular—brought chocolate back to Spain from the New World, but it was 100 years before chocolate was widely accepted. It was used as a stimulating drink, even before coffee or tea was introduced in Europe. As the Europeans discovered how the addition of milk, sugar, and other ingredients made the drink more palatable, hot cocoa and chocolate candies and desserts grew in popularity. This was especially true in cold countries, where people could afford to buy these foods and air temperatures prevented the chocolate from melting.

Today, 40 to 50 million people make their living growing cacao. The main tropical countries where the crop is grown are Brazil, Cameroon, Ecuador, Ghana, Indonesia, Ivory Coast, and Nigeria. Note that only two of these countries are in South America, the continent of cacao's origin. Some 70% of the world's crop comes from countries in western Africa, where it is a major component of the nations' economies. The countries where per capita consumption exceeds 20 pounds (44 kilograms) per year include Austria, Belgium, Denmark, Germany, Switzerland, and the United Kingdom. (Surprisingly, the United States is not on the top-consumer list!)

Cacao trees are expensive to replant and susceptible to many diseases that are difficult to manage in a climate that allows pathogens to survive throughout the year. In addition, the trees flower all year, so the valuable pods are continually produced, making many traditional disease management practices difficult to apply. Frosty pod rot and witches' broom are two fungal diseases that have caused large losses in Central and South American countries. An outbreak of witches' broom reduced Brazil's cocoa production by half during the 1990s.

Several *Phytophthora* species cause severe disease problems worldwide. One of the diseases, called black pod, destroys 40–60% of Nigeria's cocoa production each year (Figure 14.7). *Cacao swollen shoot virus* is spread by mealybug vectors and is a problem mostly in West Africa, so far. More than 160 million cacao trees have become infected by this virus and have been destroyed. The mealybugs that carry the virus from native forest reservoirs are very difficult to control.

Figure 14.7. Black pod rot of cacao caused by *Phytophthora* species.

The difficulties of producing this valuable crop should make chocolate-lovers appreciate it even more. In 2010, two U.S. chocolate manufacturers succeeded in sequencing the entire genome of *T. cacao.* This information may make it possible to determine which genes are responsible for the flavanoids in chocolate (which have the antioxidant properties thought to benefit health) and which genes are associated with resistance or susceptibility to some of the important cacao pathogens.

Threatening Diseases: Cassava, a Survival-Crop Example

Cassava (*Manihot esculenta*) is a woody, tropical crop native to South America. It is not usually familiar to people who live in temperate climates, but it is an important crop for 500 million of the world's poorest people living in Africa and South America. The starchy root, also known as manioc or yucca, is harvested and processed into many kinds of food, including the starch pearls used in tapioca pudding (Figure 14.8). The roots must be grated and washed before eating to remove compounds that release cyanide.

Cassava is an efficient carbohydrate producer, and the starchy roots can be left in the ground as a security crop for many months until needed. One goal of breeding programs is to reduce the level of toxic compounds in cassava to enhance its edibility for humans and animals, but there is some concern that this might make the plant subject to more disease and pest problems. Many insects and pathogens that attack cassava have developed cyanide detoxification mechanisms, so apparently, the cyanide serves a protective purpose. Chronic cyanide toxicity is an important health problem for those who routinely eat the crop, however.

Like the potato, cassava is a vegetatively propagated crop and subject to some threatening virus diseases. *Cassava brown streak virus,* spread by whiteflies, is a problem throughout sub-Saharan Africa and a limiting factor for food security in Tanzania (Figure 14.9). Other viruses that cause cassava mosaic disease, also spread by whiteflies, threaten this staple crop throughout Africa.

Because of the importance of cassava and the severe threats posed by its diseases, the BioCassava Plus project is being funded by the Bill and Melinda Gates Foundation as part of its Grand Challenges in Global Health project. The project's goals are to decrease the problems with and increase the food quality of cassava. In 2003, the entire cassava genome was sequenced. Researchers hope to increase the levels of protein and vitamins A and E and to increase the plant's absorption of iron and zinc from the soil.

Figure 14.8. Cassava, or manioc (*Manihot esculenta*), is an important root crop in tropical countries.

Figure 14.9. Symptoms of *Cassava brown streak virus* infection in cassava roots.

Cassava is a good source of carbohydrates, but dietary improvements could provide more complete nutrition from a meal of cassava. Scientists also hope to significantly decrease the levels of toxic compounds in the plant and to improve its resistance to the major virus pathogens through genetic engineering. Cassava rots quickly after harvest, so another goal is to extend its storage life. Through tissue culture, improved, pathogen-free planting stock could be distributed to farmers.

The International Agricultural Research Centers: Working Together

How can yields of major crop species be increased? Breeding programs have produced significant yield increases through repeated crosses and selection of promising parental lines. High-yielding cultivars of wheat and rice were first released in the 1960s.

The high-yielding rice cultivars were released from the International Rice Research Institute (IRRI) in Los Baños, Philippines, one of 15 international agricultural research centers. Like the high-yielding wheat cultivars, these rice cultivars contain a dwarfing gene that produces plants of low stature. Such cultivars can more efficiently utilize high fertilizer inputs, which greatly increase yield without causing lodging (falling over) of grain stalks during rain and wind near harvest time. The introduction of these rice cultivars contributed to increased food production and even food exports from India in the 1970s.

Although there is no doubt that total food production increased when these new cultivars were made available, there also were accompanying problems. Critics pointed to the social upheaval that resulted, in which small farmers were sometimes at a disadvantage because they were unable to obtain credit to purchase the fertilizers necessary for the higher yields. In many areas, the "land races" of wheat and rice, which had been selected by farmers over many centuries for their adaptation to local requirements, were lost, causing a significant decrease in the genetic diversity of the crops. Besides genetic vulnerability, the new cultivars were susceptible to both traditional and novel pathogen and pest problems, leading to significant losses in some areas and increased pesticide use in others.

Despite the media's reporting of a "Green Revolution," the scientists who created these "miracle" grains never considered the early releases to be the final answer to world food production. On the contrary, breeding programs are ongoing, resulting in the continual release of new cultivars with improved resistance to many of the pathogens and pests that attacked the earlier releases (Figure 14.10). Also, no plant geneticist or pathologist would state that the use of one or a few cultivars of any important crop is a reliable or desirable situation. The

Figure 14.10. The "miracle" rice cultivar, IR-8, was an early release, compared to more recent cultivars with multiple resistance to insect pests and diseases. These cultivars were developed by the International Rice Research Institute in the Philippines.

tremendous yield increase produced by the use of dwarfing genes was but a first step in a program that must continue as long as the human population continues to grow and to disrupt the environment.

What is the future of the Green Revolution? Both its successes and the problems that accompany them clearly illustrate that improvements in the agricultural sciences can be successfully implemented only after the appropriate economic, sociological, and political problems have been solved. However, agricultural scientists must develop the most productive and sustainable production methods to provide the foundation on which other disciplines can build policies to ensure that the new cultivars and production methods will be successfully implemented.

The 15 international agricultural research centers were established in 1971—many of them in the centers of origin of important food crops (Figure 14.11). These research centers grew out of previous Rockefeller Foundation and Ford Foundation projects, and today, they are supported by the Consultative Group on International Agricultural Research (CGIAR). CGIAR sponsors these research centers to foster the improvement of food crops of world importance. Scientists working at the centers not only select and develop new cultivars, but they also are active in the collection, preservation, and exploitation of the diverse species and cultivars of these species.

Many of the research centers are in the tropics. Most of the world's important food crops originated in the tropics, which is not surprising, since the greatest concentration of genetic diversity is present there. Researchers can collect germplasm from many of the nearby wild relatives of important food crops. Research centers that concentrate on a wide variety of tropical food crops include the organizations listed in Table 14.1.

Worldwide support for the international agricultural centers has dwindled in recent years. As demonstrated by many of the examples presented in previous chapters, pathogens do not respect political boundaries. International cooperation is needed for pathogen and pest monitoring, responses to global climate change, and efforts to breed improved crops.

Scientists work together and share resources at the international agricultural research centers. For example, two CGIAR centers (CIMMYT and ICARDA), in collaboration with

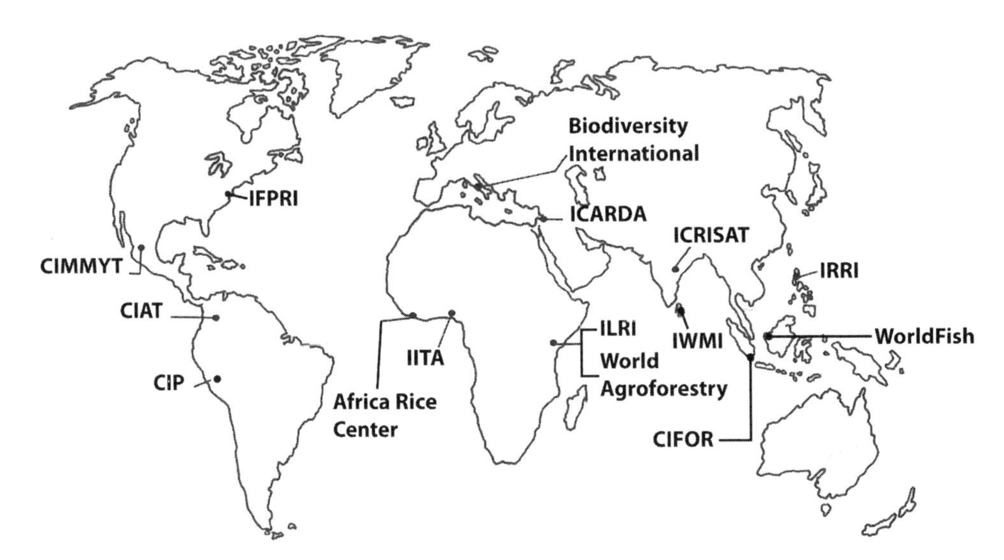

Figure 14.11. Sites of the international agricultural research centers. (See Table 14.1 for the names of the centers and the primary crops they focus on.)

national research centers of countries under threat, have developed high-yielding wheat cultivars with resistance to the Ug99 strain of wheat stem rust. Researchers are multiplying the seed and distributing it in the most threatened areas.

Perhaps the success of modern agriculture has caused people in industrial nations to forget how dependent we all are on photosynthesizing plants, which form the foundation of our food supply. Globally, research and development for agriculture was down one-third in the late 1990s from its level a decade earlier. Agricultural research in Africa from all sources is only $1.5 billion annually. The U.S. Agency for International Development (AID) cut its agricultural staff by more than two-thirds from its peak in 1990. The budget of the entire CGIAR system is about $500 million a year, and the organization's leaders hope to find donors to double that amount in the next 5 to 10 years. Compare these numbers to those of recent budgets of some U.S. institutions: the National Institutes of Health, $31.2 billion (2009); the National Aeronautic and Space Administration (NASA), $17.2 billion (2009); the National Science Foundation, $6.87 billion (2010); and the Agricultural Research Service, 1.2 billion (2010).

Every country must decide how to spend its limited financial resources. Unfortunately, agricultural research may not be well funded until world food shortages become common and widespread, which will be too late to benefit many people.

Table 14.1. International Agricultural Research Centers Supported by the Consultative Group on International Agricultural Research (CGIAR)

Abbreviation	Full Name and Location (primary crops)
—	Africa Rice Center; Cotonou, Benin
—	Biodiversity International; Rome, Italy
CIAT	Centro Internacional de Agricultura Tropical; Cali, Colombia (cassava, common bean, rice, tropical forages)
CIFOR	Center for International Forestry Research; Bogor, Indonesia
CIMMYT	Centro Internacional de Mejoramiento de Maiz y Trigo; Mexico City, Mexico (maize and wheat)
CIP	Centro Internacional de la Papa; Lima, Peru (potato)
ICARDA	International Center for Agricultural Research in the Dry Areas; Aleppo, Syrian Arab Republic (barley, faba bean, kabuli chickpea, lentil, wheat, pasture and forage species)
ICRISAT	International Crops Research Institute for the Semi-Arid Tropics; Patancheru, India (chickpea, groundnut, pigeonpea, pearl millet, small millets, sorghum)
IFPRI	International Food Policy Research Institute; Washington, DC, USA
IITA	International Institute of Tropical Agriculture; Ibadan, Nigeria (banana/plantain, cassava, cowpea, maize, soybean, yam)
ILRI	International Livestock Research Institute; Nairobi, Kenya
IRRI	International Rice Research Institute; Los Baños, Philippines
IWMI	International Water Management Institute; Colombo, Sri Lanka
—	World Agroforestry Center; Nairobi, Kenya
—	WorldFish Center; Penang, Malaysia

Biodiversity: Ecosystem Stability

Ask anyone on the street how he or she feels about genetic erosion or the loss of species diversity, and you will probably not get much of a response. The United Nations General Assembly declared 2010 as the International Year of Biodiversity, but many people think biodiversity is the domain of environmentalists. Although the economic benefits of conserving important agricultural species are obvious, a more critical question remains: What level of biological diversity is necessary to maintain a stable world ecosystem?

According to estimates, we have lost 75% of the genetic diversity of crop plants in the past century. Part of this loss has resulted from the production of seed by large companies, which offer only certain kinds of plant species and cultivars. There is growing interest among gardeners and small farmers in preserving our crops' genetic heritage by maintaining so-called heirloom varieties. The Seed Savers Exchange facilitates exchange of these genetic resources. Plants are not only important for food but also for fiber, fuel, and pharmaceutical chemicals and as the foundation of our natural ecosystems.

Each lost plant species takes other species with it. This may be easier to understand after reading the previous chapters. Bacteria, fungi, nematodes, viruses, and other organisms (including insects) often depend on specific plants for their survival. A variety of organisms is necessary for ecosystem stability. The genes that have evolved in each species provide the evolutionary potential for adaptation to a changing environment.

No one knows how many organisms exist. Most of them are in the tropics, where inventories are incomplete, especially for fungi and microorganisms. By 2015, we could easily lose one-quarter of all species. One estimate is that we are losing 27,000 species a year, three every hour. If you live in a temperate climate, this projection may seem unbelievable, but tropical environments are very diverse. Many species live in small, specialized habitats that can be completely eliminated with the destruction of a few acres of rain forest. If land is continually converted to farmland and living areas to support the human population, this destruction will accelerate.

Collections of genetic resources of crop species exist throughout the world. In the United States, the National Center for Genetic Resources Preservation is in Fort Collins, Colorado. As part of the National Plant Germplasm System, specialized collections exist in other areas, such as the apple and grape germplasm collection in Geneva, New York. The international agricultural research centers all collect and preserve germplasm.

Figure 14.12. Bananas in La Lima, Honduras, showing genetic diversity.

Unfortunately, economic constraints, political unrest, and natural disasters can threaten these valuable collections. In 1971, an earthquake destroyed the national seed bank of Nicaragua. In 1998, Hurricane Mitch destroyed the national seed bank of Honduras, and in 2006, a typhoon flooded a seed bank in the Philippines. In 2003, the national seed bank of Iraq—which contained seeds of chickpeas, lentils, wheat, and other crops—was destroyed in the U.S. invasion. More recently, a seed bank in Afghanistan was destroyed when looters stole the jars that contained the seeds of cereal crops, as well as almonds, melons, pistachios, and pomegranates. The seeds were not stolen, but their histories were lost and they were so mixed up that they are now useless.

Clearly, redundant collections are needed. To help meet this need, the Svalbard Global Seed Vault was established in Norway in 2008. It is located in a cold, remote part of the world, where international genetic resources of important plants can be safely stored.

These germplasm collections are important to maintain crop diversity and to provide sources of materials for plant breeders (Figure 14.12) Yet it is also critical to maintain the ecosystems where these species evolved. In natural ecosystems, species continue to evolve with other organisms—including pathogens and pests—adapting to both biological changes and long-term climate changes. Wild versions of crop plants and their relatives, as well as potential food plants that have not yet been exploited, are our agricultural future (Figure 14.13).

Figure 14.13. Exploring for wild potato species in the Andes Mountains of South America.

A "Perfect Storm"?

In January 2010, the Council for Agricultural Science and Technology (CAST) published a paper entitled "Agricultural Productivity Strategies for the Future: Addressing U.S. and Global Challenges," from which a number of facts were taken for this final chapter. The authors of the paper warn that we face an unprecedented confluence (a "perfect storm") of important factors that will challenge our ability to maintain the standard of living in developed countries and to improve the lives of people in developing countries.

These factors include a growing population, a lack of new agricultural land, climate change, a decrease in the water supply, an uncertain energy supply, and a continual loss of biodiversity. It has been estimated that worldwide, only about 0.55 acre (0.22 hectare) of farmland is allocated per capita, and the trend is downward. In addition, people around the world expect a better standard of living and will have greater energy needs. Few people in the United States understand the production and constraints of their food supply.

Although modern agriculture has its critics, some recent studies suggest that highly productive farming cannot only increase the food supply but also can mitigate CO_2 emissions. Less productive agriculture requires more land, destroys biodiversity, and increases greenhouse gas production by opening up land to cultivation. The connection between crop yields and the loss of wild lands was first proposed by Norman Borlaug, "father" of the Green Revolution, and has been called the Borlaug hypothesis.

One recent study determined that agricultural advances from 1961 to 2005 have saved an area the size of Russia from being converted into farmland and reduced greenhouse emissions equivalent to one-third of the total produced since the Industrial Revolution. During this time period, the world's population increased 111%. However, yields increased by 135% while global cropland only increased by 27%.

It also has been determined that the agricultural research that led to these increased yields reduced carbon emissions at a cost of less than one-quarter the price of permits for carbon trading in Europe. Another recent study found that increasing yields, instead of increasing farmland, could reduce CO_2 emissions as much as wind and solar power technologies could. It has been predicted that if yields cannot be increased to create the equivalent of a second Green Revolution, 5 billion acres (2 billion hectares) of additional farmland may be needed. This can be avoided if the productivity of current farmland can be increased, especially in the developing world.

The ecological impact of the human population on the world is challenging not only the sustainability of our environment but the very survival of life on Earth. The political, social, economic, and ecological choices that we make in the coming years will have far-reaching effects on the environment. These decisions will determine whether life becomes more uncertain in an increasingly inhospitable environment or whether we can find an ecological niche that provides a reliable and sustainable quality of life for both the human population and other members of our ecosystem. The science of plant pathology will continue to play a critical role in this quest.

In addition to hard-working scientists around the world, ingenious people with different areas of expertise will be needed to address and help solve these complex problems. We are optimistic that educated citizens can make a difference, and we hope that this book has contributed to your appreciation and understanding of the challenges we must address to feed all of the people on our hungry planet.

Glossary

Note: A similar glossary with images (diagrams and photographs) and pronunciation guides for many of the terms is available online at www.apsnet.org/edcenter/illglossary/.

A

abiotic pertaining to the absence of life; a disease not caused by living organisms but by chemical and physical factors (*see also* noninfectious) (*contrasts with* biotic, infectious)

acceptable daily intake (acronym ADI) a measure of the level of a chemical, i.e., a pesticide residue, that is believed to be able to be consumed on a daily basis over a lifetime without harm; used by the U.S. government to establish safe levels of potentially hazardous substances in food

acid having a pH less than 7 (*contrasts with* alkaline)

acid rain precipitation with a low pH due to the presence of nitric and sulfuric acid formed by the combination of air pollutants (NO_x and SO_2) with water

acute pertaining to symptoms that develop suddenly

acute toxicity ability of a single dose of a compound to poison (*contrasts with* chronic toxicity)

ADI (acronym for acceptable daily intake) a measure of the level of a chemical, i.e., a pesticide residue, that is believed to be able to be consumed on a daily basis over a lifetime without harm; used by the U.S. government to establish safe levels of potentially hazardous substances in food

aecial host the host plant on which a heteroecious rust fungus produces aeciospores and pycniospores (*contrasts with* telial host)

aeciospore the dikaryotic spore of a rust fungus produced in an aecium; in heteroecious rusts, a spore stage that infects the alternate host

aerobic living only in the presence of oxygen (*contrasts with* anaerobic)

aflatoxin a chemical by-product from *Aspergillus flavus* and *A. parasiticus* that is harmful to humans and other animals

agar a gelatinlike material derived from algae and used to solidify liquid culture media; term also applied to the medium itself

air pollution any air contaminant that causes undesirable effects on living organisms or materials

alkaline having basic (nonacidic) properties; pH greater than 7 (*contrasts with* acid)

alkaloid a nitrogen-containing ring compound produced by plants that causes physiological effects in animals

allele any of one or more alternative forms of a gene

alternate host one of two kinds of plants required by a heteroecious rust fungus to complete its life cycle; sometimes used as a general term for the aecial host or for the economically less important host

amino acid an organic nitrogenous acid from which protein molecules are constructed

ammonification the conversion of organic matter during decay by bacteria, fungi, and some other organisms into ammonia and ultimately ammonium, NH_4, which can be absorbed by plants

anaerobic living in the absence of oxygen (*contrasts with* aerobic)

annual a plant that completes its life cycle and dies within one year (*contrasts with* perennial)

annual ring a single-year growth of xylem in a woody stem

antagonism a general term for interference between organisms that may include antibiosis or competition for nutrients or space; action of two or more pesticides that reduces the effectiveness of one or all

antheridium (pl. antheridia) a male sexual organ (male gametangium) found in some fungi and funguslike organisms

antibiotic a chemical compound produced by one microorganism that inhibits growth of or kills other living organisms

antibody a specific protein formed in the blood of animals in response to the presence of an antigen

antigen any foreign chemical (normally a protein) that induces antibody formation in animals

antiserum (pl. antisera) blood serum containing antibodies

apex (pl. apices; adj. apical) the tip of a root or shoot containing the apical meristem

aphid a small, sucking insect of the family Aphididae (order Hemiptera) that produces honeydew; large populations can injure plants

appressorium (pl. appressoria) the swollen, flattened portion of a fungal filament that adheres to the surface of a higher plant, providing anchorage for invasion by a fungus

aquifer an area in which groundwater accumulates

arable able to be cultivated for agriculture

arthropod a member of the phylum Arthropoda, which consists of animals with articulated bodies and limbs and which includes insects, arachnids, and crustaceans

ascomycete an informal term for a member of the Ascomycota

Ascomycota a group of fungi that produce sexual spores (ascospores) within a saclike structure called an ascus; ascomycetes; sac fungi

ascospore a haploid (1N) sexual spore borne in an ascus

ascus (pl. asci) a saclike structure containing ascospores (typically eight) and usually borne in a fungal fruiting body

asexual vegetative; without sex organs, gametes, or sexual spores; the imperfect or anamorphic stage of a fungus

asexual reproduction any type of reproduction not involving the union of gametes and meiosis (*contrasts with* sexual reproduction)

autoecious in reference to rust fungi, producing all spore forms on one species of host plant (*contrasts with* heteroecious)

autotroph an organism that synthesizes its nutritive substances from inorganic molecules; e.g., plants capable of photosynthesis (*contrasts with* heterotroph)

auxin a plant hormone (growth regulator) that influences growth through cell elongation

avoidance a principle of plant disease management in which plants are grown at times or locations in which the pathogen is inactive or not present

B

bacilliform shaped like short rods with rounded ends

bacterium (pl. bacteria) a prokaryotic, microscopic, single-celled organism that reproduces by binary fission

bark all tissues outside the vascular cambium of a woody plant, generally including the cork layers, cork cambium (phellogen), and phloem

basidiomycete an informal term for a member of the Basidiomycota

Basidiomycota a group of fungi that produce sexual spores (basidiospores) externally on a structure called a basidium; basidiomycetes; mushroom fungi

basidiospore a usually haploid (1N) sexual spore produced on a basidium

beetle any insect of the order Coleoptera characterized by elytra (thickened outer wings), chewing mouth parts, and complete metamorphosis

berry a fleshy fruit containing seeds

binary fission a type of asexual reproduction in which two cells, usually of similar size and shape, are formed by the growth and division of one cell

binomial, Latin the scientific name of an organism; composed of two names, the first designating the genus and the second designating the specific epithet, together making the species name

bioassay any test (assay) using a living organism

biocide a compound toxic to all forms of life

biocontrol or **biological control** the exploitation by humans of the natural competition, parasitism, and/or antagonism of organisms for management of pests and pathogens

biodiversity or **biological diversity** the existing genetic variability among living organisms

biolistic transformation a method used for genetic engineering in which plant cells are bombarded with metal particles coated with foreign genes using a "gene gun"

biotechnology the development of genetically modified organisms through the use of modern technology and processes, including genetic engineering

biotic pertaining to life; a disease caused by a living organism (*see also* infectious) (*contrasts with* abiotic, noninfectious)

blight the sudden, severe, and extensive spotting, discoloration, wilting, or destruction of leaves, flowers, stems, or entire plants

breaking a disease symptom, usually caused by a virus, involving addition or loss of flower color to create a variegated pattern

brown rot (of wood) a type of wood decay resulting from selective removal of cellulose and hemicellulose, leaving a brown amorphous residue that usually cracks into cubical blocks and consists largely of slightly modified lignin (*contrasts with* white rot)

bud a terminal or axillary structure on a stem consisting of a small mass of meristematic tissue; generally covered wholly or in part by modified scale leaves

bug any insect of the order Hemiptera characterized in part by piercing-sucking mouth parts, a triangular scutellum, two pairs of wings, and gradual metamorphosis

C

callus the specialized tissues that form over a wound or cut in a plant; cork cambium may form and the cells produced will gradually seal the wound

canker a plant disease characterized (in woody plants) by the death of cambium tissue and loss and/or malformation of bark, or (in nonwoody plants) by the formation of sharply delineated, dry, necrotic, localized lesions on the stem; the term may also be used to refer to the lesion itself, particularly in woody plants

canopy the expanded leafy top of a plant or plants

capsid the protective layer of protein surrounding the nucleic acid core of a virus; the protein molecules that make up this layer (*see also* coat protein)

carbon cycle the continuous circulation of carbon atoms from inorganic carbon dioxide (CO_2) to organic molecules and back to CO_2

carcinogen a substance or agent that causes cancer

Casparian strips material between the cells of the root endodermis that blocks the passive movement of solutes and water

causal agent an organism or agent that incites and governs disease or injury

cell membrane the structure that bounds a cell and helps control the movement of substances into and out of the cell

cell wall the protective, resistant, but permeable structure secreted externally to the cell membrane in plants, bacteria, fungi, and certain other organisms

cellulose a carbohydrate polysaccharide composed of glucose units; a major component of a plant primary cell wall

center of origin the geographical area where a plant or other organism originated

certification describes seeds, propagative plant material, or nursery stock produced and sold under inspection to maintain genetic identity and purity and freedom from harmful pathogens, insect pests, and weed seeds; it is approved and certified by an official certifying agency

chitin a complex polysaccharide carbohydrate in fungal cell walls, animal exoskeletons, and nematode egg shells

chloroplast a disklike organelle containing chlorophyll in which photosynthesis occurs in the cells of green plants

chlorosis the failure of chlorophyll development, caused by disease or a nutritional disturbance; fading of green plant color to light green, yellow, or white

chromosome the structure, composed of DNA, that contains the genes of an organism; in eukaryotes, chromosomes are in the nucleus and can be visualized with an optical microscope as threads or rods during meiosis and mitosis; in bacteria, the chromosome is usually a single circle of DNA that cannot be visualized with an optical microscope

chronic toxicity poisoning resulting from exposure to low levels of exposure to a compound over a period of time (*contrasts with* acute toxicity)

clamp connection a bridge or buckle protrusion found at the septa of hyphae in basidiomycetous fungi and associated with cell division

coat protein the protective layer of protein surrounding the nucleic acid core of a virus; the protein molecules that make up this layer (*see also* capsid)

coccus (pl. cocci) a spherical (or near-spherical) bacterial cell

colony a growth of a microorganism in mass, especially as a pure culture

compartmentalization the isolation of a specific tissue area by host barrier tissues

compost a mixture of organic residues and soil that is allowed to decompose biologically

conidium (pl. conidia) an asexual, nonmotile fungal spore that develops externally or is liberated from the cell that formed it

conk a shelflike, typically hardened basidiocarp of a wood-decaying fungus, usually a polypore

contact fungicide a fungicide that remains on the surface where it is applied; has no after-infection activity (*see also* protectant fungicide) (*contrasts with* systemic fungicide)

cork the external, protective tissue of a stem or root, impermeable to water and gasses; the primary component of bark

cork cambium a cylinder of meristematic cells (lateral meristem) in the stems of woody plants that produces cork (bark)

cotyledon a seed leaf, one in moncots and two in dicots; the primary embryonic leaf within the seed in which nutrients for the new plant are stored

crop rotation the successive planting of different crop species; often used to improve soil fertility or to reduce disease and pest problems

cross-protection the process by which a normally susceptible host is infected with a less virulent pathogen (usually a virus) and thereby becomes resistant to infection by a second, usually related, more virulent pathogen

cultivar (abbr. cv.) a plant type within a species, resulting from deliberate genetic manipulation, which has recognizable characteristics (color, shape of flowers, fruits, seeds, and height or form) (*see also* variety)

cultural practices the manner in which plants are grown, including factors such as application of nutrients, irrigation practices, and type of cultivation; may be used for disease management

culture the growth and propagation of microorganisms on nutrient media; the growth and propagation of living plants

cuticle (adj. cuticular) the noncellular outer layer of an insect or a nematode; the water-repellent, waxy layer of epidermal cells of plant parts, such as leaves, stems, and fruit

cv. (abbr. for cultivar) a plant type within a species, resulting from deliberate genetic manipulation, which has recognizable characteristics (color, shape of flowers, fruits, seeds, and height or form)

cyst in fungi, a resting structure in a protective membrane or shell-like enclosure; in nematode females, the egg-laden carcass of a female nematode; in bacteria, a specialized type of bacterial cell enclosed in a thick wall, often dormant and resistant to environmental conditions

cytokinin a plant hormone (growth regulator) that controls cell division and is important for shoot stimulation of callus in tissue culture

cytoplasm the living protoplasm in a cell, except the nucleus

cytoplasmic inheritance the inheritance of genes not located in the nucleus, i.e., those in mitochondria and chloroplasts

D

damping-off the death of a seedling before or shortly after emergence resulting from decomposition of the root and/or lower stem; it is common to distinguish between preemergence damping-off and postemergence damping-off

fungistasis the inhibition of fungal growth, sporulation, or spore germination but not death; used to describe the nonspecific phenomenon in natural soils where spore germination is inhibited, often overcome by rhizosphere nutrients

fungus (pl. fungi) a eukaryotic organism that is usually filamentous (forming a mycelium) and heterotrophic, has cell walls composed of chitin, and reproduces by sexual and/or asexual spores

G

gall an abnormal swelling or localized outgrowth, often roughly spherical, produced by a plant as a result of attack by a fungus, bacterium, nematode, insect, or other organism (*see also* knot, tumor)

gallery an insect tunnel in bark and wood

gene a unit within an organism that controls heritable characteristics; genes are organized on chromosomes

gene gun a device used to bombard plant cells with metal particles coated with foreign genes to accomplish genetic engineering (*see also* biolistic transformation)

gene silencing the switching off of a gene by a process other than genetic modification; one example is the disabling of a gene by small interfering RNAs (siRNAs) that target messenger RNA for destruction

gene-for-gene hypothesis the hypothesis that corresponding genes for resistance and virulence exist in the host and pathogen, respectively

genetic code the system of triplet codons composed of nucleotides of DNA or RNA that determines the amino acid sequence of a protein

genetic engineering the transfer of specific genes between organisms using enzymes and laboratory techniques rather than biological hybridization

genetically modified organism (acronym GMO) an organism whose genetic makeup has been altered using techniques of genetic engineering (*see also* transgenic)

genetics the study of heredity and variation in organisms

genome the complete genetic information of an organism or virus

genotype the genetic constitution of an individual or group; a class or group of individuals sharing a specific genetic makeup (*contrasts with* phenotype)

genus (pl. genera) a taxonomic category that includes a group of closely related (structurally or phylogenetically) species; the genus or generic name is the first name in a Latin binomial

germ theory the theory that infectious or contagious diseases are caused by microorganisms (germs)

germ tube a hypha resulting from an outgrowth of the spore wall and cytoplasm after germination

germplasm the bearer of heredity material; often loosely applied to cultivars and breeding lines

giant cell an enlarged, multinucleate cell formed in roots by repeated nuclear division without cell division; induced by secretions of certain sedentary plant-parasitic nematodes

gibberellin a plant hormone (growth regulator) that affects stem elongation

gill a thin, radial membrane producing basidiospores in the cap of certain basidiomycetous fungi (e.g., mushrooms)

glucoside a substance that, on decomposition, yields glucose and certain other compounds; some glucosides are defense compounds produced by plants, e.g., cyanogenic or phenolic glucosides

GMO (acronym for genetically modified organism) an organism possessing a gene from another species; used to describe organisms that have been the subject of genetic engineering (*see also* transgenic)

graft the transfer of aerial parts of one plant (e.g., buds or twigs—the scion) into close cambial contact with the root or trunk (the rootstock) of a different plant; a method of plant propagation; the joining of cut surfaces or growing roots of two plants to form a living union

Gram-negative pertaining to bacteria staining red or pink after treatment with Gram stain

Gram-positive pertaining to bacteria staining violet or purple after treatment with Gram stain

Gram stain a procedure used for identification of bacteria in which crystal violet stain, Gram's iodine, ethyl alcohol, and safranin stain are applied in succession to cells of the bacteria

greenhouse effect the heating of the atmosphere as radiation from the sun is converted to infrared radiation but is trapped by so-called greenhouse gases (e.g., water vapor, CO_2, methane, ozone), preventing its release; this is similar to a greenhouse, in which glass traps air heated by infrared radiation

groundwater the water contained in aquifers in the soil, sometimes in underground rivers but more often in small accumulations mixed with sand

growth regulator a chemical substance produced in one part of an organism and transported in minute quantities to induce a growth response in another part, e.g., in plants, auxins, cytokinins, and gibberellins (*see also* hormone)

H

haploid having a single complete set of chromosomes (1N chromosomes) (*contrasts with* diploid)

haustorium (pl. haustoria) a specialized branch of a parasite formed inside host cells to absorb nutrients

heartwood the central cylinder of nonfunctional xylem in a woody stem (*contrasts with* sapwood)

herbicide a chemical used for killing plants or inhibiting plant growth, e.g., a weed or grass killer

hermaphrodite an organism having both male and female reproductive organs

heteroecious pertaining to a rust fungus requiring two unrelated host plants for completion of its life cycle (*contrasts with* autoecious)

heterotroph an organism that obtains nourishment from outside sources and must obtain its carbon from organic carbon compounds (*contrasts with* autotroph)

heterozygous having alternate forms (alleles) of a gene on homologous chromosomes (*contrasts with* homozygous)

homozygous having the same form (allele) of a gene on homologous chromosomes (*contrasts with* heterozygous)

honeydew a sugary ooze or exudate, often from aphids, and a characteristic symptom of ergot

hormone a chemical substance produced in one part of a an organism and transported in minute quantities to induce a growth response in another part, e.g., in plants, auxins, cytokinins, and gibberellins (*see also* growth regulator)

host range the range of plants on which an organism, particularly a parasite, feeds

hybrid vigor the increased vigor of hybrid offspring when compared to either parent; heterosis

hybridoma a cell produced by the fusion of an antibody-producing cell and a lymphoma (cancer) cell for production of monoclonal antibodies

hydathode an epidermal leaf structure specialized for secretion or exudation of water; a leaf opening at the terminus of a vein

hydrogen fluoride (HF) a primary pollutant from volcanoes and several industrial processes, e.g., brick making, aluminum smelting, and phosphate fertilizer production

hydrological cycle the cycling of water molecules among gaseous form (water vapor), liquid, and ice

hyperparasite an organism that is parasitic upon another parasite

hypha (pl. hyphae) a single, tubular filament of a fungal thallus or mycelium; the basic structural unit of a fungus

hypovirulence the reduced ability to cause disease

I

icosahedral having 20 faces, as a polyhedral virus particle (*see also* isometric)

immunoassay a detection method based on antibodies specifically selected to react with the substance to be detected (the antigen)

incomplete dominance the interaction of alleles of a gene that produces an intermediate phenotype, as in the production of pink flowers when red and white alleles are present in a heterozygous individual

indexing testing of a plant for infection, often by mechanical transmission or by grafting tissue from it to an indicator plant

indicator plant a plant that reacts to a pathogen or an environmental factor with specific symptoms and is used to detect or identify the pathogen or determine the effects of the environmental factor

indirect penetration the penetration of plant tissues by a pathogen through natural openings (e.g., stomata) or wounds

infect to enter, invade, or penetrate and establish a parasitic relationship with a host plant

infection peg the specialized, narrow hyphal strand on the underside of an appressorium that penetrates host cells; penetration peg

infection period the time required for infection to occur under conducive environmental conditions (usually hours of leaf wetness and temperature for foliar pathogens)

infectious pertaining to a disease that is caused by a biotic agent capable of spreading from plant to plant (*see also* biotic) (*contrasts with* abiotic, noninfectious)

infest to attack as a pest (used especially of insects and nematodes); to contaminate, as with microorganisms; to be present in large numbers

injury the damage caused by transitory interaction with an agent such as an insect, chemical, or unfavorable environmental condition

inoculate to place inoculum in an infection court; to insert a pathogen into healthy tissue

inoculum (pl. inocula) a pathogen or its parts that are capable of causing infection when transferred to a favorable location

inorganic not derived from living material

insect a member of the class Hexapoda (phylum Arthropoda) possessing three sets of limbs attached to a central body segment

insecticide a chemical or physical agent that kills insects

integrated pest management (acronym IPM) a combination of strategies to reduce losses caused by pests and pathogens based on environmental and economic considerations

IPM (acronym for integrated pest management) a combination of strategies to reduce losses caused by pests and pathogens based on environmental and economic considerations

isolate (n.) a culture or subpopulation of a microorganism separated from its parent population and maintained in some sort of controlled circumstance; (v.) to remove from soil or host material and grow in pure culture

isometric usually used to describe virus particles that are icosahedral in structure and appear approximately round

J

juvenile an immature form that appears similar to but usually smaller than the adult and is not sexually mature; e.g., insects with gradual metamorphosis, nematodes

K

karyogamy the fusion of nuclei

knot an abnormal swelling or localized outgrowth, often roughly spherical, produced by a plant as a result of attack by a fungus, bacterium, nematode, insect, or other organism (*see also* gall, tumor)

Koch's postulates the procedure used to prove the pathogenicity of an organism, i.e., its role as the causal agent of a disease mobiles and other internal combustion engines is in the presence of sunlight

ozone layer a protective layer of ozone in the upper atmosphere that reduces ultraviolet radiation

P

palisade parenchyma the tissue found beneath the upper epidermis of a leaf; composed of elongate, tubular cells arranged upright in the manner of posts in a palisade fortification

PAN (acronym for peroxyacetyl nitrate) a secondary air pollutant formed when the exhaust of automobiles and other internal combustion engines is in the presence of sunlight

pandemic a widespread and destructive outbreak of disease occurring simultaneously in several countries

parasite an organism that lives in intimate association with another organism on which it depends for its nutrition; not necessarily a pathogen (*contrasts with* saprophyte)

parenchyma the soft tissue of living plant cells with undifferentiated, thin, cellulose walls

parthenogenesis (adj. parthenogenetic) reproduction by the development of an unfertilized egg

pasteurization a process using heat to free a material, usually a liquid, of selected harmful microorganisms

pathogen (adj. pathogenic) a disease-producing organism or biotic agent

pathogenesis the production and development of disease

pathogenicity the ability to cause disease

pathology the study of disease

pathovar (abbr. pv.) a subdivision of a plant-pathogenic bacterial species defined by host range; pathovar for bacteria is equivalent to *forma specialis* for fungi

PCR (acronym for polymerase chain reaction) a technique used to amplify the number of copies of a specific region of DNA to produce enough of the DNA for use in various applications, e.g., identification and cloning

pectin a methylated polymer of galacturonic acid found in the middle lamella and primary cell wall of plants; the jelly-forming substance found in fruit

penetration the initial invasion of a host by a pathogen

perennial something that occurs year after year; a plant that survives for several to many years (*contrasts with* annual)

peroxyacetal nitrate (acronym PAN) a secondary air pollutant formed when the exhaust of automobiles and other internal combustion engines is in the presence of sunlight

persistent transmission a type of virus transmission in which the virus is acquired and transmitted by the vector after a relatively long feeding time and remains transmissible for a prolonged period while in association with its vector; circulative transmission

pesticide a chemical used to control pests

pesticide label a legal document approved by the U.S. Environmental Protection Agency that defines uses and provides other safety information for a commercial pesticide

pH the negative logarithm of the effective hydrogen ion concentration; a measure of acidity (pH 7 is neutral; values less than pH 7 are acidic; values greater than pH 7 are alkaline)

phenotype the expressed characteristics of an organism determined by the interaction of its genotype with the environment (*contrasts with* genotype)

phloem the food-conducting, food-storing tissue in the vascular system of roots, stems, and leaves

photochemical reaction a chemical reaction that requires sunlight

photosynthate a product of photosynthesis

photosynthesis the manufacture of carbohydrates from carbon dioxide and water in the presence of chlorophyll(s), using light energy and releasing oxygen

phytopathology the study of plant diseases (*see also* plant pathology)

phytoplasma a plant-parasitic mollicute (prokaryote with no cell wall) found in phloem tissue; cannot yet be grown on artificial nutrient media

phytotoxic harmful to plants; usually used to describe chemicals

plant pathology the study of plant diseases (*see also* phytopathology)

plasmid a circular, self-replicating hereditary element that is not part of a chromosome; plasmids are used in recombinant DNA experiments as acceptors and vectors of foreign DNA

plasmodesma (pl. plasmodesmata) the cytoplasmic strands that connect living plant cells

plasmogamy the fusion of two sex cells

polyclonal antibodies a mixture of antibodies produced by different antibody-producing cells (*contrasts with* monoclonal antibodies)

polycyclic having several to many disease cycles in a growing season (*contrasts with* monocyclic)

polygenic pertaining to or governed by many genes

polymerase chain reaction (acronym PCR) a technique used to amplify the number of copies of a specific region of DNA to produce enough of the DNA for use in various applications, e.g., identification and cloning

powdery mildew a common name for a disease caused by a white, powdery, superficial ascomycetous fungus that is an obligate parasite

primary inoculum the inoculum, usually from an overwintering source, that initiates disease in the field, as opposed to inoculum that spreads disease during the season; initial inoculum (*contrasts with* secondary inoculum)

primary pollutant an air pollutant that is released directly into the atmosphere and is harmful to plants, e.g., SO_2 and NO_x

prokaryote an organism without internal membrane-bound organelles; lacking a distinct nucleus, e.g., bacteria and mollicutes (*contrasts with* eukaryote)

propagule any part of an organism capable of independent growth

protectant fungicide a fungicide that remains on the surface where it is applied; no after-infection activity (*see also* contact fungicide) (*contrasts with* systemic fungicide)

protection the various methods of plant disease management, including cultural practices that create barriers or reduce the chance of infection, chemical protection, methods of biological control that protect plants, and genetic resistance

protein a nitrogen-containing organic compound composed of units called amino acids

protoplast a living cell exclusive of a wall

Puccinia pathway the region through which rust urediniospores move each season from southern areas through all grain-producing areas of the United States to Canada

pustule a small, blisterlike elevation of epidermis formed as spores emerge

pv. (abbr. for pathovar) a subdivision of a plant-pathogenic bacterial species defined by host range; pathovar for bacteria is equivalent to *forma specialis* for fungi

pycniospore a haploid, sexually derived spore formed in a pycnium of rust fungi; spermatium

pyramiding the addition, through plant breeding or genetic engineering, of several resistance genes into a single plant cultivar

Q

quarantine the legislative control of the transport of plants or plant parts to prevent the spread of pathogens or pests

quorum sensing the ability of bacteria to interact with each other through a variety of mechanisms; allows a population of bacteria to behave more like a multicellular organism

R

receptive hypha the part of a rust fungus pycnium (spermogonium) that receives the nucleus of a pycniospore (spermatium)

recessive describes a phenotypic trait that is expressed in diploid organisms only if both parents contribute the trait to the progeny (*contrasts with* dominant)

recombinant DNA DNA molecules in which sequences that are not normally contiguous have been placed next to each other by laboratory methods

replication the process by which a virus particle induces the host cell to reproduce the virus; copying or multiplication of DNA

resistant (n. resistance) possessing properties that prevent or impede disease development (*contrasts with* susceptible)

respiration the series of chemical reactions that makes energy available through oxidation of carbohydrates and fat

restriction endonuclease an enzyme that cleaves DNA at a particular base sequence; sometimes informally referred to as a restriction enzyme

reverse transcriptase an enzyme used to make complementary DNA from a piece of RNA, e.g., a plant virus

rhizomorph a macroscopic, ropelike strand of compacted tissue formed by certain fungi

rhizosphere the microenvironment in the soil immediately around plant roots

rhizosphere competent used to describe microorganisms adapted to living in the rhizosphere of a plant

ribonucleic acid (acronym RNA) any of several nucleic acids composed of repeating units of ribose (a sugar), a phosphate group, and a purine (adenine or guanine) or a pyrimidine (uracil or cytosine) base; transcribed from DNA and involved in translation to proteins

ribosome a subcellular protoplasmic particle, made up of one or more RNA molecules and several proteins, involved in protein synthesis

ringspot a disease symptom characterized by yellowish or necrotic rings enclosing green tissue, as in some plant diseases caused by viruses

RNA (acronym for ribonucleic acid) any of several nucleic acids composed of repeating units of ribose (a sugar), a phosphate group, and a purine (adenine or guanine) or a pyrimidine (uracil or cytosine) base; transcribed from DNA and involved in translation to proteins

RNA interference (acronym RNAi) a process within living cells in which a double-stranded complementary RNA targets a specific messenger RNA for destruction, blocking the function of (silencing) the gene from which the mRNA was transcribed

rogue to remove and destroy individual plants that are diseased, infested by insects, or otherwise undesirable

root cap a group of cells on a root that protects the growing tip

root exudates the various compounds that leak from growing and expanding sections of roots, as well as from broken cells at exit points of lateral roots

root graft the fusion of roots of two adjacent plants so that their water- and food-conducting (vascular) systems become joined

root hair a threadlike, single-celled outgrowth from a root epidermal cell

rot the softening, discoloration, and often disintegration of plant tissue as a result of fungal or bacterial infection

rotation the growth of different kinds of crops in succession in the same field

roundworm a nematode

rust a disease caused by a specialized group in the Basidiomycota (the basidiomycetes) that often produces spores of a rusty color

S

sanitation the destruction or removal of infected and infested plants or plant parts; the decontamination of tools, equipment, containers, work space, hands, etc.

sap transmission the transmission of pathogens, usually viruses, by rubbing sap from an infected plant onto a healthy plant to cause infection

saprophyte an organism that obtains nourishment from nonliving organic matter (*contrasts with* parasite)

sapwood the physiologically active zone of wood contiguous to cambium (*contrasts with* heartwood)

SAR (acronym for systemic acquired resistance) the reduced disease symptoms on a portion of a plant distant from the area where a hypersensitive response occurred or other stimulus was applied; a rapid and coordinated defense response against a variety of pathogens as a signal travels throughout the plant

scab a rough, crustlike diseased area on the surface of a plant organ

sclerotium (pl. sclerotia) a vegetative resting body of a fungus, composed of a compact mass of hyphae with or without host tissue, usually with a darkened rind

scorch any symptom that resembles the result of flame or fire on the affected part; often seen at the margins of leaves

secondary growth the growth in vascular plants from lateral meristems, resulting in wider stems and branches

secondary inoculum the inoculum produced by infections that occurred during the same growing season (*contrasts with* primary inoculum)

secondary metabolite a compound produced in microbes (e.g., mycotoxins, syringomycins) or plants (e.g., caffeine or nicotine) that is not necessary for normal growth and development

secondary phloem the phloem produced by the vascular cambium in stems and branches of woody plants

secondary pollutant an air pollutant that must be chemically produced from other air pollutants; e.g., ozone (O_3), which is a product of a photochemical reaction of exhaust products from combustion engines in the atmosphere

secondary xylem the xylem produced by the vascular cambium in stems and branches of woody plants; also known as wood

sedentary remaining in a fixed location (*contrasts with* migratory)

seed bank an organization that preserves and maintains species and cultivars of plants for use by plant breeders

seedborne carried on or in a seed

selective medium a culture medium containing substances that specifically inhibit or promote the growth of some organisms over others

septate with cross-walls; having septa

septum (pl. septa) a dividing wall; in fungi, a cross-wall

serology a method using the specificity of the antigen/antibody reaction for the detection and identification of antigenic substances and the organisms that carry them

sexual reproduction reproduction involving fusion of two haploid nuclei (karyogamy) to form a diploid nucleus followed by meiosis (reduction division) back to haploid nuclei at some point in the life cycle, resulting in genetic recombination (*contrasts with* asexual reproduction)

sieve element or **sieve tube element** a tube-shaped living cell in the phloem functioning in the transport of dissolved organic substances in the plant

sign an indication of disease from direct observation of a pathogen or its parts (*contrasts with* symptom)

smut a disease caused by a smut fungus (Ustilaginomycotina) in the Basidiomycota (the basidiomycetes) or the fungus itself; characterized by masses of dark brown or black, dusty to greasy masses of teliospores that generally accumulate in black, powdery sori

soft rot the softening, discoloration, and often disintegration of plant tissue as a result of fungal or bacterial infection

soilborne carried on or beneath the soil surface

soil inhabitant an organism that maintains its population in soil over a period of time

soil invader an organism whose population in soil diminishes in several months to years

soil pasteurization the process used to free soil of selected harmful microorganisms using heat

soil sterilization the process used to free soil of all microorganisms

solarization a disease control practice in which soil is covered with polyethylene sheeting and exposed to sunlight, thereby heating the soil and inhibiting or killing soilborne plant pathogens

sp. (abbr. for species; pl. spp.) when a genus name is followed by sp., it means that the particular species is undetermined; spp. after a genus name means that several species are being referred to

species any one kind of life subordinate to a genus but above a race; a group of closely related individuals of the same ancestry, resembling one another in certain inherited characteristics of structure and behavior and relative stability in nature; the individuals of a species ordinarily interbreed freely and maintain themselves and their characteristics in nature

specific epithet the second word in a Latin binomial

spicule the copulatory organ of male nematode

spiroplasma a spiral-shaped, plant-pathogenic mollicute (a prokaryote without a cell wall)

spontaneous generation, theory of the theory, now known to be invalid, that plants, animals, and microorganisms arose suddenly from nonliving materials under certain environmental conditions

sporangium (pl. sporangia) a saclike structure in which the entire contents are converted into asexual spores in certain fungi and funguslike organisms

spore a reproductive structure of fungi and some other organisms, containing one or more cells; a bacterial cell modified to survive an adverse environment

sporulate to produce spores

spot a symptom of disease characterized by a limited necrotic area, as on leaves, flowers, and stems

stele the central cylinder of vascular tissue (especially in roots)

sterile unable to reproduce sexually; to be free of living microorganisms

sterilization (adj. sterilized) the total destruction of living organisms by various means, including heat, chemicals or irradiation

stolon a slender, horizontal stem that grows close to the soil surface; in fungi, a hypha that grows horizontally along the surface

stoma (pl. stomata; adj. stomatal; also stomate) a structure composed of two guard cells and the opening between them in the epidermis of a leaf or stem, functioning in gas exchange

stone fruit a fruit with a stony endocarp, e.g., cherry, peach, plum

strain a distinct form of an organism or virus within a species; differs from other forms of the species biologically, physically, or chemically

stunting the reduction in height of a vertical axis resulting from a progressive reduction in the length of successive internodes or a decrease in their number

stylet the stiff, slender, hollow feeding organ of plant-parasitic nematodes or sap-sucking insects, e.g., aphids and leafhoppers

substrate the substance on which an organism lives or from which it obtains nutrients; a chemical substance acted upon, often by an enzyme

sulfur dioxide (SO_2) a primary air pollutant produced in industrial processes and coal burning that causes interveinal necrosis on broadleaf plants and tip necrosis on conifers

sunburn or **sunscald** the injury of plant tissues burned or scorched by direct sun

susceptible prone to develop disease when infected by a pathogen (*contrasts with* resistant)

symbiosis the living together of two different kinds of organisms that may, but does not necessarily, benefit each organism

symptom an indication of disease by reaction of the host, e.g., canker, leaf spot, wilt (*contrasts with* sign)

synergism (adj. synergistic) a greater-than-additive effect of interacting factors

synnema (pl. synnemata) compact or fused, generally upright conidiophores, with branches and spores forming a headlike cluster

systematics the study of the kinds of organisms and the relationships between them

systemic pertaining to a disease in which the pathogen (or a single infection) spreads generally throughout the plant; pertaining to chemicals that spread internally through the plant

systemic acquired resistance (acronym SAR) the reduced disease symptoms on a portion of a plant distant from the area where a hypersensitive response occurred or other stimulus was applied; a rapid and coordinated defense response against a variety of pathogens as a signal travels throughout the plant

systemic fungicide a fungicide that is absorbed into plant tissue and may offer some curative or after-infection activity; includes fungicides that are locally systemic, xylem-mobile (upward moving), and amphimobile (move in phloem upward as well as downward in the plant) (*contrasts with* contact or protectant fungicide)

T

taxonomy (adj. taxonomic) the science dealing with classifying organisms

telial host the host plant on which a heteroecious rust fungus produces teliospores and sometimes urediniospores (*contrasts with* aecial host)

teliospore a thick-walled resting or overwintering spore produced by the rust fungi (Pucciniales) and smut fungi (Ustilaginomycotina) in which karyogamy occurs; it germinates to form a promycelium (basidium) in which meiosis occurs

teratogen a chemical that causes malformations in the fetus

tillage the process of turning or stirring the soil

tissue culture a laboratory method of propagating healthy cells from plant tissues

titer the concentration of a virus

tolerance the ability of a plant to endure an infectious or noninfectious disease, adverse conditions, or chemical injury without serious damage or yield loss; in terms of pesticides, the amount of chemical residue legally permitted on an agricultural product entering commercial channels and usually measured in parts per million (ppm)

totipotency the potential ability of a single cell to be regenerated into a whole organism because each nucleus contains the full genome

toxicity the capacity of a substance to interfere with the vital processes of an organism

toxin a poisonous substance of biological origin

tracheid an elongated conducting cell of the xylem, with tapering or oblique end walls and pitted walls

transcription the production of a complementary strand of RNA from a segment of DNA

transfer RNA (acronym tRNA) the RNA that moves amino acids to the ribosome to be placed in the order prescribed by the messenger RNA

transformation the transfer of genetic materials from one organism to another by humans (genetic engineering); a means of genetic variation in bacteria by absorption and incorporation of DNA from another bacterial cell

transgenic possessing a gene from another species; used to describe the organisms that have been the subject of genetic engineering (*see also* genetically modified organism, GMO)

translation the assembling of amino acids into a protein using messenger RNA, ribosomes, and transfer RNA

transpiration the water loss by evaporation from leaf surfaces and through stomata

transposon a piece of DNA capable of moving to a different location in the genome

triplet codon a set of three nucleotide bases in DNA or RNA that code for an amino acid

tRNA (acronym for transfer RNA) the RNA that moves amino acids to the ribosome to be placed in the order prescribed by the messenger RNA

tuber an underground stem adapted for storage; typically produced at the end of a stolon

tumor an abnormal swelling or localized outgrowth, often roughly spherical, produced by a plant as a result of attack by a fungus, bacterium, nematode, insect, or other organism (*see also* gall, knot)

tylosis (pl. tyloses) a balloonlike extrusion of a parenchyma cell into the lumina of a contiguous vessel that partially or completely blocks it

U

urediniospore the asexual, dikaryotic, often rusty-colored spore of a rust fungus, produced in a structure called a uredinium; the "repeating stage" of a heteroecious rust fungus, i.e., capable of infecting the host plant on which it is produced

V

variety a plant type within a species that is true to type and has recognizable characteristics, described by a third italicized word in a scientific name (*see also* cultivar)

vascular pertaining to fluid-conducting (xylem and phloem) tissues in plants

vascular bundle a strand of conductive tissue, usually composed of xylem and phloem; in leaves, small bundles are called veins

vascular cambium a cylinder of meristematic cells (lateral meristem) that produces secondary phloem to the outside and secondary xylem (wood) to the inside of a branch or trunk of a woody plant

vascular wilt disease a xylem disease that disrupts the normal uptake of water and minerals, resulting in wilting and yellowing of foliage

vector a living organism (e.g., insect, mite, bird, higher animal, nematode, parasitic plant, human) able to carry and transmit a pathogen and disseminate disease; in genetic engineering, a vector or cloning vehicle is a self-replicating DNA molecule, e.g., a plasmid or virus, used to introduce a fragment of foreign DNA into a host cell

vegetative referring to the somatic or asexual parts of a plant, which are not involved in sexual reproduction

vegetative propagation a form of asexual reproduction in plants in which cuttings, bulbs, tubers, or other vegetative plant parts are used to grow new plants

vermiform worm-shaped

vessel a water-conducting structure of xylem tissue with openings in the end walls

virion a complete virus particle

viroid an infectious, nonencapsidated (naked) circular, single-stranded RNA

virulence the degree or measure of pathogenicity; the relative capacity to cause disease

virus a submicroscopic, intracellular, obligate parasite consisting of a core of infectious nucleic acid (either RNA or DNA) usually surrounded by a protein coat

volunteer a self-set plant; a plant seeded by chance

W

white rot (of wood) a type of wood decay resulting from enzymatic action of fungi; it degrades all components of wood, including lignin, leaving the wood light colored and spongy or stringy (*contrasts with* brown rot)

wilt the drooping of leaves and stems from lack of water (i.e,. inadequate water supply or excessive transpiration); a vascular disease that interrupts normal water uptake

winterburn the foliar necrosis, often marginal, of plants that retain their leaves in winter; caused by water deficiency because plants cannot take up water from frozen soils

witches' broom a disease symptom characterized by an abnormal, massed, brushlike development of many weak shoots arising at or close to the same point

wood the secondary xylem produced by the vascular cambium in stems and branches of woody plants

wound an injury to plant tissue that often breaches barriers (i.e., cuticle, bark, cell walls) that might otherwise exclude a pathogen; some pathogens (e.g., viruses) can enter plants only through a wound; wounds may occur from natural growth processes, physical and chemical agents, animals (especially insects), and many human agricultural activities (e.g., pruning)

X

xylem the water- and mineral-conducting, food-storing, supporting tissue of a plant

xylem-limited fastidious bacteria (acronym XLB) the bacterial pathogens of plants found only in the xylem, causing wilt, scorch, and stunting symptoms; vectored by insects that feed on xylem fluid; not sap transmissible; require complex culture media

Y

yeast a unicellular ascomycetous fungus that reproduces asexually by budding

yellows a disease characterized by chlorosis and stunting of the host plant

Z

zoospore a spore of a fungus or funguslike organism with flagella capable of locomotion in water

Figure Credits

Chapter 1

Figures 1.1, 1.4, and 1.9. Drawings by Nancy Haver.

Figures 1.2 and 1.13. Used by permission of H. D. Thurston, Cornell University, Ithaca, NY.

Figure 1.3. Courtesy H. Murphy.

Figure 1.5. Drawing by Vickie Brewster. Reprinted from Schumann, G. L., and D'Arcy, C. J. 2010. Essential Plant Pathology, 2nd ed. American Phytopathological Society, St. Paul, MN.

Figure 1.6. Used by permission of S. Johnson, University of Maine, Orono.

Figure 1.7. Courtesy W. E. Fry. Reprinted from 2002, Fundamental Fungi Image Collection, American Phytopathological Society, St. Paul, MN.

Figure 1.8. Left, Courtesy R. V. James. **Right,** Courtesy B. G. Turgeon. Reprinted from Stevenson, W. R., Loria, R., Franc, G. D., and Weingartner, D. P. 2001. Compendium of Potato Diseases, 2nd ed. American Phytopathological Society, St. Paul, MN.

Figure 1.10. From Woodham-Smith, C. 1962. The Great Hunger. Harper and Row, New York.

Figure 1.11. Cartoon by Thomas Nast, circa 1880. © Bettmann/CORBIS.

Figure 1.12. Courtesy American Phytopathological Society.

Chapter 2

Figures 2.1, 2.2, 2.4, 2.6, and 2.10. Drawings by Vickie Brewster.

Figure 2.3. Courtesy J. M. F. Yuen. Reprinted from Diseases of Herbaceous Ornamentals and Roses Digital Image Collection. 2001. American Phytopathological Society, St. Paul, MN.

Figures 2.5, 2.7, 2.8, and 2.9. Drawings by Nancy Haver.

Figure 2.11. Reprinted from Zentmyer, G. A. 1980. *Phytophthora cinnamomi* and the Diseases It Causes. Fig. 27. American Phytopathological Society, St. Paul, MN.

Figure 2.12. Courtesy J. Parke. Reprinted from Parke, J. L., and S. Lucas. 2008. Sudden oak death and ramorum blight. The Plant Health Instructor, DOI: 10.1094/PHI-I-2008-0227-01.

Figure 2.13. Courtesy USDA-ARS. Reprinted from Fundamental Fungi Image Collection and Teaching Resource. 2002. American Phytopathological Society, St. Paul, MN.

Chapter 3

Figure 3.1. Courtesy G. L. Schumann.

Figure 3.2. Reprinted, by permission, from Schieber, E., and Zentmyer, G. A. 1984. Coffee rust in the Western Hemisphere. Plant Disease 68:89–93. Fig. 1.

Figure 3.3. Drawing by Vickie Brewster. Reprinted from Arneson, P. A. 2000. Coffee rust. The Plant Health Instructor, DOI: 10.1094/PHI-I-2000-0718-02.

Figure 3.4. Reprinted, by permission, from Oliver, F. W. 1913. Makers of British Botany. Cambridge University Press, New York.

Figures 3.5, 3.6, 3.7, 3.8, and 3.9. Used by permission of H. D. Thurston, Cornell University, Ithaca, NY.

Figure 3.10. Drawing by Vickie Brewster.

Figure 3.11. Drawing by Vickie Brewster. Reprinted from Schumann, G. L., and D'Arcy, C. J. 2010. Essential Plant Pathology, 2nd ed. American Phytopathological Society, St. Paul, MN.

Figure 3.12. Courtesy USDA-APHIS, copyright-free.

Figure 3.13. Used by permission of C. Krebs, Issaquah, WA.

Chapter 4

Figures 4.1, 4.2, 4.3, 4.4, 4.5, 4.6, 4.7, 4.9, and 4.11. Drawings by Nancy Haver.

Figure 4.8. Courtesy N. Mattson.

Figure 4.10. Courtesy J. C. Wynne. Reprinted from Diseases of Legumes Digital Image Collection. 2001. American Phytopathological Society, St. Paul, MN.

Figure 4.12. Photo by Bruce Fritz. Courtesy Agricultural Research Service, USDA.

Chapter 5

Figure 5.1. This illustration was published in Plant Pathology, 3rd ed., by G. N. Agrios, Figure 1-2, p. 7, Copyright Elsevier 1988. Drawing by Nancy Haver.

Figure 5.2. Courtesy D. L. Charbonneau and W. C. Ghiorse. Reprinted from Smiley, R. W., Dernoeden, P. H., and Clarke, B. B. 2005. Compendium of Turfgrass Diseases. 3rd ed. American Phytopathological Society, St. Paul, MN.

Figure 5.3. Courtesy University Archives, University of Illinois. Reprinted from Campbell, C. L., Peterson, P. D., and Griffith, C. S. 1999. The Formative Years of Plant Pathology in the United States. American Phytopathological Society, St. Paul, MN.

Figure 5.4. Courtesy S. V. Beer.

Figures 5.5 and 5.6. Drawings by Nancy Haver.

Figure 5.7. Drawing by Vickie Brewster. Adapted from Johnson, K. B. 2000. Fire blight of apple and pear. The Plant Health Instructor, DOI: 10.1094/PHI-I-2000-0726-01.

Figure 5.8. Courtesy of the Florida Department of Agriculture and Consumer Services, Gainesville.

Figure 5.9. Top, Courtesy R. McCoy. **Bottom,** Courtesy R. E. Davis, copyright-free.

Figure 5.10. Courtesy R. McCoy.

Figure 5.11. Courtesy J. R. Hartman. Reprinted from Gould, A. B. and Lashomb, J. H. 2007. Bacterial leaf scorch (BLS) of shade trees. The Plant Health Instructor, DOI: 10.1094/PHI-I-2007-0403-07.

Figure 5.12. Courtesy R. L. Forster. Reprinted from Kado, C. I. 2002. Crown gall. The Plant Health Instructor, DOI: 10.1094/PHI-I-2002-1118-01.

Chapter 6

Figure 6.1. Drawing by Vickie Brewster.

Figures 6.2, 6.3, 6.4, and 6.5. Drawings by Nancy Haver.

Figure 6.6. Used, with permission, from Priestley, R. H. 1978. Plant Disease Epidemiology. P. R. Scott, and A. Bainbridge, eds. Blackwell Scientific Publications, Oxford. Fig. 1. Adapted from the original.

Figure 6.7. Used by permission of H. D. Thurston, Cornell University, Ithaca, NY.

Figure 6.8. Courtesy M. S. Mount.

Figure 6.9. Used by permission of K. Wood, Promega, Madison, WI.

Figure 6.10. Used, by permission, from Genetic Engineering: A National Science. Monsanto Company, St. Louis, MO. Redrawn from the originals.

Figure 6.11. Used by permission of Monsanto Company, St. Louis, MO.

Chapter 7

Figure 7.1. Courtesy R. Kirkby, Centro Internactional de Agricultural Tropical, Debre Zeit, Ethiopia.

Figure 7.2. Courtesy J. Galindo, Centro Agronomico Tropical Investigacion y Ensenanza, Turrialba, Costa Rica.

Figure 7.3. Drawing by Nancy Haver.

Figure 7.4. Courtesy North Dakota State University, Fargo.

Figure 7.5. Courtesy K. F. Baker, Corvallis, OR. Redrawn from the original.

Figure 7.6. This illustration was published in Disease Resistance in Plants, 2nd ed., by J. E. Vanderplank, Figure 13.7, p. 142, Copyright Elsevier 1984.

Figure 7.7. Reproduced with permission of Agriculture and Agri-Food Canada, from W. C. James, A Manual of Assessment Keys for Plant Diseases, Agriculture Canada Publication no. 1458 (1971), Keys 3.1.1 and 1.6.1. This reproduction is a copy of an official work published by the Government of Canada and has not been produced in affiliation with, or with the endorsement of, the Government of Canada.

Figure 7.8. Drawing by Vickie Brewster.

Figure 7.9. Courtesy L. Farrar, Auburn University, Auburn, AL.

Figures 7.10 and 7.11. Courtesy W. F. Moore, Extension Plant Pathology, Mississippi State University.

Figure 7.12. Courtesy D. Groth. Reprinted from Webster, R. K. and Gunnell, P. S. 1992. Compendium of Rice Diseases. American Phytopathological Society Press, St. Paul, MN.

Figure 7.13. Reprinted from Cook, R. J., and Veseth, R. J. 1991. Wheat Health Management. American Phytopathological Society Press, St. Paul, MN.

Chapter 8

Figure 8.1. Courtesy Cornell University, Ithaca, NY.

Figure 8.2. Courtesy Fungicide Resistance Action Committee. Reprinted from Schumann, G. L., and D'Arcy, C. J. 2000. Late blight of potato and tomato. The Plant Health Instructor, DOI: 10.1094/PHI-I-2000-0724-01.

Figure 8.3. Courtesy G. Johnson and M. J. Jackson, Montana State University Cooperative Extension, Bozeman.

Figure 8.4. Courtesy Cornell University Cooperative Extension, Ithaca, NY.

Figure 8.5. Drawing by Nancy Haver.

Figure 8.6. Based on US Mortality Data, 1960 to 2006, US Mortality Volumes, 1930 to 1959, National Center for Health Statistics, Centers for Disease Control and Prevention, 2009.

Figure 8.7. Courtesy U.S. Food and Drug Administration.

Figure 8.8. Courtesy W. E. MacHardy. Reprinted from Vaillancourt, L. J., and Hartman, J. R. 2000. Apple scab. The Plant Health Instructor, DOI: 10.1094/PHI-I-2000-1005-01.

Figure 8.9. Courtesy J. R. Hartman. Reprinted from Vaillancourt, L. J., and Hartman, J. R. 2000. Apple scab. The Plant Health Instructor, DOI: 10.1094/PHI-I-2000-1005-01.

Figure 8.10. Reprinted from Vaillancourt, L. J., and Hartman, J. R. 2000. Apple scab. The Plant Health Instructor, DOI: 10.1094/PHI-I-2000-1005-01.

Figure 8.11. Courtesy Neogen Corporation, East Lansing, MI.

Chapter 9

Figures 9.1, 9.2, 9.4, 9.5, and 9.11. Drawings by Nancy Haver.

Figure 9.3. Courtesy B. H. Waite, American Phytopathological Society Slide Set 15, Slide 32 (1980).

Figures 9.6 and 9.7. Courtesy R. S. Hussey. Used by permission of Society of Nematologists.

Figure 9.8. Courtesy Cornell University, Ithaca, NY.

Figure 9.9. Courtesy B. M. Zuckerman, University of Massachusetts, Amherst.

Figure 9.10. Courtesy G. L. Barron. Used by permission of Society of Nematologists.

Chapter 10

Figure 10.1. Courtesy of Staatliche Graphische Sammlung München, Munich, Germany. Redrawn from the original.

Figure 10.2. Courtesy Department of Plant Pathology, North Dakota State University, Fargo.

Figures 10.3, 10.8, and 10.9. Drawings by Nancy Haver.

Figure 10.4. Courtesy C. R. Funk, Rutgers University, New Brunswick, NJ.

Figure 10.5. Used by permission of M. F. Brown and H. G. Brotzman.

Figure 10.6. Reprinted from M.C. Shurtleff, ed. 1980. Compendium of Corn Diseases, 2nd ed. American Phytopathological Society, St. Paul, MN.

Figure 10.7. Courtesy of R. Duran, Washington State University, and R. Cruz, Broomfield, CO.

Figure 10.10. Courtesy J. L. Maas. Reprinted from Maas, J. L. 1998. Compendium of Strawberry Diseases. 2nd ed. American Phytopathological Society, St. Paul, MN.

Figure 10.11. Courtesy of S. Colleen O'Keefe-Safir, Chateau Grand Traverse, Ltd., Traverse City, MI.

Chapter 11

Figure 11.1. Reprinted by permission, from: Zadoks. J. 1982. Cereal rusts, dogs, and stars in antiquity. Garcia de Orta, Serie de Estudos Agronomicos, Lisboa 9(1-2):13-20.

Figure 11.2. Courtesy of the U.S. Department of Agriculture, Soil Conservation Service.

Figure 11.3. Left, Courtesy B. Steffenson. Reprinted from Schumann, G. L. and K. J. Leonard. 2000. Stem rust of wheat (black rust). The Plant Health Instructor. DOI: 10.1094/PHI-I-2000-0721-01. **Right,** Courtesy D. L. Long, U.S. Department of Agriculture.

Figures 11.4 and 11.8B. Used by permission of M. F. Brown and H. G. Brotzman.

Figures 11.5 and 11.10. Drawings by Nancy Haver.

Figure 11.6. Courtesy Cornell University, Ithaca, NY.

Figure 11.7. Drawing by Vickie Brewster.

Figure 11.8A. Photo by Mr. Eugene Herrling, Department of Plant Pathology, University of Wisconsin–Madison.

Figure 11.9. Used by permission of M. Daughtrey.

Figure 11.11. Courtesy E. G. Kuhlman, U.S. Department of Agriculture, Forest Service.

Chapter 12

Figure 12.1. Drawing by Vickie Brewster.

Figure 12.2. Courtesy G. L. Schumann.

Figures 12.3, 12.6, 12.8, 12.9, and 12.10. Drawings by Nancy Haver.

Figure 12.4. Courtesy D. W. French.

Figure 12.5. Courtesy The American Chestnut Foundation. **Top,** First appeared in "Scenes in Fairmount Park," The Art Journal, Vol. 4, 1878, Appleton and Co., New York. **Bottom,** First appeared in American Lumberman, 1910.

Figure 12.7. Used by permission of G. Hudler, Cornell University, Ithaca, NY.

Figure 12.11. Courtesy U.S. Department of Agriculture, Animal and Plant Health Inspection Service, Science and Technology.

Chapter 13

Figure 13.1. Courtesy H. D. Shew. Reprinted from Scholthof, K-B.G. 2000. Tobacco mosaic virus. The Plant Health Instructor, DOI: 10.1094/PHI-I-2000-1010-01.

Figure 13.2. Image copyright © The Metropolitan Museum of Art / Art Resource, NY.

Figure 13.3. Top and Center, Courtesy H. Israel. **Bottom,** Courtesy American Phytopathological Society, St. Paul, MN.

Figure 13.4. Drawing by Vickie Brewster. Reprinted from Scholthof, K-B.G. 2000. Tobacco mosaic virus. The Plant Health Instructor, DOI: 10.1094/PHI-I-2000-1010-01.

Figure 13.5. Top, Courtesy L. Nault, American Phytopathological Society Slide Set 21, slide 20 (1978). **Bottom,** Courtesy J. E. Duffus. Reprinted from Whitney, E. D., and Duffus, J. E. 1991. Compendium of Beet Diseases and Insects. American Phytopathological Society, St. Paul, MN.

Figure 13.6. Used by permission of H. D. Thurston, Cornell University, Ithaca, NY.

Figure 13.7. Original drawing by Bunji Tagawa. Used courtesy Donald Garber.

Figure 13.8. Redrawn, by permission, from Clark, M. F., and Adams, A. 1977. Characteristics of the microplate method of enzyme-linked immunosorbent assay for the detection of plant viruses. Journal of General Virology 34:475-483.

Figure 13.9. Drawing by Vickie Brewster. Reprinted from Schumann, G. L., and D'Arcy, C. J. 2010. Essential Plant Pathology, 2nd ed. American Phytopathological Society, St. Paul, MN.

Figure 13.10. Courtesy G. L. Schumann.

Figure 13.11. Courtesy T. M. Nourse, Nourse Farms, Inc., Deerfield, MA.

Figure 13.12. Courtesy K. Maramorosch, Rutgers University, New Brunswick, NJ.

Figure 13.13. Courtesy T. O. Diener, University of Maryland, College Park.

Chapter 14

Figure 14.1. Courtesy Norman E. Borlaug.

Figures 14.2, 14.7, 14.8, 14.10, and 14.12. Used by permission of H. D. Thurston, Cornell University, Ithaca, NY.

Figures 14.3 and 14.6. U.S. Department of Agriculture.

Figures 14.4 and 14.11. Drawings by Nancy Haver.

Figure 14.5. Courtesy W. J. Manning, University of Massachusetts, Amherst.

Figure 14.9. Used by permission of Dr. Rory Hillocks, Natural Resources Institute. Copyright University of Greenwich, 2010.

Figure 14.13. Reprinted, by permission, from Potatoes for the Developing World. 1984. International Potato Center, Lima Peru, p. 56.

Index